... to
Elec... ...ations

Technician's Guide to Electronic Communications

Frederick L. Gould

McGraw-Hill

New York San Francisco Washington, D.C.
Auckland Bogotá Caracas Lisbon London
Madrid Mexico City Milan Montreal New Delhi
San Juan Singapore Sydney Tokyo Toronto

Library of Congress Cataloging-in-Publication Data

Gould, Frederick L.
Technician's guide to electronic communications / Frederick L. Gould.
p. cm.
Includes index.
ISBN 0-07-024536-3. — ISBN 0-07-024537-1 (pbk.)
1. Telecommunication systems. 2. Electronic circuits. I. Title.
TK5105.G68 1997
621.382—dc21 97-8074
CIP

McGraw-Hill

A Division of The ***McGraw-Hill*** *Companies*

1 2 3 4 5 6 7 8 9 0 DOC/DOC 9 0 2 1 0 9 8 7

ISBN 0-07-024536-3 (hc)
0-07-024537-1 (pbk)

The sponsoring editor for this book was Scott Grillo, the editing supervisor was Scott Amerman, and the production supervisor was Tina Cameron. It was set in Century Light. It was composed in Hightstown, N.J.

Printed and bound by R.R. Donnelley & Sons Company.

I dedicate this book to my best friend, life's partner, and love of my life—my beautiful wife, Suzanne.

Contents

Introduction

While electronics encompasses many areas, one of the most absorbing is communications. Modern communications is the nervous system that binds our complex society together, transforming the earth into a global village. Without it, commerce, entertainment, defense, medicine, and space exploration would be severely limited, if not impossible. Today we think nothing of communicating over thousands of miles to conduct business and obtain information at the speed of light. Just a few short decades ago, such activities were found only in the realm of science fiction. Communications systems are available in many different forms. A communications link can be as simple as walkie-talkies used by hikers, or as complex as a satellite system supporting international business and diplomacy. Information that is exchanged can be digital, code, voice, facsimile, or video. It is a modern-day miracle that is often taken for granted.

Communication is the act of giving or exchanging information, usually by gesturing, speaking, writing, signaling, or transmitting. The need for long-range communication is as old as the family of man. Earlier civilizations used gestures, speech, and symbols to communicate with one another. When our ancestors consisted of wandering bands of hunter-gatherers, speech, hand, and smoke signals were sufficient means to link the bands together. However, as we gained in capabilities and ranged farther afield, the only reliable method of long-range communications was the use of human messengers. As they could move only as fast as horses, or later, sailing ships, the time required for the receipt of a message and a reply was often measured in days, months, or even years. A good example is the War of 1812 between the United States and Great Britain. The Treaty of Ghent, signed on December 24, 1814 and ending the war, did not reach New York until February 11, 1815. Because of the lack of rapid communications, the war continued for six weeks after it could have ended. Due to the painfully slow communications links, the unnecessary Battle of New Orleans was fought on January 8, 1815.

For the intent of this book, discussion of communications will be limited to the use of electromagnetic waves in the radio frequency range. The waves are used to transmit electronic signals through space without the use of wires, cables, waveguides, or any other interconnecting conductors. The rapid evolution of electromagnetic communications systems has been an electronics revolution that has touched all of our lives in ways that would have been seen as science fiction only a short 50 years ago. In 1950, a typical American family owned an AM radio and had access to a telephone—usually a party line at that. The passage of time and rapid adoption of advanced technology has changed that forever.

The wide acceptance and use of inexpensive electronic components, coupled with advanced techniques and designs, has fueled an explosion of communications systems. Radio communications for information now consist of AM, FM, and shortwave broadcasts. Personal two-way radio communications has become CB, amateur radio, GMRS, and the emerging two-way systems called family radio. Television, in its infancy in the late 1940s, has now grown to include over-the-air broadcasts, cable TV, and several varieties of satellite TV systems. The simple telephone—once always in black—has now become a designer product. The well-connected individual owns a pager, cellular phone, and cordless telephones. It has now become difficult not to be informed and electronically bound to the rest of the world.

Communications changed little throughout our history until the eighteenth century. Human beings' harnessing of the electromagnetic spectrum would answer our communications needs. From the first halting steps with the telegraph to the sophisticated personal communications systems of today has only been 200 years. Our global village is tightly bound by an invisible web of communications links. The highly touted and yet-to-be-fully-exploited information superhighway is only the latest link, and one of many links. Virtually everyone has access to radio and television transmissions. Telephone systems are now common throughout the world. Now more than just a simple means of two-way communications, the telephone system offers many services, including an entry point to the Internet. By adapting computers, digital electronics, and satellites, point-to-point communications is possible anywhere in the world under virtually any conditions. An area long driven by the research and developmental budgets of the armed forces, communications allows our complex business and personal world to function at virtually the speed of light. As the complexity and flexibility of these systems has grown, equipment size has dramatically shrunk. Circuitry and hardware developed for the space and defense programs have gained widespread adaptation in the civilian markets. With the demise of the Cold War, forcing decline in military research dollars, the concept of COTS (civilian off-the-shelf) is gaining in popularity with the military community. This has resulted in the military leading in some areas of technology, and civilian consumer corporations in others.

This book is written for technicians who already have a basic understanding of electronics theory and would like to expand upon that knowledge. Its purpose is to introduce the new technician to communications theory as it is known today, and to provide a review for the experienced individual. Information introduced in this book will begin with the history of electromagnetic communications. Visionary scientists who had an impact on physics and mathematics led the way in this field. A presentation of the electromagnetic spectrum will show how much of it is currently being exploited. The first chapter will then be rounded out with a block diagram discussion of a representative communications system.

Further chapters will continue with a more in-depth study of electronic basics, antenna theory, and RF energy propagation. The communications block diagram will be explored in greater detail, presenting how modern communications systems function. Safety and important maintenance techniques will each be discussed in separate chapters. When maintenance is being performed on electronic systems that radiate energy, safety cannot be overemphasized. The final chapter is an examination of what the future might hold for the field.

This book will provide communications information to the new technician. The almost frantic advances seen in communications technology will doubtless continue unabated for years to come. This once-small area of the electronics field that was at one time the home of the eccentric experimenter and theoretical researcher has become fascinating, critical to our lives, and a driving force behind much of the economy. With its scope and constant advances, communications is more than enough to fill an entire career.

Abbreviations and Acronyms

A/D	analog-to-digital
AF	audio frequency
AM	amplitude modulation
CMRS	Cellular Mobile Radio Telephone
COMSAT	communications satellite
COTS	commercial off-the-shelf
CW	continuous wave
dB	decibel
dBm	decibel as referenced to 1 mW
dBW	decibel, as referenced to 1 watt
DSB	double sideband
DSBAM	double sideband amplitude modulation
EHF	extremely high frequency
EIA	Electronics Industry Association
EL	elevation
ELF	extremely low frequency
ELOS	extended line-of-sight
EM	electromagnetic
FCC	Federal Communications Commission
FET	field effect transistor
FM	frequency modulation
GPS	global positioning system
HERF	hazard to electromagnetic radiation to fuel

HERO	hazard to electromagnetic radiation to ordinance
HERP	hazard to electromagnetic radiation to personnel
HF	high frequency
HPA	high-powered amplifier
Hz	hertz
IF	intermediate frequency
ISB	independent sideband
LF	low frequency
LOS	line-of-sight
LSB	lower sideband
MF	medium frequency
MIL-STD	military standard
MOF	maximum observable frequency
MOSFET	metal oxide semiconductor field effect transistor
MTBF	mean time between failures
MTSO	Mobile Telecommunications Switching Office
MTTR	mean time to repair
NCS	Network Control Switch
OWF	optimum working frequency
PLL	phase-locked loop
RADHAZ	radiation hazard
RF	radio frequency
RFI	radio frequency interference
RT	receiver/transmitter
SHF	super high frequency
SNR	signal-to-noise ratio
SSB	single sideband
SSBAM	single sideband amplitude modulation
TRF	tuned radio frequency

UHF	ultra high frequency
USB	upper sideband
VCO	voltage controlled oscillator
VF	voice frequency
VHF	very high frequency
VLF	very low frequency
VSWR	voltage standing wave ratio

Electronic Communications for Technicians

BECAUSE OF THE UNIVERSAL NATURE OF RESEARCH, IT IS difficult to mention all the advances and individuals who have had an impact on electronics and communications. Often, similar research and conclusions took place in several different locations, at the same time, without researchers knowing of one another. However, there are many who will always be remembered for their efforts. This list of notable electronics inventors, experimenters, and pioneers is long, consisting of many distinguished intellectual explorers. It is the efforts of these rare individuals that still have an impact on our world today; they have written the history of electronics.

The History of Electromagnetic Communications

From its beginnings, electromagnetic communications has been an effort in innovation. Due to the inquisitiveness of the human mind, research has not been based upon the talents or studies of one nation. The new ideas and experiments over the past 200 years have come from the efforts of many individuals, collaborations, and organizations in many countries. Because of the sum of the many collective efforts, present-day electromagnetic communications systems are among the most complex, extensive, and costly of civilization's technological creations. From the first theoretical experiments on the nature of magnetism in the 1700s to today's universal systems has been a rapid evolution in capabilities and acceptance. It has evolved into an invisible, instantaneous web that links virtually every point in the globe. Telecommunications, computer networks, radio, and television links have become indispensable. If industry is the backbone of modern civilization, then the electromagnetic communications system is the nervous system. Without it, our complex civilization simply would not be possible. We have become so dependent upon it that, were it suddenly to disappear, commerce, diplomacy, defense, and entertainment would return to only what was possible in the 18th century.

Communications between wide-ranging groups has always been a problem. The first methods in recorded history were limited to messengers, signals, smoke, and flags. While the concept of messengers was effective, it did suffer from a major limitation: time. The message could travel no faster than the means of conveyance. It could take months or even years for the original message to be sent and the reply received. Until the establishment of the intercontinental telegraph networks, communications across continents took weeks over land and months by sea. Something better was required if society was to advance. Another problem was ensuring a continuous means of long-range communications. Any crude links that existed were ad hoc in nature and not meant for long-term use. The first successful answer was the European semaphore system.

It was the violence of the French Revolution that gave impetus to the first long-range communications system. With the rapid changes in the volatile French political landscape, reliable and swift communications were mandatory throughout the strife-torn nation. Designed and constructed by the brilliant engineer Claude Chappe, it was placed in service in the 1770s. When completed, it consisted of over 500 stations stretching over 5000 kilometers throughout the country. Each station consisted of a tower mounting manually movable arms or apertures with covers. To increase the range between each station, the towers were mounted on hilltops with a clear field of view.

As communications technology does not acknowledge international boundaries, Britain recognized the importance that a semaphore communications system could have to a complex society. The British system was smaller in scope, linking only the cities of London, Portsmouth, and Deal with a network of 15 stations. Even with the limitations of slow rate of data transmission and the requirement for good visibility, it was a very effective start.

Based upon the experiments and theories of several physicists and mathematicians, research was progressing in several countries to develop the first electromagnetic communications system. The work of Allesandro Volta, Michael Faraday, and Hans Christrian Oersted promised the possibility of communications using electricity and magnetism. The device was much sought after, as it would greatly increase the speed of data transmission and free the system from the effects of weather. As the industrial revolution was in full swing, commerce had become worldwide. Reliable and rapid communications would greatly facilitate the economic growth that industrialized nations needed. In the 1800s, attempts to fabricate a practical system were noted in Britain, Bavaria (Germany), Russia, Switzerland, and the United States, to name a few.

The required breakthroughs occurred in Britain and the United States. The British team of Charles Wheatstone and William Cooke developed the first practical electromagnetic communications system. Based on the experiments of themselves and others, the team took out their first patent on July 10, 1837. It was a hard-wired system consisting of five magnetic needles. Through the application of electrical current, any combination of needles could be deflected. Through the use of a simple code, 20 letters of the alphabet could be represented. For a first effort, this was an ingenious answer to a critical problem. Its practical application gave other experimenters a reference point from which to build.

The next leap in communications technology occurred in the United States. It was software related, rather than a hardware innovation. Samuel B. F. Morse made a long-lasting contribution to the field of electromagnetic communications. A student of electricity in college, it became a hobby for him. In pursuit of his interest, he attended lectures and spoke with any electricity experimenters he encountered. His first attempt at a communications system was a printing telegraph that he first demonstrated in 1835. In the system, a key switch controlled current applied to an electromagnet with a pencil attached to it. The result was a mark on the paper any time the key was activated. The length of the mark was determined by the amount of time that the key was closed, completing the circuit. Initially, the "electric pencil" could have been just an experimenter's oddity, but for the foresight, talents, and efforts of many individuals.

The name Samuel F. B. Morse and electromagnetic communications will be forever linked. Morse's contribution to technology is the code that bears his name. For more than a century it was the method of choice for the electronic conveyance of information. Morse code is simple to learn and use, and is presented in Figure 1-1. A short mark is called a dot. A long mark, equal in time to two to three dots, is a dash. Different combinations of dots and dashes are used to represent letters, numerals, and punctuation marks. Morse's design of the code was actually a common sense approach. He simply counted the number of each letter in a printer's type box. The more numerous a particular letter was, the more frequently it would be used in print. Therefore, he assigned the most common letters the shortest code representations. Hence, "E" is a single dot and "T" is a single dash. As letters such as "X" and "Z" are the least common, they are assigned the longer codes of "dot-dash-dash-dot" and "dash-dash-dot-dot." As operators became familiar with the system, rather than read the marks on a paper tape, they listened to the sounds of the clicking electromagnet. That was how the least complex and most flexible of all telegraph systems was developed—a key, a power source, an interconnecting conductor, and a sounder. Even though his code and simple hard-wire communications system was an obvious improvement, he had a very difficult time in obtaining funding. It took until 1843 before Congress funded a small system to link Washington with Baltimore as a practical demonstration. It was a resounding success, resulting in the globe becoming interconnected by telegraph systems in the following decades. So long-lasting was his contribution that if one scans the shortwave radio bands (3 MHz to 30 MHz) today, its rhythmic sounds can still be heard. However, new technology is replacing it. On January 31, 1997, the French Coast Guard had its last CW transmission. The message was "Calling all. This is our last cry before silence."

Morse's simple-but-effective code for quickly and accurately transmitting complex data was revolutionary, rather than evolutionary. Time was no longer a factor in providing communications links between two interconnected points. Locations were no longer limited to hilltops with good fields of view. All that was needed was a very narrow pathway wide enough to place the poles carrying the wires. Its rapid adaptation in overland communications sparked interest in developing underwater cables to bridge the oceans and thereby to provide the first effective intercontinental communications systems. Other inventors, in their drive to improve upon the system, were determined to provide mobile communications, for example providing ships at sea with communications links to shore bases. The simple little code ultimately led to radio communications, television, and facsimile (fax).

Morse's idea was improved upon by the development and adaptation of duplex, quadruplex, and time-division multiplex telegraphy. The technique was to use one conductor to carry multiple communications links. It was a highly desirable and much-sought-after

INTERNATIONAL RADIO ALPHABET AND MORSE CODE

A: Alpha	. _	W: Whiskey	. _ _
B: Bravo	_ ...	X: X-ray	_ .. _
C: Charlie	_ . _ .	Y: Yankee	_ . _ _
D: Delta	_ ..	Z: Zulu	_ _ ..
E: Echo	.	1:	. _ _ _ _
F: Foxtrot	.. _ .	2:	.. _ _ _
G: Golf	_ _ .	3:	... _ _
H: Hotel	....	4:	.... _
I: India	..	5:	.....
J: Juliet	. _ _ _	6:	_ ....
K: Kilo	_ . _	7:	_ _ ...
L: Lima (leema)	. _ ..	8:	_ _ _ ..
M: Mike	_ _	9:	_ _ _ _ .
N: November	_ .	10:	_ _ _ _ _
O: Oscar	_ _ _	period:	. _ . _ . _
P: Papa	. _ _ .	comma:	_ _ .. _ _
Q: Quebec (kaybec)	_ _ . _	question mark:	.. _ _ ..
R: Romeo	. _ .	semicolon:	_ . _ . _ .
S: Sierra	...	colon:	_ _ _ ...
T: Tango	_	hyphen:	_ .... _
U: Uniform	.. _	apostrophe:	._ _ _ _ .
V: Victor	... _		

■ **1-1** *Morse code.*

technique, as more message traffic could be handled by each line. Increased traffic per line decreased costs and increased profits for the communications companies. The first viable use was demonstrated in 1853, when Dr. Gintle's duplex system simultaneously transmitted and received messages over only one conductor. Demonstrated in Vienna, Austria, his system used a balanced bridge with an artificial line to prevent the two messages from interfering with one another. Building upon Gintle's work, Thomas Edison obtained a patent award in 1874 to allow quadruplex messages to be sent over a single wire.

The next improvement was the adaptation of time-division multiplexing (TDM) developed by Emile Baudot. An official in the French Telegraph Service, his concept allowed for the transmission of many messages at the same time, a great savings in equipment. His system consisted of synchronized rotating commutators with a five-unit code. For transmission, a keyboard with five keys was used. The message was printed out at the receiving station with a tape printer. His invention, TDM, can be considered to be the basis of modern digital techniques.

The next great hurdle that challenged the radio experimenters were the barren wastes of the oceans separating the continents. The first ocean to be crossed by cable would be the Atlantic. It is a tale of epic proportions that required courage, ingenuity, vision, and perseverance. The successful completion of the project required the contributions of two very different, but brilliant, men—industrialist Cyrus W. Field, and eminent scientist Lord Kelvin. The first electromagnetic link was attempted between Newfoundland and Valentia, Ireland, in 1857. Because the cable-laying vessels were U.S. and British warships rather than specialized vessels, it was a failure. Due to improper cable-laying techniques and materials, the cable suffered numerous breaks and was a complete loss. A second attempt in 1858 was initally effective. The communications link allowed President James Buchanan and Queen Victoria to exchange messages on August 14, 1858. However, the cable was in operation for less than a month. In an attempt to improve signal quality, an overzealous telegraph operator boosted the voltage applied to the line to such an excessive level that it caused an insulation breakdown, leading to cable failure.

Due to the American Civil War, a third attempt was not undertaken until 1865. Leaving little to chance, the American Telegraph Company began by utilizing Lord Kelvin's expertise in the redesign of the cable to improve performance. A larger vessel, the USS Great Eastern (a failed passenger liner) was employed as the cable layer. Modified with improved equipment and a specialized crew, it was the only ship in service at the time that had the internal volume to store and handle the entire length of cable to cross the Atlantic. Even with the improved materials and equipment it still took two attempts to bridge the ocean. On July 27, 1866, the North American and European continents were electromagnetically linked, and have been so ever since. So successful was the project that by 1875, a worldwide network of underwater and land-based telegraph cables provided an electronic nervous system interconnecting the nations of the world.

Lord Kelvin's technical advice was critical due to changes in cable design. Electrically, a cable functions as a reactive component, specifically a capacitor. The longer the cable, the greater the distributed capacitance. The causes and effects will be more fully explored when antenna theory is addressed. Briefly, the effect of the distributed capacitance is to limit the rate of change of current flow through the conductor. As Morse Code consists of changes in current flow (ON and OFF), any reduction in the rate of current flow change reduces the speed at which messages can be transmitted over the cable. As communications is a profit-driven endeavor, a reduction in message traffic results in reduced revenue.

Therefore, the utilization of Lord Kelvin as a scientific advisor was a sound business decision.

The research that ultimately resulted in radio communications was first demonstrated in theory by the studies of Clerk Maxwell, a Scottish mathematician, and in experiment by Heinrich Hertz, a German physicist. In 1886, Hertz's goal was to prove the existence of radio waves through the use of a crude spark generator. In his work, sparks were generated by the discharge of an induction coil across a gap between a pair of round electrodes. To prove that the "spark" actually traveled through the atmosphere, he used a simple detector that was nothing more than a wire loop with a small gap. By placing the detector in close proximity to the spark gap, radio wave action could be observed. When a spark was created in the "transmitter," one would then be observed in the "detector." Through the careful placement of various objects, Hertz determined that electromagnetic radio waves passed through insulators, were reflected by metallic surfaces, and could be refracted by prisms. Scientific discussion had reached the point that, by the 1890s, it was thought that "Hertzian waves" would provide the means to free reliable communications from the limits imposed through the use of interconnecting cables.

Although much experimentation occurred throughout the world, the one individual given the most credit for the advancement of radio communications is Guglielmo Marconi. Rather than a scientist or physicist searching for answers, Marconi was an inventor whose goal was to exploit the electromagnetic spectrum for communications. His early experiments in his home laboratory were similar to those of his contemporaries. His simple equipment consisted of a battery power supply, an induction coil, and a simple antenna known as a Hertzian dipole radiator. Transmission consisted of generating high voltage sparks across the air gap in the dipole. The receiver equipment also used a radiator as the antenna, and a coherer as the radio wave detector. A coherer is a vacuum tube with two metal plugs at either end; it electrically acts as a capacitor. Metal filings are placed between the plugs. When charged, the filings cling (or *cohere*) together. The phenomenon was noted and improved upon by several different researchers. Professor E. Branly of Catholic University in Paris, France, noted and explored its actions in depth. His studies resulted in an improved coherer. In 1894, Sir Oliver Lodge completed a series of experiments utilizing a Branly coherer as a radio wave detector. He proved that a coherer was a more efficient detector than a gapped wire loop. When the receiver detected an electromagnetic wave, the metal filings cohered, decreasing the resistance between the plugs. That in turn allowed current to flow in the battery circuit. After signal reception, the coherer had to be mechanically tapped to return it to normal sensitivity. The required tapping mechanism was devised by a Russian experimenter, Alexander Popov.

After observing the efforts of other experimenters, Marconi changed the antenna design to improve performance. He devised an elevated aerial with a metal plate on the ground. The same design was used for both the transmitter and receiver. Additional design changes included the use of a coherer for the receiver detector. The result of all the changes was a greatly improved range of about a mile and a half. This simple technological oddity (by today's standards), with a range measured in hundreds of yards, would ultimately be considered the first usable communications system.

Successful commercial exploitation would take several more years due to disinterest in the new communications medium. Marconi's first attempt to gain backing was with the Italian government, who showed very little interest in his proposals. Undeterred, he selected Great Britain for his next demonstration. As Great Britain had the largest merchant fleet and navy, he felt that there would be the interest he needed to advance commercial communications. As a nation with worldwide defense commitments and economic interests, a reliable communications system freed from the restraints of interconnecting cables would be an important improvement. Reliable long-range systems would allow a more effective control of Great Britain's vast colonies. To facilitate acceptance, he contacted Sir William Preece, the Chief Engineer of the British Post Office, to offer a demonstration. The tests were held in two separate locations in 1896—in London, and on Salisbury Plain. In both experiments, a range of one and three-quarters miles was achieved. Sir Preece lectured on Marconi's experiments, not as an avid supporter, but as a rival. The British Post Office had been granted a monopoly for telephone and telegraph communications within the United Kingdom. With foresight, he considered the new technology of wireless communications as the ultimate replacement for telegraph cables. By ensuring widespread knowledge of the Marconi's achievements, he felt that the Post Office would gain governmental backing to exploit the new medium.

Undaunted by the lack of acceptance, Marconi continued his experiments. He was issued the first patent from the British government for wireless communications in 1896. It was, in part, based on the application of increased antenna height to improve communications ranges. In 1897, through the use of antennas supported by balloons, a range of four miles was achieved. Encouraged, he founded the Marconi Company to continue experimentation and possible commercial exploitation.

With improved equipment and techniques, ranges gradually increased. Figure 1-2 is a schematic diagram of his early transmitters that were used from about 1897 to 1900. As can be seen, it is a very simple design, consisting only of a code key, induction coil, spark gap, antenna, and power source. The receiver, illustrated in Figure 1-3, is a little more complex. In this design, the antenna provides an input to the coherer. The resulting energy is coupled via RF chokes to a relay. A Morse code printer is used to provide a hard copy of the received intelligence. With the modest technological improvements, in one demonstration contact was made between an Italian cruiser and the Spezia dockyard, separated by 11 miles. The test was so successful that the Italian Navy adopted his system for wireless communications. The Alum Bay demonstration in 1898 was a watershed in that constant communications were maintained with commercial passenger steamers over a distance of 18 miles. That was rapidly followed by Marconi achieving a major technological feat—the first wireless communications across the English Channel.

For years, all early transmitters operated on virtually the same low frequency. That was because they were a spark gap design that radiated over a broad band of low frequencies. As long as spark gap transmitters were widely separated in distance and rare, there was

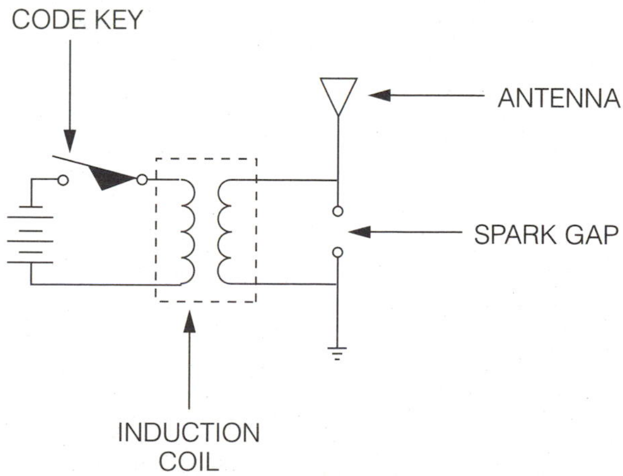

■ **1-2** *Marconi spark gap transmitter.*

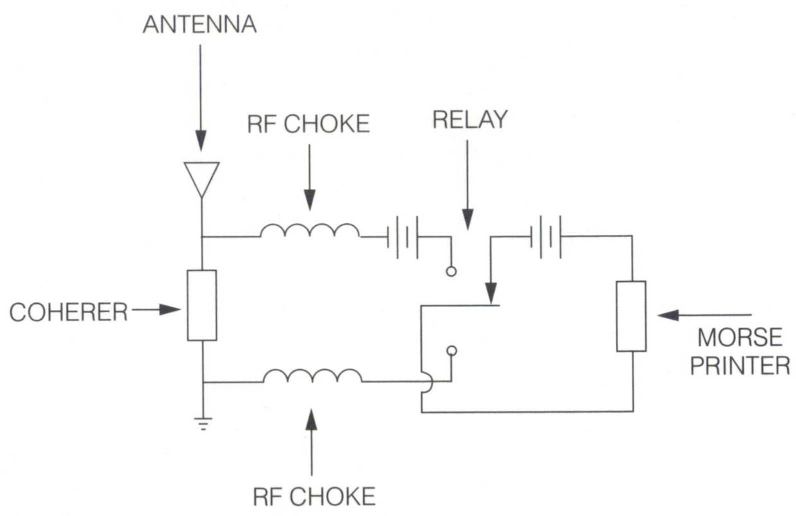

■ **1-3** *Marconi coherer receiver.*

very little observed interference. As designers increased transmitter power to obtain greater operational ranges, problems began to appear. Greater operational ranges meant that more ships and shore stations would be equipped with the new, more powerful spark gap transmitters. The results from more transmissions would be predictable with our experiences in modern communications. As the number of transmitters and output power increased, the weaker ones would be masked, or jammed, by the stronger ones. Because the equipment was meant to provide ships with emergency communications, the results could be disastrous. Using a technique called syntony, developed by Sir Lodge, Marconi obtained a patent in 1900 for a circuit of his design that would revolutionize the fledgling medium—tuning. Syntony, by definition, is a condition that exists when two or more oscillators have precisely the same resonant frequency. In his system, the antenna was connected to the transmitter with an inductor. By changing the value of the inductor, the transmitter frequency could be varied. The receiver used a high-frequency transformer, a tapped inductor, and a variable capacitor. The benefit of the variable components was that the receiver's resonance could be varied to select between several different transmitters.

The final technological chasm to be bridged by Marconi was the Atlantic Ocean. Prevailing scientific opinion was that wireless communications would never improve to the point where it could be used to connect North America and Europe. Marconi was motivated by the desire to prove the scientific community wrong and to challenge the submarine cable companies. The locations selected were to be as physically close together as possible. In Great Britain, Cornwall, England was to be the British site for the transmitter. Cape Cod in the United States was to be the North American terminus.

Location was not going to be the only aspect of the experiment that received his personal attention. The equipment was to be revolutionary, and the finest available. The spark gap transmitter was to be energized by a 25 KW alternator. It would be keyed by short-circuiting the inductance in the alternator circuit. The transmitting and receiving antennas were identical in construction. Each consisted of a 200 foot diameter circle consisting of 20 masts, 200 feet high. However, before the system could be tested, storms in September of 1901 damaged both antenna arrays.

Undeterred, Marconi had his assistant repair the British array. The result was an inverted triangle of 50 copper wires, 160 feet high. Moving the North American installation to Canada, Marconi set up the receiving station on Signal Hill, located in St John's, Newfoundland. The site was also the location for the landing of the first transatlantic telegraph cable. The receiving antenna was somewhat less elaborate than first envisioned. As a suitable mast was not available, the antenna was a 500 foot long wire supported by a kite.

The first transatlantic signal, consisting of the letter "S" in Morse code, was received on December 12, 1901. As the event was not observed by any independent observers, there was considerable skepticism toward the claim. Also, physicists still felt that the alleged achievement was an impossible goal. Further experiments were successful; however, to compound matters, the phenomenon was not always repeatable. Anyone familiar with the problems encountered in long-range communications caused by atmospheric conditions can relate to the problem. Further experimentation determined that signal propagation was affected by the time. During daylight, range was limited to 700 miles. At night, due to the effects of the upper atmosphere, range increased to over 2000 miles. That indicated that North Atlantic shipping could be in communications virtually all of the time.

By 1914, Marconi's private company had begun commercial wireless communications from Glace Bay, Canada to Carnavon, Wales. Long-range plans were for an Imperial Wireless Chain to link Britain, India, Australia, Canada, Egypt, and other smaller nations. Realizing the advantages that reliable communications links provided to Great Britain, other European nations, such as

France and Germany, began to construct their own systems. However, technology was to overtake the low-frequency communications systems.

From Marconi's experiments in 1902, the advances in communications only increased in number and frequency. The first types of transmitters, known as spark gap transmitters, were low in power and frequency. With new advances, transmitter power and usable frequencies increased. Fueled by both commercial interests and enthusiastic amateurs, the field expanded.

A noticeable improvement in the science of communications was the carrier modulation, rather than the old spark gaps. The advantage was a narrower, cleaner signal that was usable over a wider range of frequencies. Continuous wave (CW) was the first type of modulation to be employed. Increasing the operating frequency of the transmitters and receivers resulted in smaller antenna systems. As an example, an early wireless station operating on a low frequency of 16 kHz required an antenna farm extending over a full square mile. The array was supported by 12 masts, each one over 800 feet high.

Antenna and transmitter designs were not the only areas to advance. The earliest receivers used a crude component called a crystal to extract the intelligence from the transmitted signal. The crystal functioned as a rectifier and was made of either carborundum or galena. A moveable wire called a cat's whisker was used for tuning. The nature of a crystal allowed current to flow in one direction, which allowed only the positive portion of the modulated wave to pass into the receiver. Reactive components, such as capacitors, would then be used to filter out the unwanted high-frequency carrier, leaving only the audio intelligence. The receiver, as it was not equipped with a speaker, required headphones. The crystal receiver used the incoming signal as its only source of power. That limited reception to only high-powered signals. With a good ground, antenna, and strong signal reception, ranges of up to 1000 miles were routine.

The answer to how to increase operating frequency was provided by the vacuum tube for use in both transmitters and receivers. The diode (two-element) tube was invented in 1905 by Sir Ambrose Fleming. As can be seen in Figure 1-4, it is an odd looking component by today's standards. The grid structure was very large, permitting close examination. As crude as we would think of the device, it permitted more efficient and higher power-supply voltages and the detection of high-frequency radio waves. Application of the diode improved communications to the point where, in 1906, human speech was first successfully transmitted and received by Reginald Arbrey Fessenden. A physicist, Fessenden used radio speech communications to contact ships in the Atlantic from a station located at Brant Rock, Massachusetts. Building upon the success of the diode, Lee De Forest invented the triode (three-element) vacuum tube in 1907. Figure 1-5 is a schematic diagram of how it would be interconnected. The battery symbols represented true batteries, not conventional DC power supplies. Many early radios were battery operated. To provide proper operating voltages, several different batteries were required. The addition of the third element, or grid, permitted the amplification of radio waves and audio signals.

Experiments yielded improvements in transmitters that increased in operating power from a few watts to ultimately thousands of watts by the mid-1920s. In 1924, using his yacht Elettra as a mobile receiver base, Marconi proved that the new technology of vacuum tubes could provide many advantages. He selected a wavelength of 100 meters, a frequency of about 3 MHz. The transmitting equipment consisted of a 10 kW transmitter, using a parabolic antenna

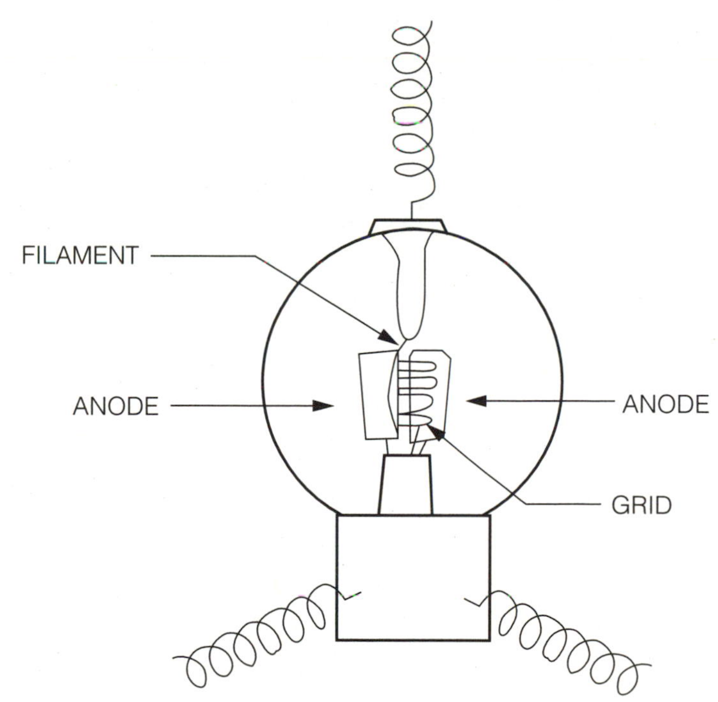

■ **1-4** *Early thermionic tube.*

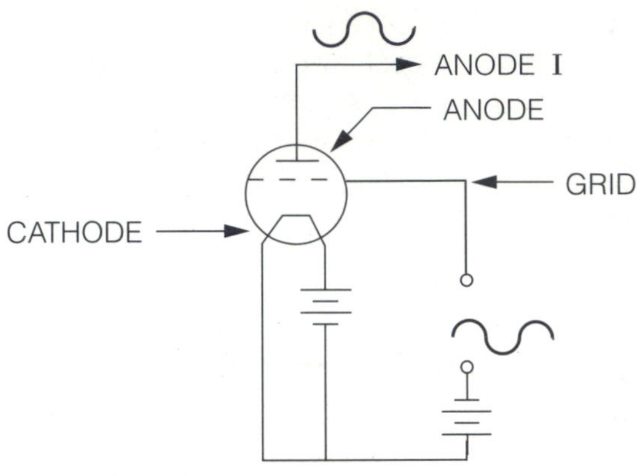

■ **1-5** *Early triode diagram.*

mounted on a 325-foot-high mast at the Poldhu, England base station. From the South Atlantic, ranges of over 1200 miles during the day and 2200 miles at night were achieved. After increasing transmitter power output to 20 kW, good nighttime reception was reported from New York. Commercial transatlantic communications were to quickly become routine events using long-wave frequencies by today's standards—60 kHz.

Receiver advances kept pace with the improvements in transmitters and antennas. New receiver designs, such as the tuned-radio-frequency (TRF) design, used triodes to improve sensitivity, selectivity, and audio output. Plagued with tuning difficulties, the TRF was rapidly superseded by the superheterodyne receiver circuit. The new design was first demonstrated by Major Edwin Armstrong in 1912. Encouraged by the now-routine use of voice communication, the idea of transmitting music was first proposed by David Sarnoff in 1916. At the time, he called it a "Music Box."

As often happens, conflict provided the impetus that resulted in many technological advances in electronic communications. World War I was a titanic struggle that affected almost all of Europe. The control of tremendous numbers of combat troops in the field, continental-sized combat zones, and large numbers of ships at sea all required accurate and timely means of communications. Military aviation, also a new technology, used air-to-ground communications to provide reconnaissance reports and artillery target information. Research and applications, devised for war use, fueled peacetime advances. The research high point of the war years was W. Schottky's experiments that led to the development of the tetrode.

During the 1920s, radio communications flourished. In 1919, Dr. H. W. Eccles's research lead to the introduction of the temsdiode and further applications for the triode. Domestic radio service expanded rapidly in the United States, filling an entertainment and information void. Westinghouse engineer Frank Conrad licensed the nation's first true entertainment radio station, KDKA in Pittsburgh, Pennsylvania. The station scheduled music, sports, and a first—the 1920 Presidential election. In 1921, Westinghouse began selling home radio receivers, with a starting price of $25. By 1924, the American radio audience numbered at least 20 million listeners, enjoying the services of over 500 AM broadcast stations.

Tube designs were also improved. The addition of more grids by Hull and Tellegen made the electronic vacuum tube more versatile. Up until the 1920s, CW and AM were the only methods of modulation. Major Edwin Armstrong began investigating the possibility of frequency modulation (FM) to eliminate unwanted noise and interference.

Radio could now be arguably categorized as domestic, international, and governmental services. Although long-wavelength communications links were in operation, by the late 1920s longwave was becoming obsolete for the level of communications that were required by society. AT&T, Western Electric, and Bell Laboratories began to experiment with shortwave frequencies for long-range communications. By 1930, an experimental transmitting site was in operation at Deal, New Jersey, and receiver stations were in Netcong and Holmedel, New Jersey. The transmitter, through the use of water-cooled output tubes, achieved an output of 60 kW. Antenna designs were also improved to take advantage of reduced interference and vastly improved gain. It has become a communications truism that improvements in antenna gain are always less expensive than boosting power to increase range. The most promising antenna design was a shortwave rhombic array, a very simple and inexpensive design that greatly enhanced performance. The first design was supported by 80-foot-tall towers. The radiating elements consisted of copper wire. A very effective design, it featured power gains on the order of 14 dB and a wide operational band—5 to 20 MHz. As each 3-dB increase in gain represents a doubling in power, the 14-dB antenna gain translates into a power increase of almost 32 times.

World War II, as with all conflicts, provided another boost to electromagnetic communications. Operational frequency, output power, and receiver sensitivity all increased. Due to design advancements, tuning and modulation techniques all improved. The result was that equipment became more common, cheaper, smaller, lighter, and easier to use. Two-way communications systems were provided to all ships, aircraft, armored vehicles, and small infantry units. The walkie-talkie, a small one-man transceiver, was very easy to carry and helped to revolutionize the concept of personal communications. It was to communications as the Jeep was to wartime transportation. As World War II was worldwide in scope, rapid long-range communications were vital to the successful completion of the conflict.

Partially based on the research of the war, and partially based on pent-up consumer demand, the late 1940s and 1950s were a golden age for electronics. All facets of communications decreased in cost and increased in scope. Transistors began to appear, offering smaller, cheaper designs. Transistors were invented in 1948, dramatically altering all phases of the electronics field. Its appearance was not a sudden technological inspiration, as it was based on years of research and work by many individuals. That crystal diode oscillations could be sustained and exploited was first recorded by Dr. H. W. Eccles in 1910. The early crystals could function as detectors to remove intelligence from the carrier frequency and to rectify AC voltage into DC. These were very important uses, but what was needed was a solid-state crystal that could be used as an amplifier.

The first "tubeless" or solid-state radio receiver was demonstrated in 1923 by Dr. Julius E. Lillienfeld in Canada. The receiver used crystal (or solid-state) devices that functioned as oscillators and amplifiers. The devices were very similar in design to the insulated gate field effect transistors in use today. Based on his work, he applied for Canadian patents covering his electronic devices in 1925 and 1928. A German engineer, O. Hell, developed a copper-oxide rectifier with a third electrode in 1934. Two Philips engineers, Hoist and van Gael, received a Dutch patent in 1936 for a bipolar solid-state device. As promising as these first efforts were, none were considered successful. The most likely cause for the failures was a lack of knowledge of semiconductor materials and the manufacturing techniques to fabricate them. The final pieces were provided by Bell Laboratories after World War II.

The scientists, Dr. John Bardeen, Dr. Walter Brattain, and Dr. William Shockley, received the 1956 Nobel Prize for physics for their efforts in fabricating the first practical solid-state crystal amplifier: the transistor. The three eminent scientists were first brought together at Bell Labs in the fall of 1945, and formed a study group along with Gerald Pearson (a physicist) and Robert Gibney (a physical chemist). The group was formed to study solids from an atomic and electronic structure standpoint. By using researchers with diverse backgrounds, it was hoped that the goal of learning how to design and fabricate new materials that could be used in the development of completely new and improved components in communications systems would be met. Building upon the wartime use of silicon and germanium as detectors in radar circuits, the first point-contact transistor was a reality in less than three years.

Frequencies continued to go higher, resulting in UHF and SHF communications links. Once more, as operating frequencies increased, antenna size decreased. However, society's appetite for communications channels was insatiable. To provide the number of communications links required by modern commerce, satellite communications networks were established. The transistor, which once replaced the vacuum tube, led to the development of an entire branch of electronics. In a few short years it would be superseded by the integrated circuit.

The integrated circuit was an electronic device that was obviously needed. The first digital computer, ENIAC of 1946 vintage, required thousands of vacuum tubes and weighed a staggering 30 tons. Tubes are bulky, require a great deal of power, and produce tremendous amounts of heat (requiring external cooling). Transistors were a vast improvement, as they consumed much less power, took up less space, and operated cooler. Heat can be a detriment, as it dramatically reduces the operational life of electronic components. The transistor had a flaw in that its leads that connected it to the circuit board were flimsy and often broke off. The answer was found by two engineers simultaneously—Robert Noyce at Fairchild Semiconductor and Jack Kilby at Texas Instruments. The simple solution that was technically difficult to produce was to make the transistor its own circuit board. Fabricated out of chips of semiconductor, entire electronic circuits could be contained within a solid piece of material.

Due to improvements in materials and manufacturing techniques, a single integrated circuit (IC) can contain up to 100,000 transistors. Its advantage is that there are no soldered wires between components, which reduces failure points, making the whole circuit very reliable. As all connections are internal to the chip, the interconnections between individual transistors or gates are very short, making circuits operate more rapidly. In today's electronic products, ICs containing 5,000 transistors produce a digital watch; 20,000, an inexpensive scientific calculator; 100,000, a small home computer microprocessor. Progress has been so rapid and successful that every year for the past 20 years engineers have virtually doubled the number of components on an individual IC chip. The result is that electronic products have become smaller, cheaper, more reliable, and more prevalent.

The superiority of ICs in many applications has vastly improved electromagnetic communications. With the adaptation of microprocessors, digital signal processing and tuning made operation much simpler. Communications techniques that were only theories a few years ago now make the limited electromagnetic spectrum more efficient. Major communications systems could now be automated and operated more efficiently. The future for electromagnetic communications is as bright as it has ever been. It is interesting to note that little of this would have come to pass if it was not for the curiosity of visionary scientists of the 18th and 19th centuries.

The Electromagnetic Spectrum

The electromagnetic spectrum is the complete range of electromagnetic radiation frequencies that extends from cosmic rays to low-frequency radio waves. As large as the region may appear, it is crowded, because modern communications needs more space to expand old services and add new ones as technology advances in an increasingly urbanized world. The region exploitable for electromagnetic communications currently ranges from 30 Hz to 300 GHz. The region is divided into several separate frequency bands, and can be identified by several different designations, depending upon format. Currently, frequency, wavelength, and letters are in use. The shorter the wavelength of the electromagnetic wave, the higher the frequency. Table 1-1 lists the current acceptable band designations.

By international agreement, the electromagnetic frequencies begin at a frequency of 0 Hz. The audio frequency band is usually classified from 0 to about 20 kHz. Currently, useful electromagnetic frequencies begin at about 30 Hz. The three lowest electromagnetic frequency bands are the Extremely Low Frequency (ELF), Very Low Frequency (VLF) and Low Frequency (LF) bands, extending from 30 Hz to 300 kHz. With current technology, the three lowest bands only have two uses: communications with

■ **Table 1-1 Electromagnetic spectrum**

Frequency Band	Frequency Range
Subaudible Frequencies	.001 to 15 Hz
Audio Frequencies	15 Hz to 20 kHz
Power Frequencies	10 Hz to 1 kHz
Video Frequencies	20 kHz to 4.5 MHz
Ultrasonic or Supersonic Frequencies	20 kHz to 2 MHz
Extremely Low Frequencies (ELF)	30 Hz to 300 Hz
Super-low Frequencies (SLF)	300 Hz to 3 kHz
Very Low Frequencies (VLF)	3 kHz to 30 kHz
Low Frequencies (LF)	30 kHz to 300 kHz
Medium Frequencies (MF)	300 kHz to 3 MHz
High Frequencies (HF)	3 MHz to 30 MHz
Very High Frequencies (VHF)	30 MHz to 300 MHz
Ultrahigh Frequencies (UHF)	300 MHz to 3 GHz
Super-high Frequencies (SHF)	3 GHz to 30 GHz
Extremely High Frequencies (EHF)	30 GHz to 300 GHz
Infrared (heat)	.3 THz to 430 THz
Visible Light	430 THz to 1 kTHz
Ultraviolet	
X-rays	
Gamma Rays	
Cosmic Rays	

submerged submarines, and utility beacons. As can be seen from the chart, the bands have severe limitations such as slow data transmission rates, antenna design difficulties, and noise. The Medium Frequency (MF) band, 300 kHz to 3 MHz, and the High Frequency (HF) band, 3 MHz to 30 MHz, are more versatile. This range is where truly global communications occur. The MF band is used for local and regional communications. A good example would be the AM broadcast stations in the United States and the slightly higher frequencies used in Europe, Asia, and South America. HF is the intercontinental long-range frequency band. Even with the advent of satellite communications, these bands are still crowded with various services. It is used for long-range shortwave broadcasts, military communications, commercial aviation, and economic communications. Very High Frequency (VHF), with a range of 30 MHz to 300 MHz, is used for line-of-sight communications. Its primary users are public safety, FM broadcast, TV channels 2 to 13, military tactical communications, radio astronomy, space telemetry, ship control close to shore, and air traffic control. Ultra-High Frequency (UHF) extends from 300 MHz to 3000 MHz.

This band is also line-of-sight. Primary uses in this range are military tactical, radar, commercial TV UHF channels, cellular phones, public safety, and satellite communications. The new public safety frequencies are in the 800 MHz bands and provide new features to the users, such as trunking to more effectively utilize band space and security codes. Many consumer products that utilize communications links are also found here, such as cordless telephones, wireless intercoms, wireless speakers, baby monitors, and remote control vehicles.

The upper reaches of the electromagnetic spectrum are also exploited. Infrared, in the frequency range of 300 GHz to 430 THz, is usually measured in length (microns), as opposed to frequency. It is currently used for detecting heat sources in environmental, military, space, and manufacturing applications. The next band of frequencies is the visible light spectrum. Above it is the ultraviolet band. It is used in security and military applications. The final exploited region of the electromagnetic frequency spectrum is the X-ray band. Its uses include medical, military, and flaw detection. Due to the pressure placed on the spectrum by applications, it is a very finite and valuable resource.

The frequency band selected for a particular service is determined by the area to be covered and the type of transmission. New technologies, techniques, and equipment designs are allowing services such as personal communications, cell phones, and pagers to be coast-to-coast systems. Public safety and commercial communications links are being assigned frequencies that only a few years ago were considered to have little economic value. These services have gradually worked up from LF, through HF, VHF, UHF, to SHF. With the market forces driving the field, change will continue to be rapid for many years.

The Representative Communications System

All communications systems must consist of a minimum number of functions in order to be useful. The transmitter, by definition, has the function to generate and amplify a radio-frequency carrier signal. The intelligence signal, which can be from a teletype, microphone, tape, or Morse key, modulates, or varies, the carrier frequency in amplitude, phase, or frequency. Modulation is required, as it is used to boost the frequency of the intelligence to one that is suitable for transmission through free space. Free space is what is between the transmitting and receiving antennas. It can be the atmosphere, outer space, or a combination of both. The modulated carrier is amplified by the RF amplifiers to a level suitable to be transmitted. The output of the transmitter is routed to the antenna, the function of which is to act as an impedance match between the output of the transmitter and free space. It has a single output from the transmitter and radiates the RF to free space.

The atmosphere and the surface of the earth are both capable of reflecting radiated RF energy. The portion of the radiated energy that follows the surface of the earth is called the ground wave. Energy that is reflected through the atmosphere is known as the sky wave. Under certain well-known conditions, the radiated RF will reflect between certain layers in the atmosphere and the earth's surface many times. The distance between "skips" in the sky wave is determined by the frequency of the radiated RF and atmospheric conditions. As this is a predictable phenomenon, operators and broadcasters take full advantage to obtain the maximum possible range. Some portion of the RF energy is intercepted by the receiver antenna. The antenna on the receiver end has the function of matching the input of the receiver to the atmosphere. A well-designed antenna will have the effect of providing some gain to the signal. The receiver then demodulates the RF. Demodulation is the action of removing the RF carrier, leaving just the original intelligence. The resulting low-frequency signal is then amplified and processed for further use.

The output of the receiver is then applied to ancillary devices for further processing and use. An ancillary device can be speakers, additional amplifiers, tape recorders, or a teletype for hard copy. All communications systems, whether a sophisticated satellite system or handheld walkie-talkies, have the same minimum functions in order to provide proper operation.

Following chapters in the book will provide a more in-depth study of how basic electronic components and circuits are interconnected to result in a complete communications system. Additionally, electrical safety and maintenance tips will be presented.

2 Reactive Components and Basic Filter Circuits

ALL TECHNICAL DISCIPLINES HAVE BASIC THEORIES AND facts that are so important as to be considered the very foundation of the specialty. The field of electromagnetic communications is no different. As communications is dependent upon stable frequencies for proper operation, reactive components are the bedrock of electronic communications design. These vital, but simple, electronic parts are used in circuits such as filters, oscillators, and amplifiers. Additionally, as antenna theory is based on resonance, it is important here as well to ensure complete understanding. When you read the chapter on antennas, you will learn how important they are to total system operation. An optimum antenna will give more system performance improvement with less effort and lower cost as compared to transmitter and receiver improvements. This chapter will present inductors and capacitors. Theory discussed will include basic component characteristics, impedance, resonance, and filter operation.

Without filters, electromagnetic communications as we know it would be impossible. A filter usually consists of a capacitor and an inductor, and is known as a tuned circuit. Through the careful selection of component values, tuned circuits can be constructed that will pass some frequencies and block others. The operation of a filter is dependent upon a phenomenon that has intrigued researchers for many years—resonance. Resonance is the condition in a tuned circuit when the capacitive reactance equals the inductive reactance. To fully understand filter operation, the voltage and current characteristics of series and parallel reactive circuits must be examined in detail.

Capacitor

The capacitor will be studied first. By definition, capacitance is the property that exists whenever two conductors are separated by an insulating material that permits the storage of electrical energy. A characteristic of the component is that it prevents a sudden change in circuit voltage by allowing only a limited rate of change. Figure 2-1 illustrates the construction of a typical capacitor. As can be seen from the diagram, it consists of two metal plates, separated by a block of material known as the *dielectric*. The plates, also known as electrodes, can be constructed from a conducting material such as aluminum, copper, gold, silver, or tin foil. The dielectric is an insulating material, such as a vacuum, air, prepared paper (oiled paper), polypropylene, glass, mica, or one of several oxides. Table 2-1 lists several common dielectrics and their permeability factors. The function of the dielectric is to store electrical energy without damage and return it to the circuit with as low a loss as possible. For the purpose of this book, losses are considered to be nonexistent.

Capacitive action is as follows, as illustrated in Figure 2-2. As the plates are separated by an insulator, current does not flow through a capacitor. If a difference in potential is placed across the device, charging action occurs. In the diagram, one plate is connected to the positive terminal, and the other to the negative terminal. The negative potential of the source repels electrons from its negative terminal to the negative plate of the capacitor. The buildup of electrons on the negative plate of the capacitor repels electrons contained within the dielectric to migrate toward the positive plate. That causes a buildup of electrons in the portion of the dielectric next to the positive plate. As that side of the dielectric has a negative charge, it drives electrons from the positive plate to the positive terminal of the power source. The result is that the voltage

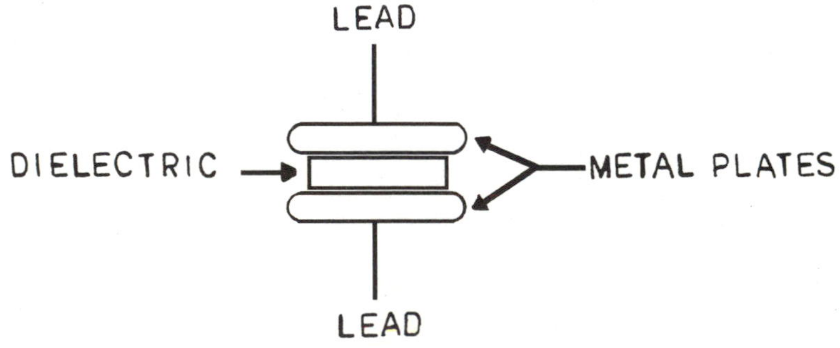

■ **2-1** *Capacitor construction.*

■ Table 2-1 Dielectric constants of common materials

Dielectric material	Dielectric constant	Dielectric strength
Air	1	20
Bakelite	4–4.7	100
Ceramics	80–1200	600–1250
Epoxy resin	3.4–3.7	130
Glass	8	330
Mica	3–8	600–1500
Paper	2–6	400–1250
Teflon	2.1	1500

source has redistributed electrons from one side of the capacitor to the other. The electron migration continues until the potential difference across the plates of the capacitor is equal to the charge of the applied voltage source. All current flow is through the circuit external to the capacitor. The current is transient in nature, as it flows only until the capacitor reaches full charge. The effect is that a capacitor stores electrical energy in its dielectric by altering the orbits of the electrons. As energy storage is based upon the movement of electrons within the dielectric, a capacitor is considered to be an electrostatic device.

The amount of charge that a capacitor stores is determined by the value of the applied voltage and the "capacitance" of the reactive component. Capacitance is a function of the physical construction of a capacitor and the insulating value of the dielectric. By referring to Table 2-1, you can see the insulating values, or *dielectric constants*, assigned to several common capacitor dielectrics. As can be seen from the table, dielectric constant values range widely. The unit of measurement of capacitance is the farad (F). A capacitor is said to have a capacitance of 1 farad if an applied voltage change of one volt per second results in an external circuit current of 1 amp. A whole farad of capacitance is far too large for practical applications. Typically, capacitors are manufactured in the microfarad (up) or picofarad (pF) range of values. The physical construction of capacitor plates in conjunction with the dielectric constant determine overall capacity. In short, a capacitor's capacitance is directly proportional to the area of the plates and inversely proportional to the distance between the plates. The larger the plates, the greater the capacitance. However, the closer the plates are together, the greater the capacitance. The formula

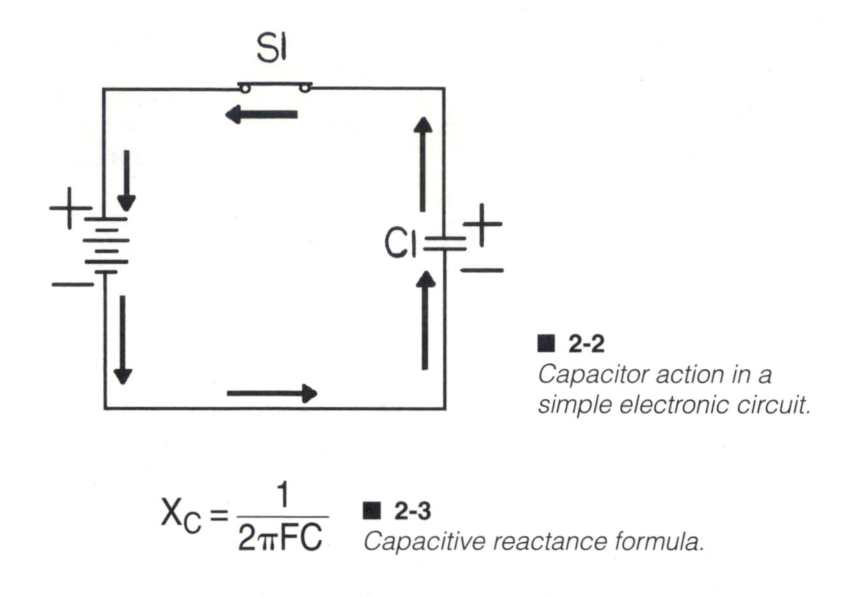

■ **2-2**
Capacitor action in a simple electronic circuit.

$$X_C = \frac{1}{2\pi FC}$$

■ **2-3**
Capacitive reactance formula.

to calculate capacitance is in Figure 2-3. As can be seen from it, capacitance is inversely proportional to distance between the plates and directly proportional to the area of the plates.

An important rating in capacitors is the working voltage DC (WVDC). WVDC is the maximum value of DC voltage that can be applied across the part without causing arc-over and subsequent damage. Due to the electrical properties of the dielectric, the plates must be separated by enough distance to prevent the insulating material from breaking down while still providing the required amount of capacitance. The breakdown voltage of a capacitor is directly related to the dielectric strength of the component. The rating is typically stated as volts per mil (V/mil). One mil is equal to .001 inches. Table 2-1 lists the dielectric strength of several common dielectrics. From the dielectric strength, constant, plate area, and separation, it can be seen that a working capacitor is the result of a complex series of considerations.

Capacitors are also affected by heat. As temperature increases, the working voltage of a capacitor decreases. Care must be taken to ensure that the WVDC is sufficient to compensate for excessive heat.

Capacitor operation is illustrated in Figure 2-2 , and is as follows. When switch S-1 is closed, source voltage is placed across the capacitor. That results in a negative potential on the lower plate and a positive potential on the upper one. Even though current does not flow through the component, a current external to the device is present. That is caused by the negative potential repelling electrons off of the plate. As they are forced toward the upper plate, that action repels electrons on the upper plate into the external circuit. External circuit current is maximum the instant the switch is closed. As the capacitor charges toward the applied voltage, current decreases. When the capacitor is fully charged, circuit current is minimum, or zero. Therefore, in a capacitive circuit, current is said to be the lead voltage. Capacitors hold energy, or a *charge*, until discharge. Discharge is accomplished by changing the potential felt on the plates. If S-1 in the schematic is placed in position 2, the capacitor is in parallel with a resistor. Current then flows from the negative plate to the positive one.

Capacitors can be connected in series or parallel to obtain desired total values and operation. Capacitors connected in parallel have a greater total capacitance. It is calculated by simply adding all the values together as in Figure 2-4a. When connected in series, total capacitance is less than that of the smallest capacitor. Figure 2-4b depicts the concept. As a review, capacitance of several capacitors in parallel is calculated the same way as resistors in series. That of

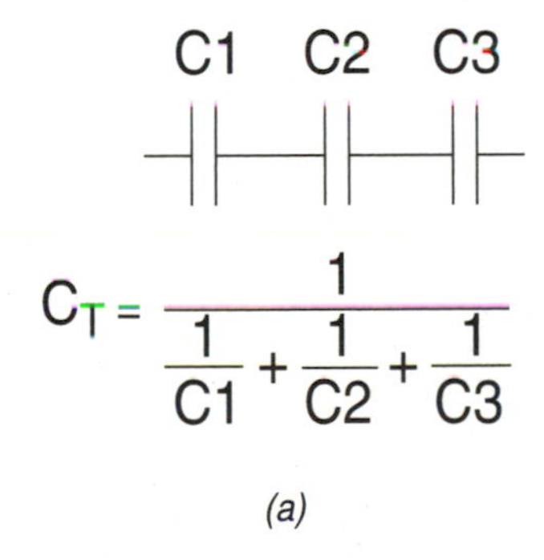

(a)

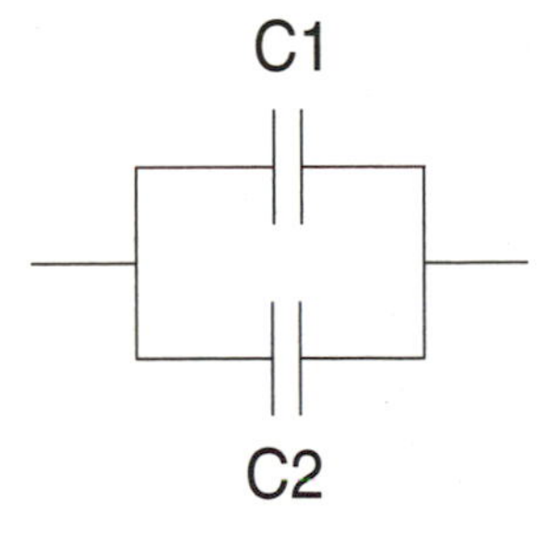

(b)

■ 2-4
Total capacitance in (a) series and (b) parallel circuits.

capacitors in series is found with the same method as that of resistors in parallel.

A capacitor opposes an alternating current flow; this opposition is known as impedance. Impedance is expressed in ohms, but is not a true loss as a resistance would be. By nature, a resistor actually limits the current flow in a circuit and causes power losses due to heat. A capacitor, by comparison, stores energy and as a result opposes any changes in voltage in the circuit. Figure 2-5 compares two series DC circuits, one resistive, one capacitive. In the resistive circuit, as voltage increases, circuit current increases—they are in phase, which means one follows the other. In the capacitive circuit, when the switch is closed, the circuit appears electrically as a short circuit, allowing maximum current to flow. That is because the capacitor is in the process of storing energy. As the capacitor charges, the voltage across it increases in a polarity that opposes the voltage source. Due to the opposition presented by the capacitor's charging action, circuit current begins to decrease. At full charge, circuit current drops to zero, as the capacitor is acting as an open circuit. In the capacitive circuit, the instant the switch closes, circuit current is maximum and the capacitor's charge is zero. As the charge increases, the current decreases until the charge reaches maximum and current reaches zero.

In an AC circuit, capacitor operation shows its true value to advanced electronics techniques. Applied voltage across the capacitor lags circuit current by 90 degrees. Also, a capacitor offers an opposition to current flow that is inversely related to the frequency of an applied AC voltage. The changing opposition presented by a capacitor to AC voltage is known as capacitive reactance. Figure 2-6 illustrates an AC sine wave. If you examine the wave, you will notice that there are points where the change in voltage as compared to time (rate of change) is greater than others. At the positive and negative peaks, the rate of change is the slowest. That is because the sine wave voltage must slow, stop increasing, and then begin

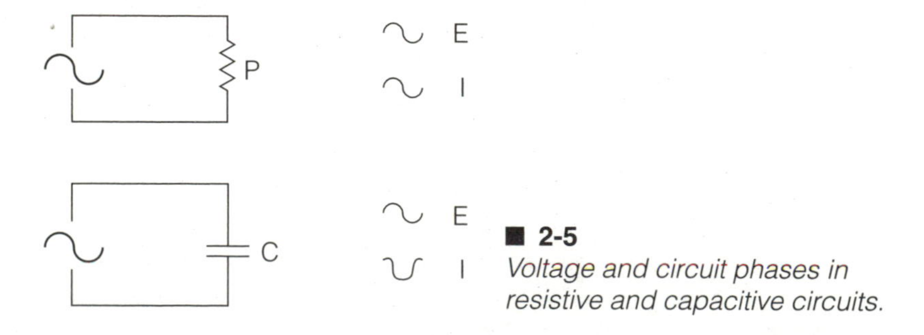

■ 2-5
Voltage and circuit phases in resistive and capacitive circuits.

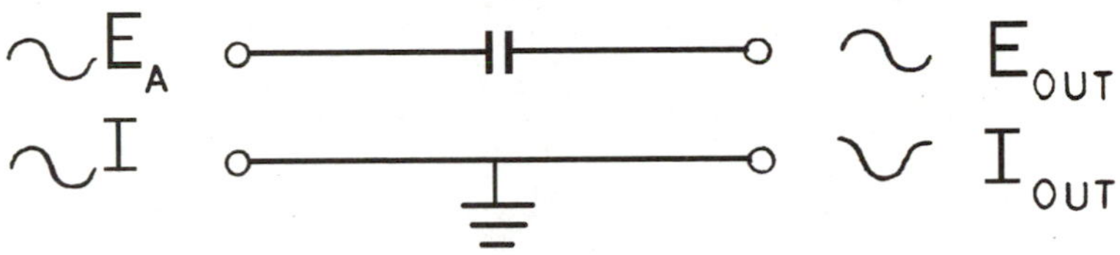

■ **2-6** *Voltage and current in a capacitive circuit.*

decreasing. When the sine wave crosses the zero axis, the rate of change is the greatest. This concept can be proven either mathematically or by observing an AC sine wave with an oscilloscope and comparing the changes in voltage amplitude to time.

Because applied voltage is an AC sine wave, the resulting circuit current is also a sine wave. Due to the nature of a capacitor, circuit current is initially maximum and decreases to zero. The resulting charge voltage starts at zero and increases as the charge reaches maximum value. That means that in a capacitive circuit, the current leads the voltage by 90 degrees. The easy way to remember this is "*ICE*," a mnemonic for *I* (current) *C* (capacitor) and *E* (voltage)

As stated earlier, a capacitor opposes (impedes) the flow of current in a circuit. The opposition, called *capacitive reactance*, is determined by the frequency of the applied voltage. That is because when a capacitor is charging, it opposes the increasing value of applied voltage by storing energy. Conversely, when applied voltage is decreasing, the capacitor attempts to maintain a constant circuit voltage by discharging energy back into the circuit. Due to the opposition to the rate of voltage change, capacitive reactance is inversely proportional to the frequency of the applied voltage. Figure 2-7 compares capacitive reactance to the frequency of the source voltage. As the frequency of the applied voltage increases, the resulting capacitive reactance decreases. The formula that calculates the impedance value of a capacitor in an AC circuit is illustrated in Figure 2-8. 2π is an electrical constant that is used in many AC calculations. It is approximately equal to 6.28 (2×3.14). It indicates the circular motion from which a sine wave is derived. As such, it has no application in DC circuits. *F* is the applied frequency in Hertz, and *C* is the capacitance in farads. The resulting impedance is expressed in ohms.

Inductor

The inductor or coil is an electronic component that exhibits the phenomenon of inductance. By nature, a component that prevents

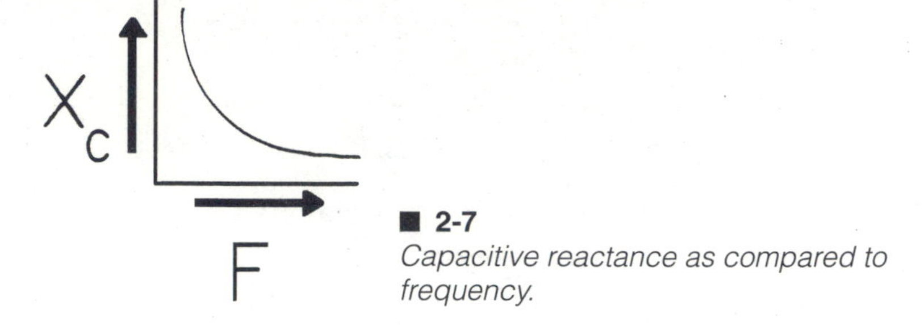

■ **2-7**
Capacitive reactance as compared to frequency.

$$X_C = 2\pi FC$$

An increase in frequency *increases* capacitive reactance.
A decrease in frequency *decreases* capacitive reactance.

An increase in capacitance *increases* capacitive reactance.
A decrease in capacitance *decreases* capacitive reactance.

■ **2-8**
Capacitive reactance formulas.

sudden or abrupt changes in circuit current through the expansion or collapse of a magnetic field possess inductance. It will also limit the rate of change in current exhibited by the circuit. As with the capacitor, an inductor can store energy. Inductors are considered to be electromagnetic devices, as they store energy in a magnetic field. Therefore, any circuit containing an inductor will not be capable of an instantaneous change in current. The construction of an inductor is quite simple, and the method has been known for over two hundred years. An inductor is a coil of wire wrapped on a cylinder, called the core. The wire is electrically insulated so that each turn of the coil will not short out to the next one; this ensures the proper effective wire length. Core construction varies, based upon the use for which the coil is designed. In its simplest design, a core might be laminated paper that forms a hollow circle. This type is called an *air core inductor*. When the coil is formed around a metal core, it is called an *iron core inductor*. Moveable cores are used to construct variable inductors. An inductor is also known as a *choke* or *coil*. That is because in the early days of electronics, it was named after its construction—a coil of wire. The term choke came about as it was often used to eliminate or choke off undesired high-frequency signal components. Whatever term it is known by, the amount of inductance exhibited by the device is measured in Henrys (H). A Henry is defined as this quantity of inductance: a 1 Henry coil

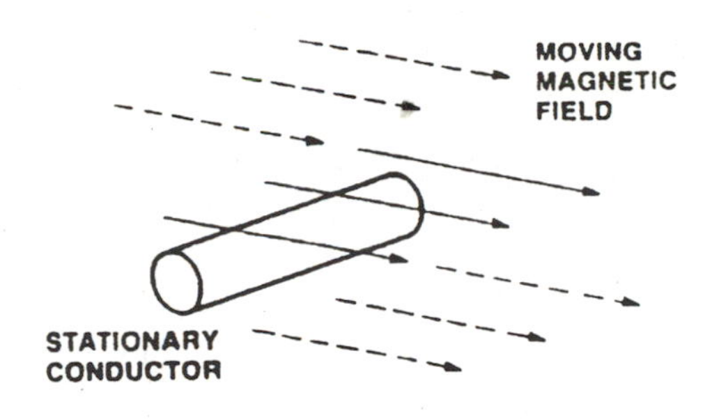

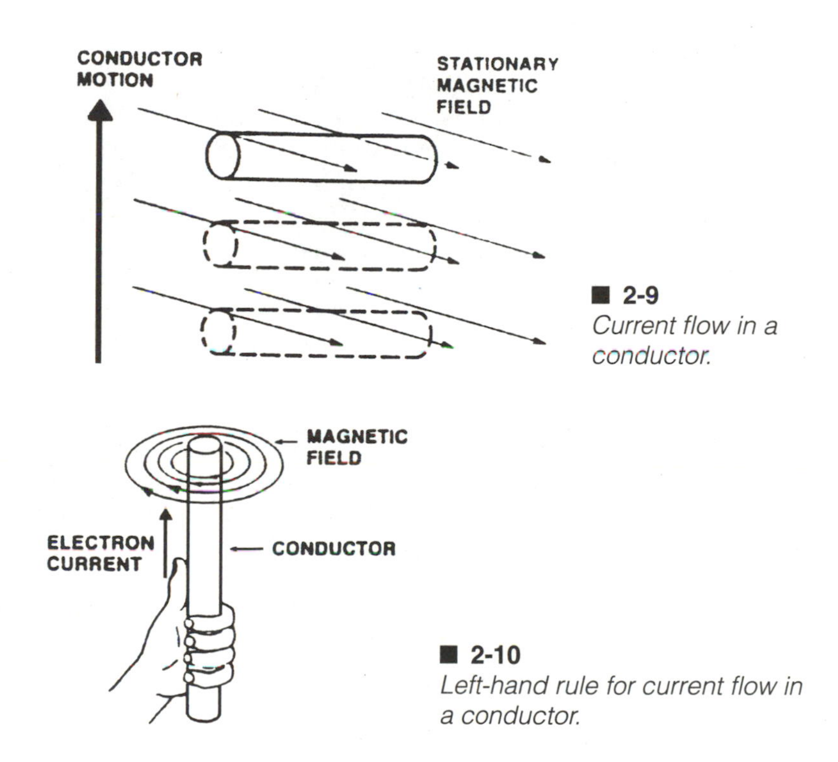

■ **2-9**
Current flow in a conductor.

■ **2-10**
Left-hand rule for current flow in a conductor.

flux emanating from one turn of wire cut through adjacent turns. The result is a cascading or multiplying effect. The more turns of wire in the coil, the greater the induced voltage within the coil. When a source provides an increased voltage, that causes an increase in opposition current, which in turn has a greater effect on the initial current flow in the circuit. The more wire contained in a coil, the greater the inductive action.

Several factors affect the value of inductance possessed by an inductor. The number of turns of wire that make up the coil affect the resulting value. The more turns of wire, the greater the inductance (and conversely, fewer turns mean less inductance). Induc-

produces 1 volt when the current flowing through it is changing at a rate of 1 amp per second.

An inductor exhibits the basic properties of electromagnetism. Figure 2-9 depicts the relationship that exists between a magnetic field, motion, and a conductor. Any time that current flows through a conductor, a magnetic field is produced that is perpendicular to the current flow. Secondly, any time a conductor passes through a magnetic field, a voltage is induced in the conductor, causing current to flow. The relationship between current flow and a magnetic field follows a predetermined pattern. Figure 2-10 illustrates the concept known as the left hand rule. Notice that the conductor is gripped by a left hand. As current flows through a conductor, the hand holds it so that the direction in which the thumb is pointing indicates the direction of current flow. As the current flows through the conductor, a magnetic field is produced. As can be seen, it is in a clockwise direction, as indicated by the fingers of the left hand. As long as the circuit current flow continues to increase, the magnetic field will continue to expand outward. If the current flow maintains a constant rate, then the magnetic field will stabilize, storing electrical energy. If current flow in the conductor begins to decrease for any reason, the magnetic field begins to collapse, feeding energy back into the circuit. In that manner, the device attempts to hold circuit current constant. Although there is some small energy loss, for the purpose of this discussion such losses are insignificant.

Inductor action is as follows and is illustrated in Figure 2-11. With the application of voltage, current flow begins through a conductor and the resulting field begins to expand, cutting through the conductor. As the field cuts through the conductor, it induces a voltage in the conductor. The polarity of the induced voltage is opposite to that of the applied voltage. The induced voltage then produces a circuit current that is opposite in direction to that of the initial current, causing an opposition. If an AC voltage is applied to the circuit, as the voltage changes, the induced voltage changes with it, and opposes any changes in circuit current. In most cases, the induced voltage is minute, measurable only with laboratory instruments. The action of the magnetic field inducing a voltage and resultant current is known as inductance.

This property becomes useful when a conductor is wound into a coil. When a conductor is in the shape of a coil, the lines of magnetic

tance is also controlled by the cross-sectional area of the core. The larger the core, the greater the inductance. The final factor is the coil length. The shorter the coil, the greater the inductance. As can be seen, the amount of inductance displayed by an inductor is the result of several different factors.

When an AC voltage is applied to an inductor, the changing voltage induces a changing magnetic field. That in turn causes a changing voltage as the magnetic field alternatively expands and collapses. The voltage that results from the changing magnetic field is known as *counter-electromotive force*, or CEMF. CEMF is the phenomenon that causes an opposition that resists any change in circuit current flow.

Figure 2-12 illustrates inductor action in a series resistive-inductive (RL) circuit with a DC power source. The switch is closed at T0, and the voltage drop across the inductor is maximum as the magnetic device electrically appears as an open. All the applied voltage is felt across the inductor as CEMF. Initially, circuit current is zero, as the complete path required for current flow is blocked by the inductor building a magnetic field. As the voltage continues to increase, current flow begins. The magnetic field expands outward. CEMF slows, but does not stop the increasing circuit current. The magnetic field will continue to expand as long as the applied voltage increases in value. At the peak of the sine wave, change slows, then stops. As the applied voltage is not increasing, the induced magnetic field stops expanding. In turn, the CEMF begins to decrease, allowing current flow to gradually increase toward maximum. As CEMF continues to decrease, current increases to the point where circuit current is maximum and CEMF is zero. As the applied AC voltage starts to decrease, the inductor produces CEMF of the opposite polarity. It begins to feed energy back into the circuit to maintain a constant current flow by collapsing the magnetic field. Because of the CEMF

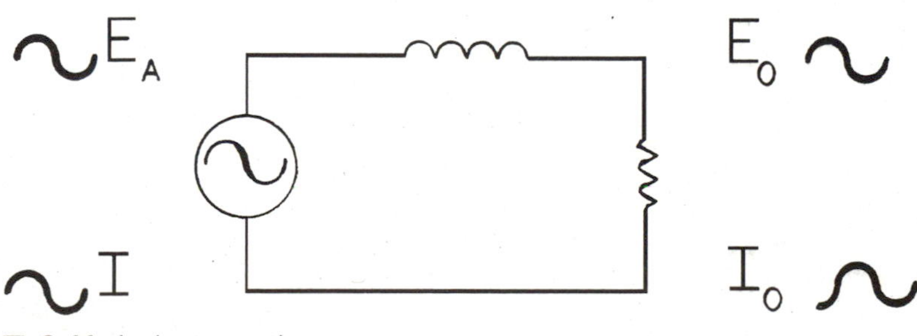

■ **2-11** *Inductor action.*

action, current lags voltage by 90 degrees. With inductance, the opposition to current flow is apparent only when current flow through the conductor is changing. A steady current flow results in no opposition to current flow. An easy rule to remember with a circuit containing inductors is that the *E* (voltage) *L* (leads) *I* (current), or *ELI*.

If several inductors are connected in a circuit, total inductance is calculated in the same manner as total resistance. When inductors are connected in series, just add them up.

$$L_1 + L_2 + L_3 = L(T)$$

Total inductance when inductors are in parallel is the reciprocal.

$$L(T) = \frac{1}{L_1} + \frac{1}{L_2} + \frac{1}{L_3}$$

Because inductors are devices that store energy electromagnetically, they can affect one another when installed in close proximity. As inductors concentrate the magnetic field naturally found in any current-carrying inductor, they can interact with one another. This property is called *mutual inductance* and is depicted in Figure 2-13. Whenever the current through an inductor is changing, the resulting magnetic field can cut across any other coil or conductor that might be installed nearby. That action will cause a voltage to be induced in the second coil. As can be seen, two inductors are connected in a series circuit with an AC source. As the current

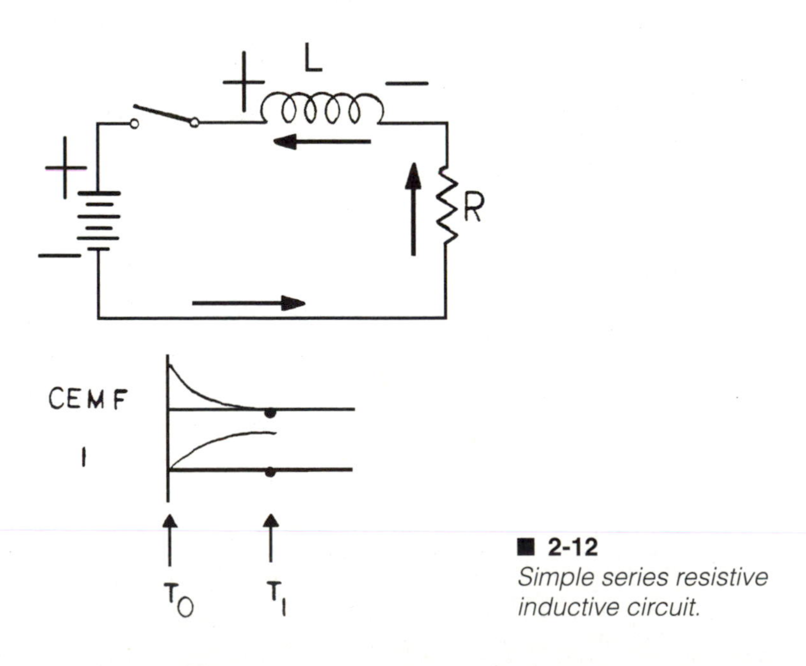

■ **2-12**
Simple series resistive inductive circuit.

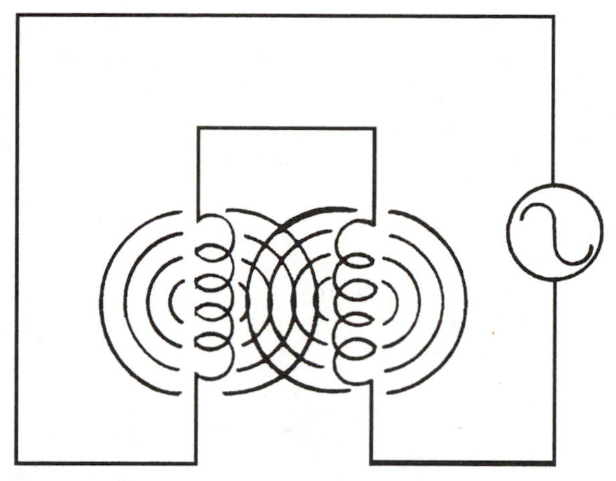

■ **2-13** *Mutual inductance.*

flow through L1 changes, it produces a varying magnetic field. As L2 is in close physical proximity, L1's magnetic field cuts through it, inducing voltage in its windings. The same circuit current is flowing through L2, causing it to produce a magnetic field that cuts through the windings of L1. Therefore, the magnetic field produced by L1 affects L2 as the magnetic field of L2 affects L1. Because of this property, it is good practice to always install new inductors in the same position as the defective ones to prevent any undesirable interactions. At RF frequencies, component placement and lead length can be critical for proper operation. Engineers often take advantage of component leads and position to provide additional circuit inductance and capacitance.

As the interaction between two coils can be beneficial in some applications, or to predict other interactions, it is desirable to be able to calculate the affect they will have on one another. That property is known as the *coefficient of coupling*. The coefficient of coupling is the fraction of total magnetic flux lines produced by both coils that are common to both. The degree of coupling is stated as a value of K. A low value of K, .1 to .3, is called *loose coupling* and indicates that relatively few magnetic flux lines are common. A high value of K, .7 to .9, is called *tight coupling* and is caused by the fact that a majority of the flux lines are common to both coils. To obtain unity coupling, or 1, where all lines are flux are shared, the coils are wound on a common iron core.

In summation, the coefficient of coupling (K) increases when the coils are placed closer together or wound on a common core. That ensures that magnetic lines of flux produced by each inductor pass through the other inductor. To lower K, the coils are located farther apart, installed perpendicular to one another, or are wound on separate cores. In most instances, K is the design consideration of the engineer, not the technician.

The effect that an inductor has on a circuit is called *inductive reactance* and is measured in ohms. Inductive reactance does provide opposition to circuit current flow. Inductive reactance, just like capacitive reactance, is affected by the frequency of the applied voltage. With inductive reactance, as the frequency of the applied voltage increases, the resulting opposition to current flow increases. The formula is as follows:

$$X(L) = 2(\pi)fL$$

L is the value of inductance in Henrys, f is the applied frequency in Hertz, and 2π is an electrical constant.

The quality, or Q, of an inductor is a very important characteristic, as it indicates the ability of a coil to develop self-induced voltage. The formula for inductive reactance includes the frequency of the current flowing through the circuit and the inductance value of the coil. The value of inductance possessed by a coil is the result of the number of turns a wire contains. Electrically, all wire has resistance, and resistance is a physical property that must be taken into consideration. The more turns a coil has, the longer the piece of wire used to fabricate it. The longer the length of wire, the greater the resistance that a coil has. Figure 2-14 consists of two diagrams. A coil is depicted on the bottom, and its actual electrical equivalent (consisting of an inductor in series with a resistor) is on the top. This diagram illustrates the concept that all inductors contain an inductive component and a resistive component. The inductive component is the result of the number of turns of wire and the core material. The resistive component is the physical length of the wire. As the two components have different electrical characteristics, energy in the circuit is affected by both. As illustrated in Figure 2-15, energy in the inductive component is

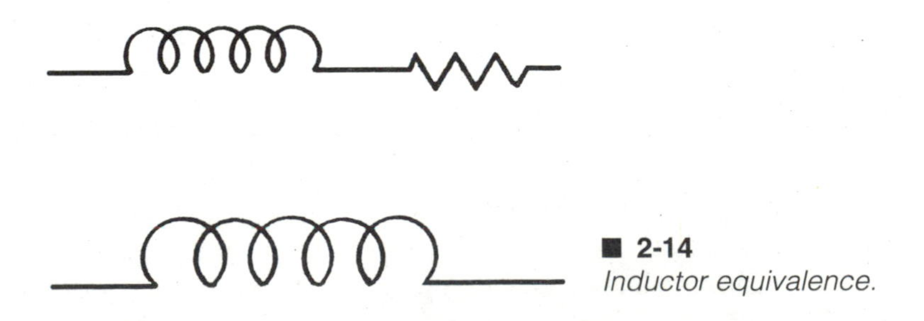

■ **2-14**
Inductor equivalence.

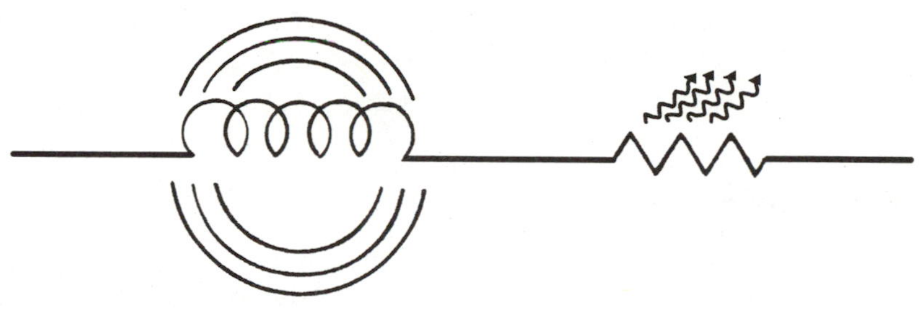

■ **2-15** *Inductor losses.*

stored in a magnetic field and is returned to the circuit when the field collapses, without significant loss. Energy utilized by the resistive component occurs in the form of heat resulting from current flow, and is lost. In effect, Q is an efficiency rating of an inductor. The higher the Q of a coil, the less energy is lost during the coil's operation; such energy is nonrecoverable.

There are several other factors that affect the overall Q of a given coil. At low frequencies, the resistance of a coil can be determined by using a multimeter to measure the wire resistance. As frequency increases, other losses come into consideration. Iron-core coils are constructed to use the magnetic properties of the metal to concentrate the strength of the magnetic lines of flux. At higher frequencies, greater losses occur due to eddy currents and hysteresis losses associated with the iron core. Any losses have the effect of increasing the resistance of the coil. An increase in resistance offsets some of the increase in inductance due to an increased operating frequency. For these reasons, iron-core coils are more common at lower operating frequencies.

At higher frequencies, air core coils are more common, as they avoid the iron-core losses. Even air-core coils have an additional loss at high frequencies: *skin effect*. Skin effect is the tendency of high-frequency currents to flow near the surface of a conductor. No matter how small a conductor, it does have a measurable cross-sectional area. At low frequencies, current flow is distributed across the entire cross section. As the frequency of the applied voltage increases, the resulting current flow tends to concentrate on the outer layers of material, or the *skin*, of the conductor. The far smaller currents that are found in the center of the conductor encounter more inductive reactance due to the concentrated lines of magnetic flux found there. The concentrated lines of flux have the effect of increasing the apparent resistance of the conductor. Hence, the same volume of electrons is moving through a smaller area of the conductor. Any increase in conductor resistance, either actual due to the type of material used in its construction, or apparent due to skin effect, decreases the inductance of the inductor.

Resonance

Resonance, a critical concept in communications theory, is defined in its simplest term as when the inductive reactance and capacitive reactive in a circuit are equal and cancel. Without resonance, many basic communications circuits could not exist, and the medium could not operate. Every facet of communications is based upon resonance and reactive components.

As was stated earlier, capacitive reactance and inductive reactance are both determined by the frequency of the applied voltage. To ease the following discussion, common and easy-to-work-with values for an inductor, capacitor, resistor, and applied frequency will be used. That will be followed by a quick review of the reactance formulas illustrates the concept of resonance and how it is derived. Figure 2-16 is a simple series LC circuit, with a variable AC voltage source. Included in the diagram is the graph of the applied AC source and the resulting impedances. As the frequency of the applied voltage increases, capacitive reactance decreases and inductive reactance increases. At only one frequency will the inductive reactance and capacitive reactance have an equal impedance. In the case of the values of 4 microfarads and 10 milliHenrys, the resonant frequency is 796 Hz, which is graphed in Figure 2-17. At that frequency the impedances of the two components are equal and opposite, and therefore cancel. Any impedance to circuit current flow is provided only by any resistors that are in the circuit.

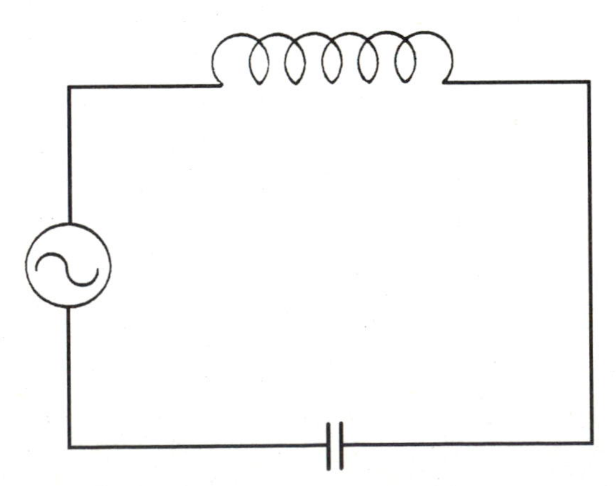

■ **2-16** *Simple LC circuit.*

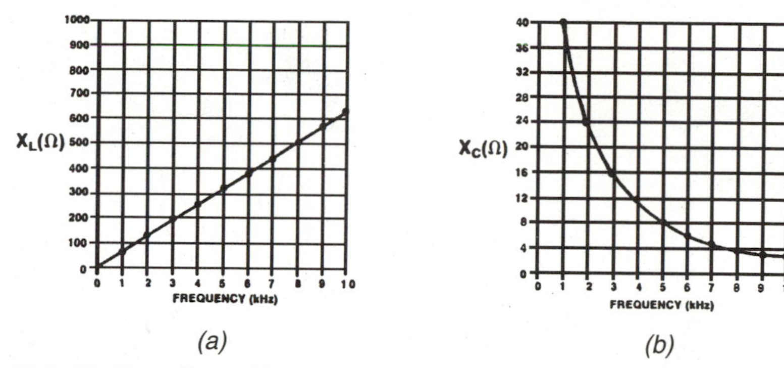

■ **2-17** *The effect of frequency on impedance.*

Figure 2-17b graphically shows the effect that the frequency of the applied voltage has on a 4-microfarad capacitor. Notice that at the low frequency of 1 kHz, the resulting impedance is 40 ohms. As the applied frequency increases, impedance decreases in an exponential manner, so that at an applied frequency of 5 kHz, impedance is now 8 ohms. Figure 2-17b is a graphic representation of the effect that frequency has on a 10 milliHenry inductor. By observing the graph, you will see that as applied frequency increases, resulting impedance increases. At a low frequency of 1 kHz, the resulting impedance is a low 80 ohms. As applied frequency increases, impedance increases in a constant manner, so that at an applied frequency of 5 kHz, impedance is now about 325 ohms.

As very few circuits contain just reactive components, RCL circuits are far more common. Figure 2-18 is a simple series RCL arrangement. When a series RCL circuit is at resonance, reactive impedance is minimum. As you remember, that is because at resonance capacitive reactance and inductive reactance are opposite and cancel one another. With the effects of the capacitance and inductance removed (at resonance), the resistance provides the only impedance to current flow. That results in maximum circuit current occurring at resonance.

Another important fact is the phase relationship that exists between the voltages and currents in a resonant circuit. Figure 2-19 depicts the relationships with vectors, or phasors. Notice that the phasors for X(*L*) and X(*C*) are 180 degrees out of phase. That is because with capacitive action, the current leads voltage. With inductive action, the voltage leads the current. The resistor phasor is identical to the voltage, indicating that it is in phase with the applied voltage and the resulting current.

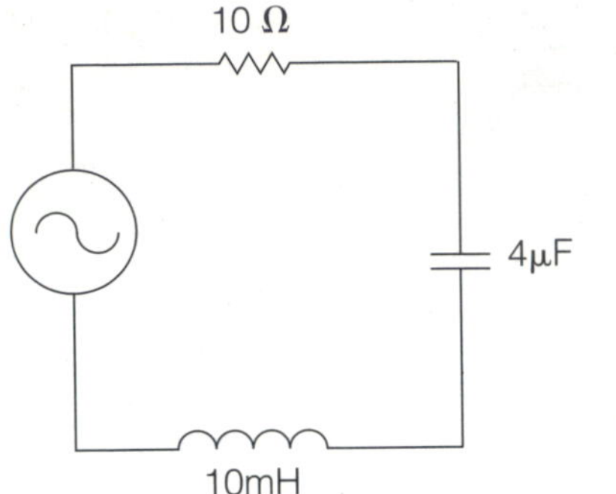

■ **2-18**
Representative RCL circuit.

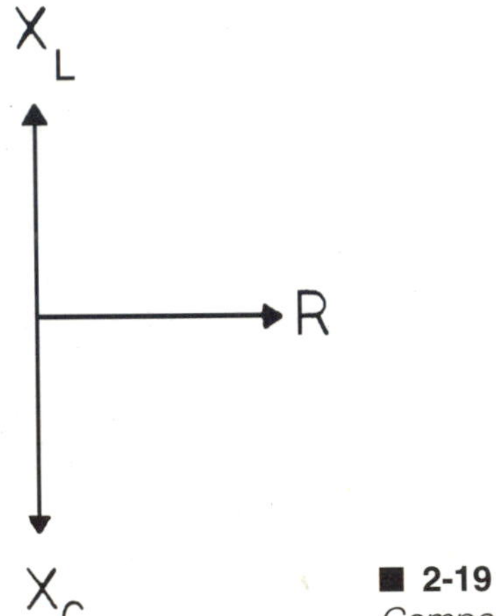

■ **2-19**
Comparison of RCL current phase vectors.

Frequency response is how a reactive circuit is influenced by a range of frequencies, as opposed to a single one as has been discussed so far. Normally it is portrayed as a graphic representation. Figure 2-20 is a graph that compares the impedance of a reactive circuit to a range of input frequencies. When the input frequency is low, impedance is high. That is because the circuit is capacitive. As the frequency increases, impedance decreases. That is because with the change in frequency, the circuit becomes less capacitive and more inductive. When impedance is at its lowest value, X(*L*) equals X(*C*) and they cancel out. As the frequency continues to increase, the circuit becomes more inductive and less capacitive. At the second maximum value of impedance, the circuit now appears all inductive, with a very small or practically nonexistent capacitive component.

Another important circuit characteristic is bandwidth. As you have seen, the effects of resonance are most evident on the exact resonant frequency. However, centered about the resonant frequency is a range of frequencies that exhibit virtually the same effects as resonance. Figure 2-21 illustrates the concept. In this example, the input frequency (dashed line) is compared to the circuit current (solid line). Notice that as the input frequency increases, impedance provided by the reactive components decreases, causing the resulting circuit current to increase. Maximum circuit current is achieved when impedance is at its minimum value, which corresponds to the resonant frequency. At that point, the impedances of the inductor and capacitor cancel one another out. This condition remains constant until the input frequency increases past resonance. At that point, circuit impedance begins to increase, which in turn decreases circuit current. The resonance effect covers the range of frequencies between the half-power points of the current waveform as illustrated in Figure 2-22. By definition, a half-power point is the frequency at which circuit current is 70.7% of maximum value, as measured at center frequency.

Bandwidth is an important circuit characteristic when working with reactive circuitry. It is the range of frequencies that are the performance limits for a filter, amplifier, or attenuator, as measured between the half-power points on a frequency response curve. The half-power point is the place on the frequency response curve where circuit current is 70.7% of maximum. The concept of bandwidth is depicted in Figure 2-23.

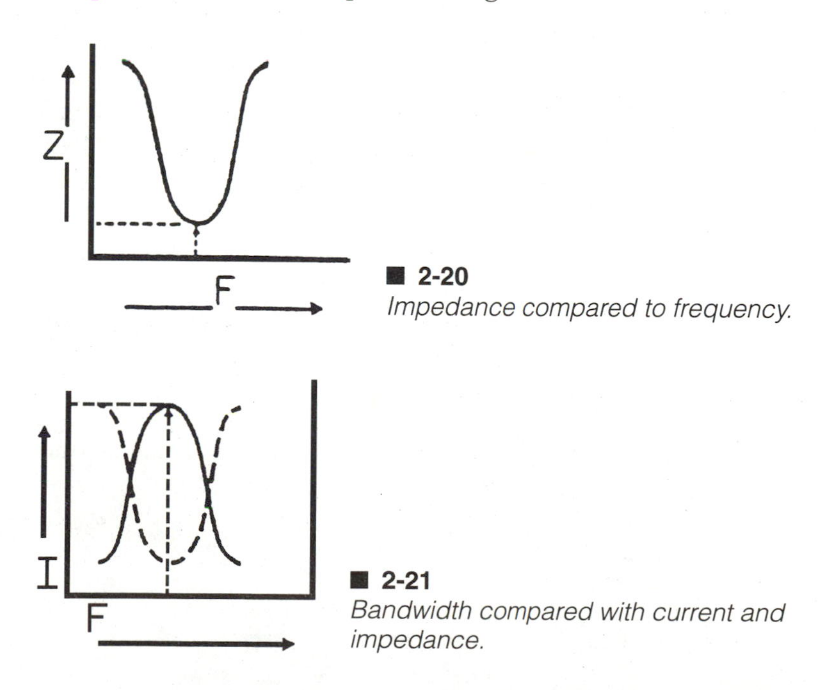

■ **2-20**
Impedance compared to frequency.

■ **2-21**
Bandwidth compared with current and impedance.

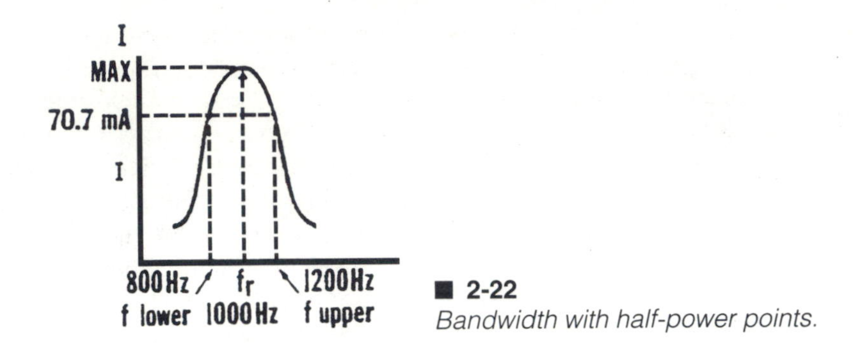

■ **2-22**
Bandwidth with half-power points.

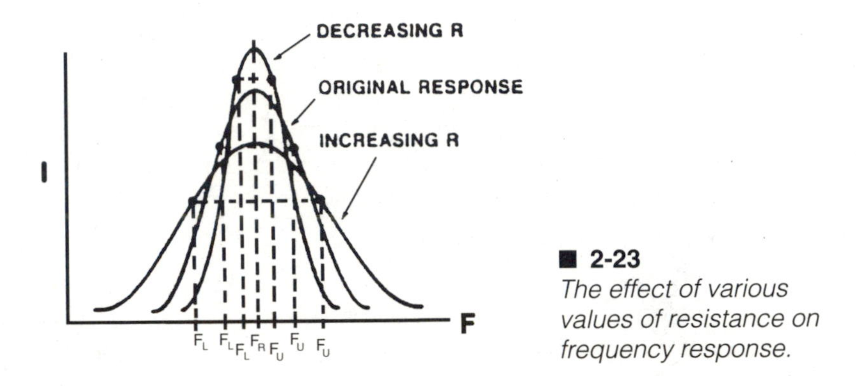

■ **2-23**
The effect of various values of resistance on frequency response.

The resonant frequency is marked as F_R. Notice that it is point of maximum circuit current, which also corresponds to minimum circuit impedance. The lower cutoff frequency is marked F_L, or frequency lower. It is the lowest input frequency that still results in a resonant effect. The highest frequency that causes a resonant effect is marked as F_U, or frequency upper. The band of frequencies that is between those two points is the circuit bandwidth. Circuit bandwidth can be calculated mathematically by subtracting the lower cutoff frequency from the upper cutoff frequency; for example:

$$BW = F_U - F_L$$

$$BW = 1200 \text{ Hz} - 800 \text{ Hz}$$

$$BW = 400 \text{ Hz}$$

In this example, the upper and lower cutoff points are 200 Hz from the resonant frequency. That is because in calculations, the components are considered to be perfect, resulting in an ideal symmetrical response curve. In actual circuits, due to differences in components, installation, and input frequency variations, resulting curves will be less than ideal. The formulas predict circuit behavior very accurately.

As was presented before, on a resonance curve, the point of 70.7% of maximum circuit current is also referred to as the half-power point. A good question is "How can 70.7% be called 'half-power?'" That is because it is the point on the response curve where real power dissipation is one-half of what it is at the resonant frequency. Real power is that which is dissipated by the resistive component of the circuit. The power used by the reactive components is classified as apparent power, as it is returned to the circuit (not lost). The series resonant circuit in Figure 2-24 represents how the half-power points are derived. In this circuit, the resistive component has a value of 10 ohms, which at resonance is the total impedance offered by the circuit to current flow. That is because at the resonant frequency the capacitive reactance and inductive reactance are equal and opposite, and cancel one another out. At any other frequency, total impedance will be greater, resulting in less total current. At the upper and lower cutoff points, due to the effects of capacitive and inductive reactance, total circuit current decreases to 70.7% of maximum. The formula for calculating power is:

$$\text{Power (Total)} = I(2)\,R$$

At the upper and lower cutoff points, you are squaring 70.7% of maximum circuit current. 70.7% times 70.7% equals 49.49%, or about one-half of maximum. That is how the half-power points were named.

There is also a relationship that exists between resonant frequency, bandwidth, and Q. The greater the Q of a reactive circuit, the sharper or narrower the response curve. If you know any two of the characteristic values, then the unknown can be easily found. The formula is:

$$\text{BW} = \frac{\text{F(r)}}{\text{Q}}$$

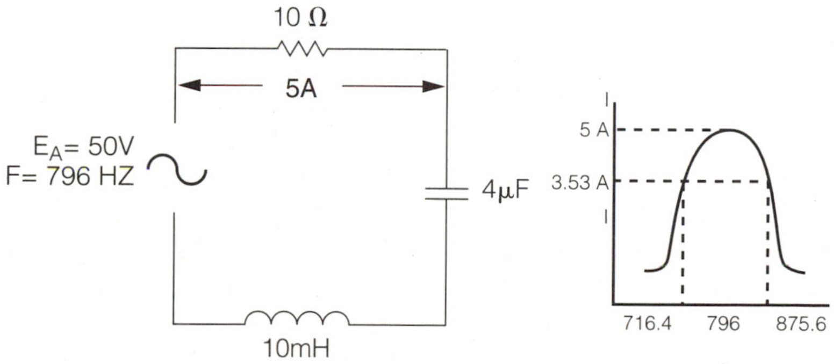

■ **2-24** *LCR circuit with frequency response curve.*

The concepts of Q and bandwidth are very important when dealing with the design and fabrication of various types of filters, as filters are one of the methods of obtaining desired circuit operation.

Frequency response is not a fixed value. By varying the component values, any required frequency response can be obtained. Due to the interrelated nature of resonant circuits, changes must be accomplished intelligently to prevent changing resonant frequency. Beginning with the bandwidth formula, you can see how all the various circuit parameters affect one another. If the values of the individual capacitor, inductor, or resistor are changed, then circuit Q, bandwidth, and frequency response of total circuit change as well.

Figure 2-25 shows several formulas that will be used in discussing how to change frequency response. Resonant frequency is determined by the capacitive and inductive values. If only one is changed, then the resonant frequency of the circuit is changed. However, resonant frequency is determined by the product of the values of the inductance and capacitance. As long as the product remains the same, then the resonant frequency will be held constant. As an example, if a circuit contains a 4 Henry inductor and a 4 farad capacitor, they will have a product of 16. Change the inductor value to 2 Henrys and the capacitor to 8 farads, then the product is still 16, which results in the same resonant frequency. Now, in order to keep the resonant frequency and product the same, the inductance has to decrease and the capacitance to increase. In actual circuits many combinations of inductance and capacitance have the same product, which results in the same resonant frequency.

By changing inductance and capacitance, resonant frequency can be held constant, but there is an effect on the frequency response of the circuit. Bandwidth is changed, affecting the curvature of the frequency response curve. When dealing with series resonant circuits, the bandwidth of a circuit is controlled by the L/C ratio, as illustrated in Figure 2-25. As an example, if the inductance is increased, the overall ratio is increased. Also referring to Figure 2-25, the increased inductance led to increased inductive reactance. Continuing on, an increased inductive reactance results in an increased Q. With Q increased, bandwidth for a given resonant frequency decreases, which results in a sharper or narrower frequency response curve.

Resonance is also found in parallel circuits constructed from reactive components. As resonance is defined as the frequency at

$$F_R = \frac{1}{2\pi\sqrt{LC}}$$

$$L/C\ RATIO = \frac{L}{C}$$

$$X_L = 2\pi FL$$

$$Q = \frac{X_L}{R}$$

$$BW = \frac{F_R}{Q}$$

■ **2-25**
Resonance formulas.

which the inductive reactance and capacitive reactance are equal and thus cancel, the components can be connected in series or parallel configuration for the same effects.

Figure 2-26 is a typical parallel RCL circuit. As can be seen, each branch consists of a single component—resistor, inductor, or capacitor. The entire arrangement is connected to an AC source. Figure 2-27 lists several important formulas associated with parallel RCL circuits. The formula for calculating resonant frequency should appear familiar, as it is also used in series resonant circuits. Total current is calculated using a more complex formula. As each branch feels the full source voltage, each one has its own value of current flow. The formula for determining total circuit current flow takes into account the characteristic differences between the components. Resistance is unaffected by frequency, therefore any voltage across it produces an in-phase current flow. The reactive branches (inductor and capacitor) are affected by frequency. If the applied voltage is at the resonant frequency of the components, then the current through them cancels. At that frequency, then, the only current flow in the circuit is that which flows through the resistor. At any other frequency, the circuit appears either capacitive or reactive. To find the reactive current, subtract the smaller one from the larger one. The resulting reactive current is squared, added to the square of the resistive current, and the square root of the sum is then taken.

Q is also an important consideration in parallel reactive circuits. When the circuit is at resonance, inductive reactance and capacitive reactance are equal. To find Q, divide the value of the resistive branch by either one. Figure 2-27 shows both.

An interesting characteristic associated with parallel reactive circuits at resonance is current magnification. This can be shown

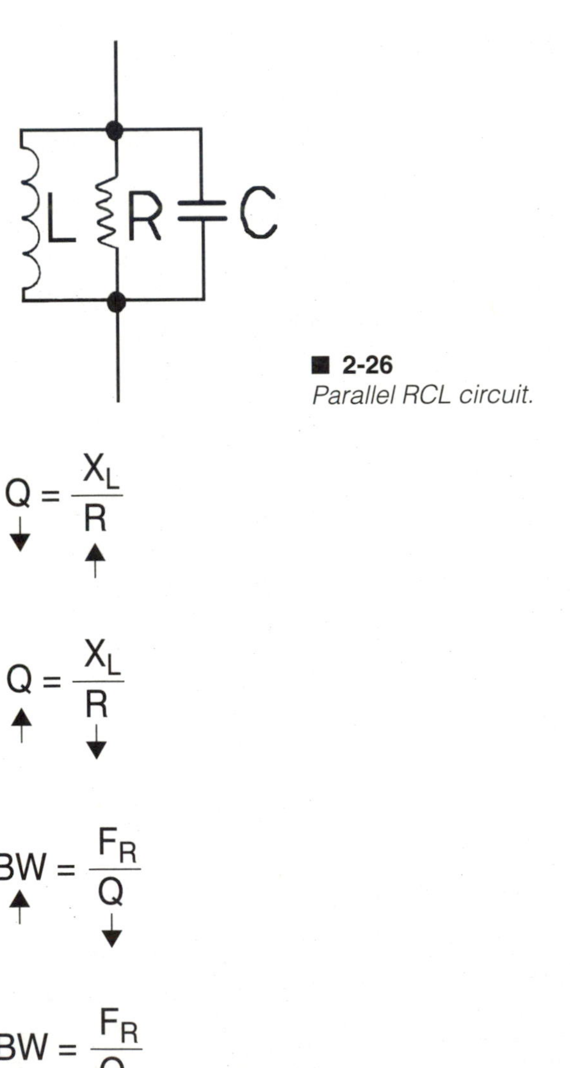

■ **2-26**
Parallel RCL circuit.

■ **2-27**
Formulas for parallel RCL circuits.

$$I_T = \sqrt{I_R^2 + (I_C - I_L)^2}$$

$$I_T = \sqrt{I_R^2 + O^2}$$

■ **2-28**
Current magnification formulas.

either mathematically or in a shop with typical test equipment. Figure 2-28 provides the formulas used in this discussion. There are two kinds of power in a reactive RCL circuit. One, *resistive power*, represents a loss due to heat caused by current flow through the resistive branch. That is known as P(R). The second, *apparent power*, P(X), represents the power of the reactive branches. As reactive components do not cause a power loss, apparent power is returned to the circuit. Circuit Q can be found by dividing the apparent power by the resistive power. As power is equal to the voltage applied to a circuit times the current through the circuit, those values can be substituted for power. By canceling the value for applied voltage as it appears on both sides of the =, you are left with just currents. At resonance, total current is equal to the current flowing through the resistive branch, so that can be substituted. Rearrange the formula by multiplying both sides by I(T), and the result is that the current flow through the inductor is equal to the Q of the circuit times the value for total current.

At resonance, capacitive and inductive current are equal. As Q can be greater than one, how do you explain that the reactive currents are greater than the total current? The answer is really quite easy and illustrated in Figure 2-29. Remember, in a capacitor, the current leads the voltage. In an inductor, the voltage leads the current. When connected in parallel, the two components become complementary and function in an interrelated fashion. As the capacitor charges, the magnetic field of the inductor collapses in an attempt to keep circuit current constant. That causes a decrease in voltage, which in turn causes the capacitor to discharge in an attempt to keep the voltage constant. This action occurs at the resonant frequency. The resulting current flow between the inductor and capacitor is known as *circulating current*. Although it is larger than the total current drawn at resonance, it is felt only between the inductor and capacitor. Because of how the two components function together, this frequency-sensitive arrangement is often called a *tank circuit*. By definition, a tank circuit is an arrangement capable for storing electrical energy of a band of frequencies.

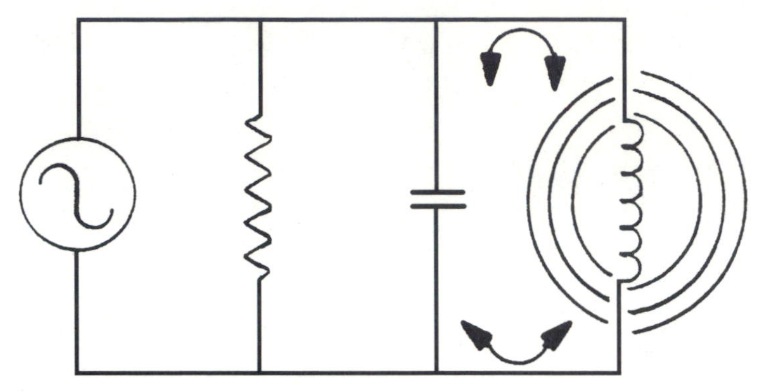

■ **2-29** *Representative tank circuit.*

Frequency response is also an important consideration in parallel reactive circuits. It is calculated exactly as was done for the series resonant circuits, as depicted in Figure 2-25. Frequency response can be modified by changing the L/C ratio. By examining the response curve in Figure 2-30, you will notice that it is the exact opposite of what happened in the series circuit. In a parallel resonant circuit, an increase in the L/C ratio causes a decrease in Q, a decrease in the sharpness of the response curve, and an increase in bandwidth. This reaction occurs in a parallel circuit because an increase in induction causes an increase in inductive reactance, which in turn decreases the inductive branch current, resulting in a decrease in the circuit Q.

Bandwidth is also affected by the value of resistance in the circuit. Although resistance is not affected by frequency, it has an impact on the response curve. Figure 2-31 illustrates the effect of changing circuit resistance on bandwidth. At resonance, if the resistance is decreased, total impedance decreases. A decrease in resistance with an unchanged inductive reactance causes circuit Q to decrease. A decrease in Q, in turn, causes overall bandwidth to increase.

Filters

A filter is a selective circuit that passes a desired frequency or band of frequencies while blocking or attenuating all others. The study of filters is a very interesting and involved area of electronics circuitry. Because of that, only the more common filter types will be presented here. The four most common types of electronic filters are the low-pass, high-pass, bandpass, and band-reject filters. Filters are very common in communications, as the field requires known, stable frequencies for proper operation. This section will begin with the low-pass filter.

A low-pass filter is a frequency-selective circuit that passes low frequencies while blocking higher ones. The definition of low and high depends on the application of the circuit. In a DC power supply, low frequency would be DC and high frequency would be 60 Hz. The function of the low-pass filter in this application would be to pass only the desired DC voltage and block the 60- or 120-cycle waveform left over from the diode rectification process. In another application, the desired low frequencies might be 50 KHz and below, while the undesired high frequencies would be 50.5 kHz to infinity. The simplest and most common low-pass filter is found in DC power supplies.

Figure 2-32 illustrates a basic DC power supply. AC is applied to the primary of T1. The function of T1 is to provide for isolation from the source and to either step up (increase) or step down (decrease) the AC voltage to the proper level required by the equipment. The diode CR1 rectifies, or converts, the AC to a pulsating DC by passing only the positive alternations of the AC signal. That is accomplished because it is forward-biased and conducts during

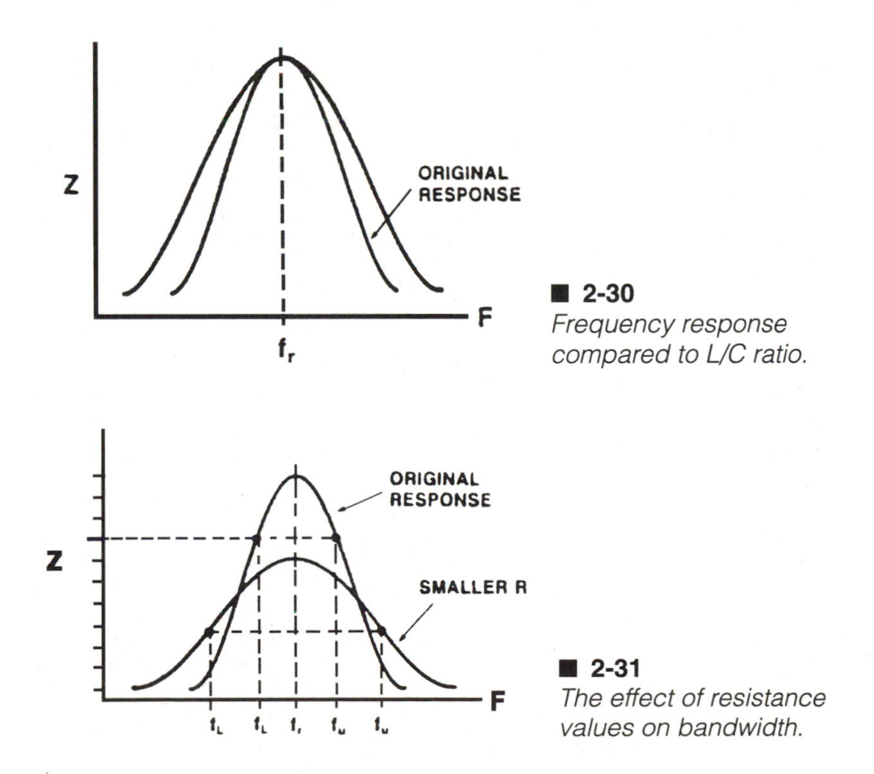

■ **2-30**
Frequency response compared to L/C ratio.

■ **2-31**
The effect of resistance values on bandwidth.

the positive alternations, and it is reverse-biased and cut off during the negative ones. Pulsating DC is not a suitable power supply for most electronic applications. The function of the filter, consisting of L1 and C1, is to smooth out the ripple left over from the rectification process. Inductors, through the expansion and collapsing of a magnetic field, attempt to keep the flow of current through the circuit a constant value. That has the effect of smoothing out the ripple from the diode. The function of the capacitor is to shunt any remaining ripple to ground. That is because a capacitor, by storing energy in the dielectric, prevents any rapid changes in voltage. This type of filter is often called an "L" filter due to the shape of the reactive components on the schematic. The resistor R(L) represents the load that uses the power supply output.

In constructing a power supply filter, the inductor must be large enough to provide a non-interrupted current flow. Any capacitors must have a reactance that is very low when compared with the inductors to provide a good path to ground for any ripple that passed through the inductor. In these applications, the inductors are usually iron-cored to provide the large magnetic field required to store the energy. Capacitors are typically electrolytic.

More sophisticated systems require ripple-free DC power supplies. In these instances, bridge rectifiers are used. A bridge provides a higher ripple frequency (120 Hz), which is easier to smooth out. Additional filtering is provided by adding components. Figure 2-33 illustrates a low-pass "Pi" filter. It so named because the two capacitors connected by the inductor resemble the shape of the Greek letter "Pi" (π). In this type of filter, most of the filtering action is accomplished by the first capacitor by shunting the AC ripple to ground. The inductor provides a smoothing action by impeding any circuit current variations. Any ripple left after passing through the inductor is shunted to ground by the second

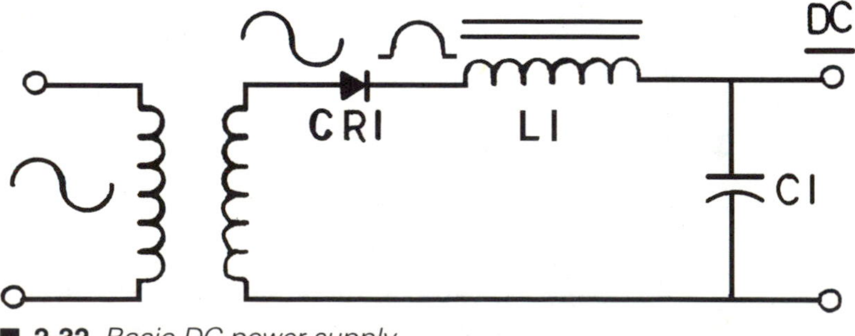

■ **2-32** *Basic DC power supply.*

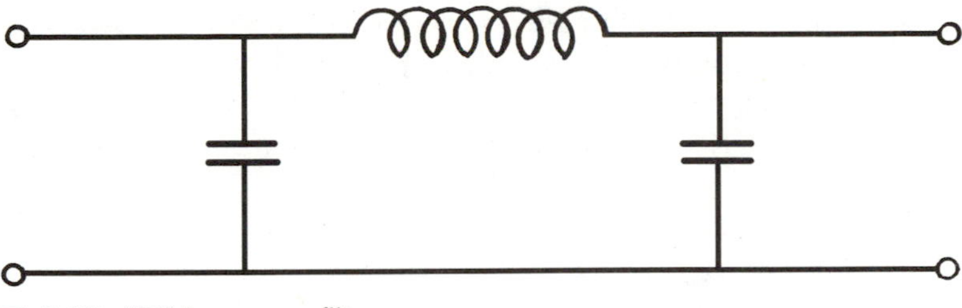

■ **2-33** *"Pi" low-pass filter.*

capacitor. That happens because the capacitor provides a low-impedance path to ground for any AC variations.

Figure 2-34 illustrates how additional "L" filter sections can dramatically reduce ripple. In Figure 2-34a, only one section is used. The reduction in ripple frequency obtained by the filter network is the reciprocal of 2π times the desired resonant frequency, and the product of the values of capacitance and inductance. As can be seen, each additional "L" section increases the ripple reduction factor dramatically.

How does a filter actually eliminate, or shunt, the undesirable higher frequencies to ground? The key is the value of the capacitance. It is selected so that at the higher frequencies it appears more and more as a short circuit. In that manner, any undesirable higher frequencies are shunted to ground.

As can be seen from Figure 2-33, a low-pass filter looks very similar to one used in a power supply. The main difference you may notice is that in this type an iron-core inductor is not used, as the current requirements are less. In this diagram, the first filter is a single section "L" filter with the resulting output. The appearance of adding a second "L" section is illustrated in Figure 2-34. As a result of the additional sections, the cutoff between the bandpass frequencies and the bandstop frequencies steepens. A variation of the PI filter is the "T," which is shown is Figure 2-35. As can be seen, it is constructed from two inductors and a capacitor.

Several types of high-pass filters are illustrated in Figure 2-36. If you compare them to Figures 2-33, 2-34, and 2-35, you will notice that these circuits are the exact opposite. The function of a high-pass filter is to block low frequencies and pass higher ones. As with the low-pass filters, the bandstop is an arbitrary frequency based on the function of the equipment in which the filter is used. In this arrangement, the inductors appear as a short to the low frequencies, shunting the undesired lower frequencies to ground,

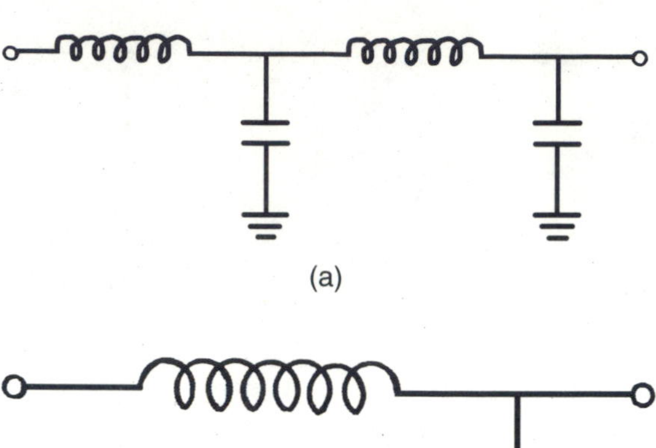

(a)

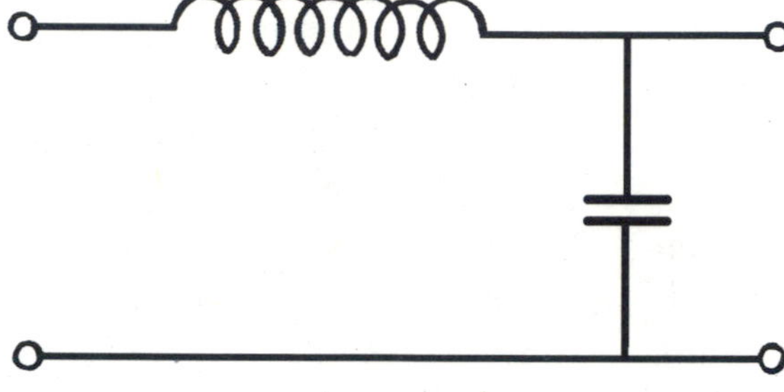

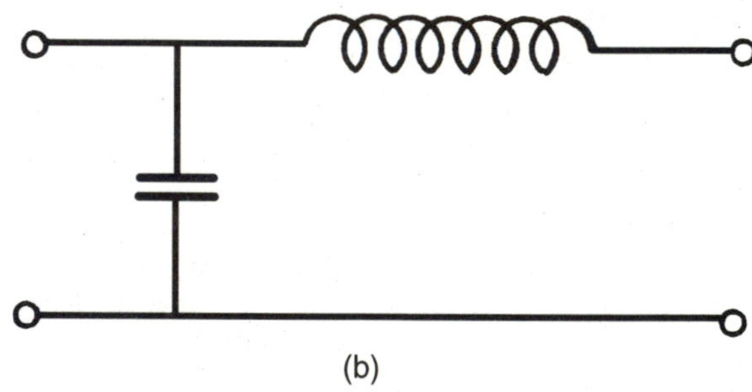

(b)

■ **2-34** *"L" low-pass filter.*

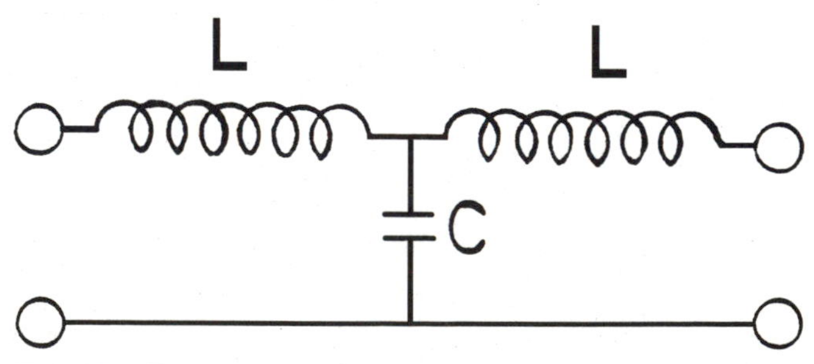

■ **2-35** *"T" low-pass filter.*

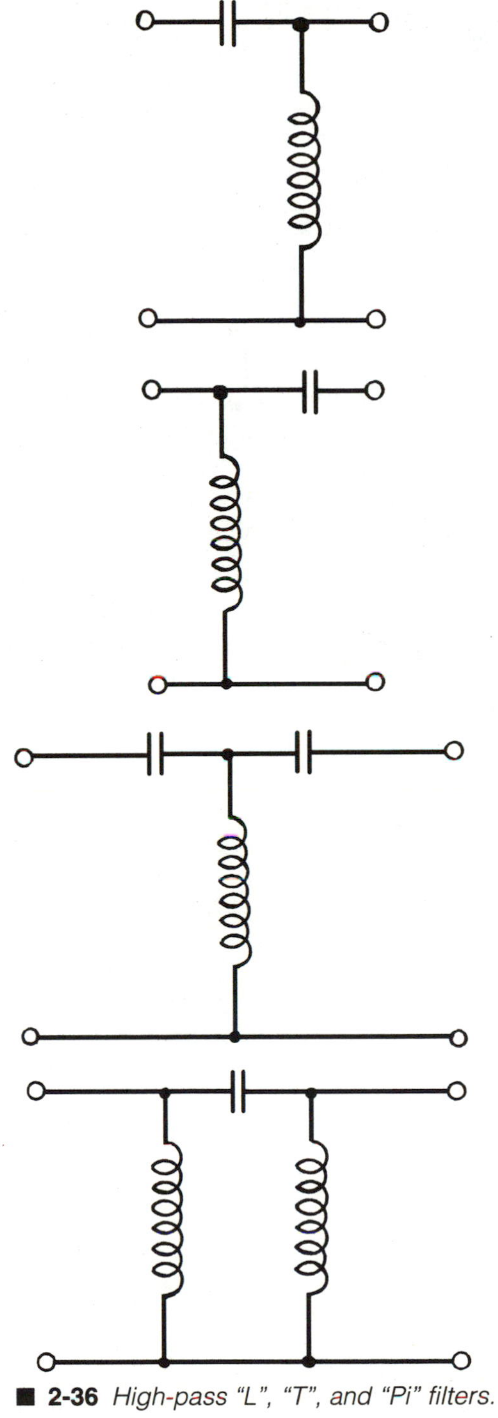

■ **2-36** *High-pass "L", "T", and "Pi" filters.*

and passing the higher ones. The function of the capacitor(s) is to represent a low impedance to high frequencies and a high impedance to low frequencies. As with the low-pass filter, "L," "Pi," and "T" filter arrangements are found as well. Also, filtering action can be improved through the addition of more filter sections. With each added filter section, the bandstop curve becomes sharper.

A bandpass filter, as shown in Figure 2-37, is very common in communications applications. This type of arrangement is designed to pass a range of designated frequencies and block any that are above or below cutoff. A good example of a bandpass filter is the band selector on an FM radio. In the United States, the FM broadcast band is 87.5 MHz to 108 MHz. Below the low cutoff of 87.5 MHz is the TV audio broadcast for channel 6. Above the high cutoff is the aircraft communications band. In actuality, then, the FM band selector on an FM radio blocks all frequencies that are outside the range of 87.5 MHz to 108 MHz.

A fourth type of filter is the bandstop, or band elimination configuration. The function of this circuit is to pass all frequencies, with the exception of a band that is not desired. A good example of its use is to eliminate interference picked up by a radio or television

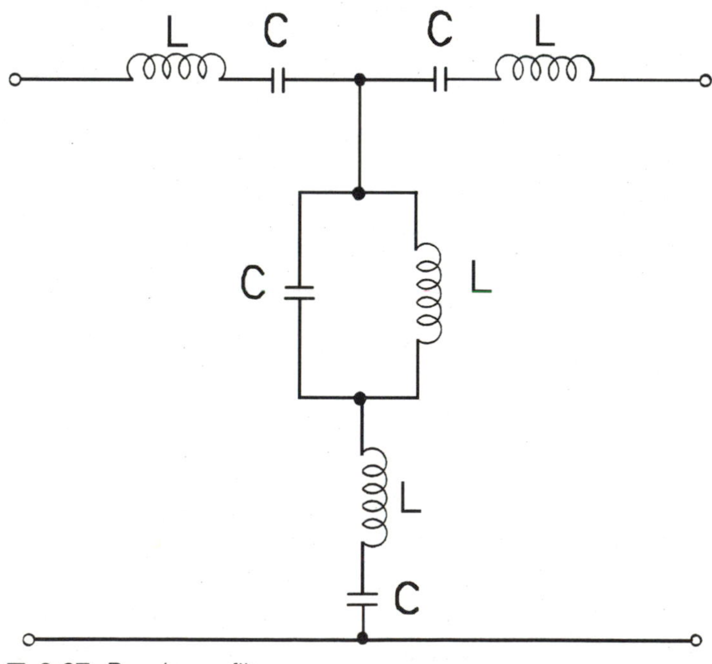

■ **2-37** *Bandpass filter.*

from a nearby transmitter. When used in that fashion it is known as a trap. A representative bandstop filter is shown is Figure 2-38.

This small chapter has covered a great deal of important information that is critical to a good foundation for the study of communications theory. This information will be built upon in following chapters, with the ultimate goal of an appreciation and understanding of the complex field that communications has become.

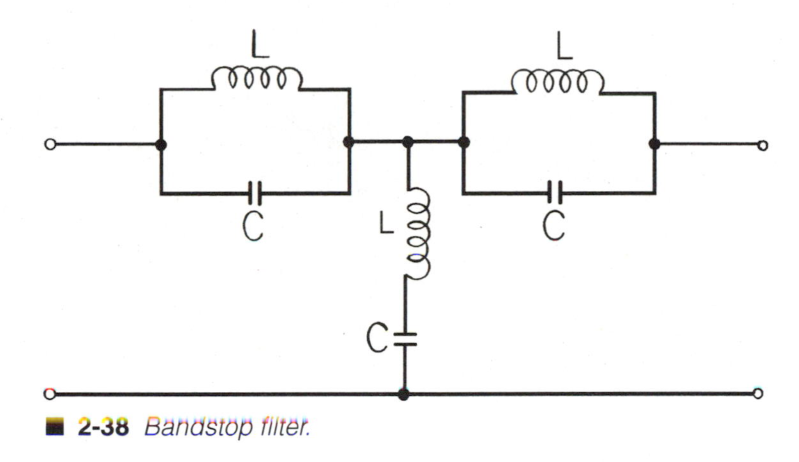

■ **2-38** *Bandstop filter.*

3

Electronic Amplifiers and Circuits

ALTHOUGH COMMUNICATION WAS POSSIBLE USING SPARK gap transmitters and coherer receivers, these devices had severe limitations. Long-range communications links were more of an art than a science. Data transmission was slow and was limited to some type of simple code, such as Morse code. To achieve full potential, the field required numerous technological breakthroughs. New components had to be developed and circuits designed to exploit their capabilities. Modern electronics is dependent upon many circuits, such as power supplies, oscillators, modulators, demodulators, and amplifiers. Possibly the most important circuit design to be developed was the one that performed amplification. Amplification is the process of increasing the strength of an electronic signal. It can be in the form of increasing the signal's current, voltage, or power level. An ideal amplifier performs the task without distortion.

The Vacuum Tube

The first amplifying device to be developed was the thermionic valve, or as we know it, the vacuum tube. Although for many years it was the mainstay of electronics, it has been superseded in many applications by the transistor and integrated circuit. Knowledge of its operation is still important, as it is still used in older equipment and high-powered applications.

Figure 3-1 is the schematic symbol for a simple thermionic electronic vacuum tube, the diode. It functions somewhat like a solid-state PN junction diode, with which you might be familiar. Vacuum tubes require two physical phenomena for proper operation. The first is the action of certain materials emitting large quantities of electrons when subjected to heat. The second is the ability to control the electron flow through the vacuum tube by the application of electric and magnetic fields. The major elements of any tube are the *cathode* (one or more electrodes to collect and control electrons) and an evacuated glass envelope. The function of the cathode is to emit electrons (current carriers), which are received by the *anode* or *plate*. Depending upon the tube design, additional electrodes or *grids* are added to control electron flow. A glass envelope is required to provide an evacuated environment for two very important reasons. The first is to prevent the heaters, metal cathode, and electrodes from rapid burnout in the presence of oxygen. The second is that an overabundance of matter within the envelope would allow for excessive electron dissipation in the forms of heat and ionization. Cathodes are fabricated from materials that emit a large quantity of electrons when heated. Anodes under normal operating conditions do not emit electrons.

Electron flow in a vacuum tube is from the cathode to the anode. To allow for the internal movement of electrons, free space within the envelope of a tube is not a perfect vacuum. That allows for

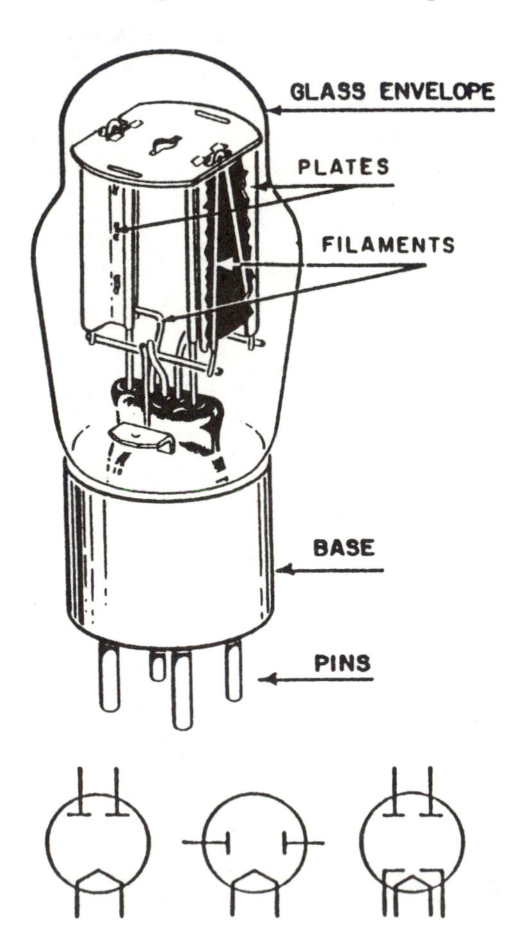

■ **3-1**
Vacuum tube construction and schematic symbol.

some matter to remain inside the tube. With the application of heat, the cathode begins to emit electrons. The phenomenon of a heated body emitting electrons is known as the Edison Effect. The electrons emitted by the heated cathode form a cloud surrounding it, called the space charge. Vacuum tube operation does not begin until an external difference in potential is applied across the tube. The application of external voltages to allow for proper operation is known as bias. In vacuum tubes, bias is considered to be an electrical force that is applied to a device to establish a desired reference level for operation. Proper bias for vacuum tube operation is when the anode is positive in respect to the cathode.

When a positive voltage is applied to the anode and a negative one to the cathode, vacuum tube operation begins. With the application of the voltages, an electric field exists between the two electrodes. The electrons forming the space cloud around the cathode are repelled by the negative potential toward the anode. As a positive potential is on the anode, or plate, the electrons are attracted to it, traveling through the tube.

Cathodes can be constructed in three different ways. The first is called a *cold cathode tube*. In this device, there is no heater. The space charge is formed from the high electron emission that naturally occurs from the cathode material. Conduction occurs when the tube is properly biased with the cathode negative and the anode positive. Usually, this type of tube has the envelope filled with an inert gas, such as argon, krypton, neon, or xenon. In this type of tube ionization is a desired characteristic, as the device is used for voltage regulation. A voltage regulator is used to maintain the DC voltage output of a power supply regardless of variations in line voltage or load.

The other two types of cathodes are heated, and the designs are illustrated in Figure 3-2. When a vacuum tube uses a heated cathode it is considered to be based on thermionic emission. Thermionic cathodes can be either directly or indirectly heated. In the directly heated cathode, the current for heating purposes flows through the emitting material of the cathode. This type of heater is commonly called a filament and resembles a very thin piece of wire. A directly heated cathode is capable of emitting large amounts of energy and is relatively efficient. A disadvantage is the small physical size of the filament wire, which allows the filament temperature to fluctuate with changes in current flow. When an AC voltage is used as the heater voltage, an undesirable hum can be introduced into the circuit. This is not normally a problem unless very low-level signals are being amplified by the circuit. To compensate for the noise problem, an indirectly heated cathode can be used. With this design, a relatively constant rate of electron emission can be obtained from the cathode in conditions of fluctuating current. As can be seen from the diagram, this type of heater is fabricated in the form of a cylinder. The heater is located in the center of the cylinder and is electrically insulated from the cathode electrode.

The simplest of tubes is the vacuum tube diode, illustrated in Figure 3-3. As can be readily determined, the actual physical construction of the tube bears little resemblance to the schematic symbol. At the center of the tube structure is the heater element. If the tube is a directly heated version as depicted in Figure 3-3a, the cathode and heater are one element. In the indirectly heated type, the heater element is surrounded by the cathode structure, as seen in Figure 3-3b. Regardless of heater type, the plate electrode surrounds the cathode and heater. Often, to save space, two diodes are mounted in the same glass envelope. In this manner, two separate components are mounted in virtually the same space at less overall cost. This design technique has been used in all types of vacuum tubes for many years. A two-diode tube is known as a duo-diode. Other, more complex tubes, such as the triode, pentode, and tetrode are also found in dual designs.

The weak link in the design of electronic vacuum tubes is the heater. Due to the tendency of tube filaments and heaters to fail by burning open, a vacuum tube has a much shorter operational life than other electronic components, such as resistors, capacitors, inductors, transformers, and semiconductor devices. To facilitate rapid replacement of electronic tubes, most are constructed with a base that serves as a plug. Tube input and output points are in the form of pins on the base of the tube. Tubes are inserted in a socket mounted on the chassis of the equipment.

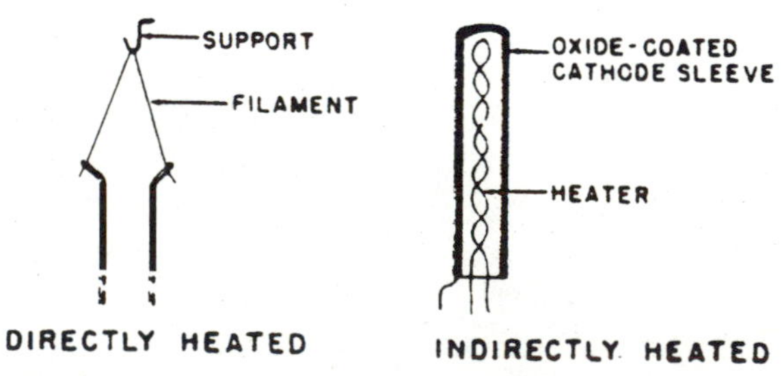

■ **3-2** *Vacuum tube heater design.*

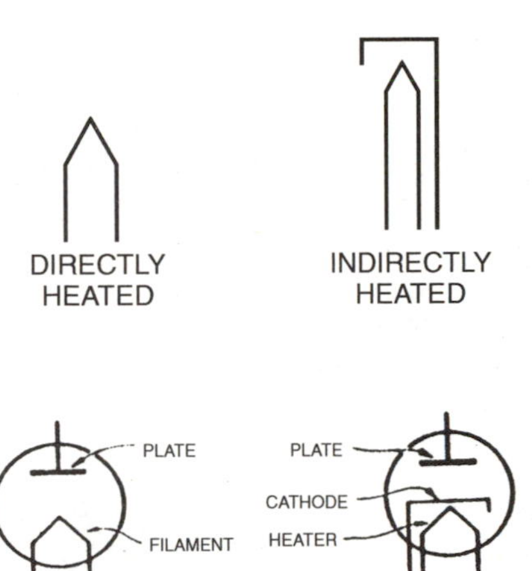

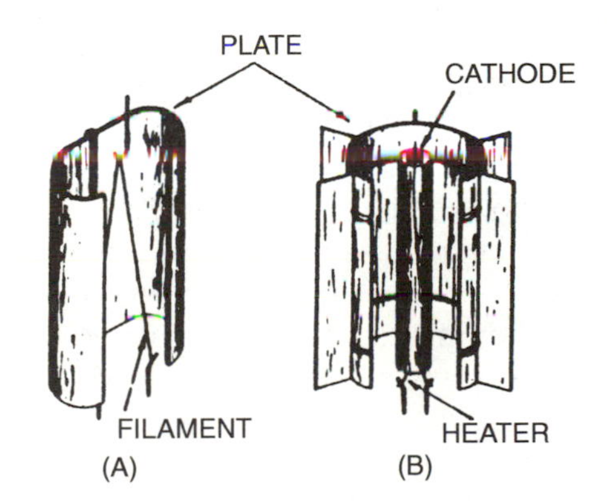

■ **3-3**
Vacuum tube diode construction and schematic symbol.

Diode operation occurs when the proper bias voltages are applied across the tube: plate positive, cathode negative. That allows electron current to flow from the cathode to the anode. The magnitude of current flow is controlled by the amplitude of the bias voltage and the internal resistance of the tube. Figure 3-4 compares the value of bias voltage across a diode vacuum tube to the resulting level of current flow. The operational characteristics of any dynamic device (tube or semiconductor) can be obtained by comparing the bias voltage to the resulting current flow through the device.

As the voltage across the tube increases, current through the tube increases. The graphing begins at the zero volt point and is slowly increased to maximum. As the voltage increases, current through the tube increases. Notice the shape of the resulting

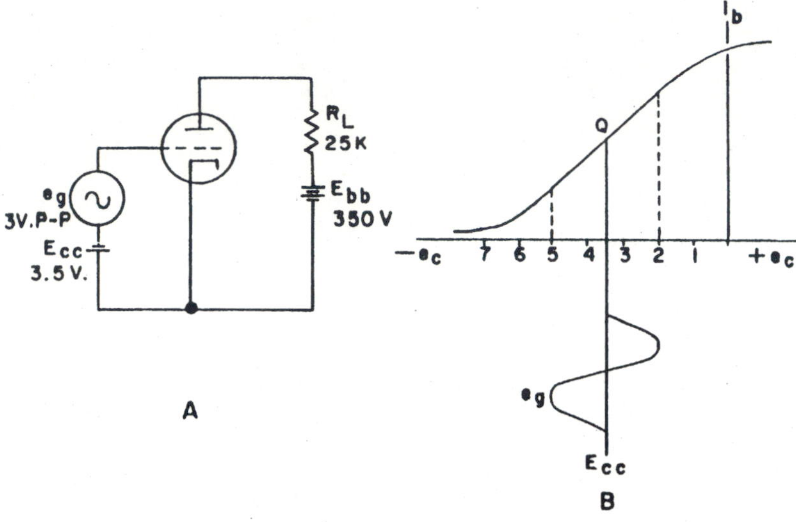

■ **3-4** *Tube plate voltage compared to tube plate current.*

voltage-versus-current curve. As can be readily seen, the curve is not linear throughout its entire range. Initially, a large increase in voltage causes a small increase in current flow as potential across the tube must overcome the internal resistance of the tube and begin attracting electrons to the plate. Then the curve is linear. As voltage across the tube increases, electron current flow through it increases proportionally. Finally, the curve becomes nonlinear again. At this point, further increases in potential across the tube do not bring about an increase in current flow. At this point, the plate is attracting all of the electrons that the cathode is capable of emitting without damage. This condition is known as *saturation* and is a characteristic of tubes and semiconductors. The internal resistance of a tube can be calculated by dividing the voltage across the tube by the resulting current flow through the tube.

All electronic components have certain ratings of current, voltage, and power that should not be exceeded to prevent damage. The most critical diode ratings are plate dissipation, maximum average current, maximum peak plate current, and peak inverse voltage. Plate dissipation, in the form of heat, is the maximum average power that the plate can dissipate without damage. The maximum average current is the greatest average value of current flow through the tube that it can handle without internal breakdown. It is affected by the tube's plate dissipation figure. Maximum peak plate current is the highest instantaneous plate current that a tube can pass repeatedly. The final rating of the peak inverse voltage (PIV) is the greatest instantaneous plate voltage that a tube can tolerate that is in the opposite polarity; that is, with the plate negative in relation to the cathode.

Diodes are very useful electronic components, as the devices are used as rectifiers, as clippers, as clampers, for signal mixing, and for detection. Figure 3-5 illustrates the most common use: voltage rectification. The figure consists of three schematics of the most common power supplies. Figure 3-5a is a half-wave rectifier, which has the function of converting the input AC power into the DC voltages used by most electronic equipment. The half-wave rectifier is forward-biased and conducting for the positive alternation of the input AC and cutoff, and not conducting for the negative alternation. As this type of circuit rectifies only half of the input AC into DC, it is very inefficient. Figure 3-5b is a full-wave rectifier. This type of circuit converts both alternations of input AC into DC. That is accomplished because the transformer secondary winding is center-tapped. That provides one-half of the input AC to each diode. Although it is more efficient, it provides a lower level of DC voltage. The most efficient rectifier circuit is the full-wave bridge illustrated in Figure 3-5c. This circuit uses both input AC alternations. Notice

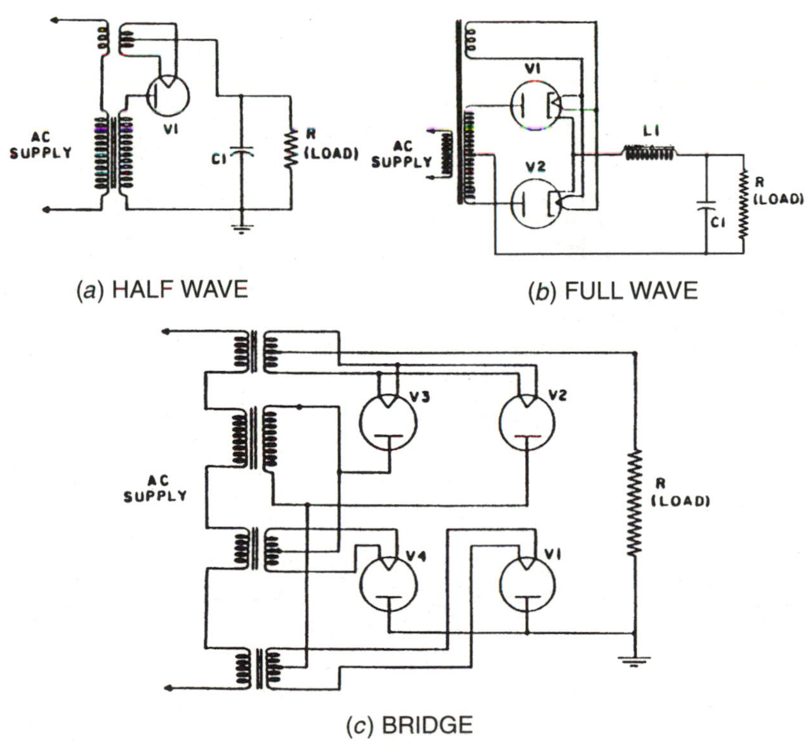

■ **3-5** *Vacuum tube rectifier circuits.*

that the transformer secondary is not center-tapped. That action provides full input voltage across each diode, resulting in a higher DC output. To complete any rectifier circuit, filters are added to smooth the output to a pure DC.

As important as the vacuum tube diodes are, electronic communications required amplifiers to expand beyond the spark gap transmitter stage. By adding a third electrode to a tube, amplification was possible. Figure 3-6 is the schematic representation of a triode and a drawing of its physical construction. The device is nothing more than a diode with a third electrode added between the cathode and plate. Any additional grids that are added to a tube are indicated as a dashed line. As can be seen from Figure 3-7, the control grid consists of very fine wire wound on support rods. Two of the most common designs are the elliptical helix on the left and the ladder type on the right. The wire selected for grid construction has a very low level of electron emissions. The function of any grid is to control electron movement, not initiate it. The spacing of grid wire is very large when compared to the size of an electron. That allows an unobstructed path of current flow from the cathode to the grid.

A control grid has only one function in a vacuum tube—to control the flow of plate current. Figure 3-8 illustrates the electric field that exists within a triode. In this example, the plate is positive in respect to the cathode. The control grid and cathode are at the same potential. Due to the difference in potential across the triode, electric field lines of force originate at the plate and terminate on the cathode. Some of the lines of force terminate on the wires of the control grid. As the grid consists of widely spaced wires, only a few lines of force terminate there.

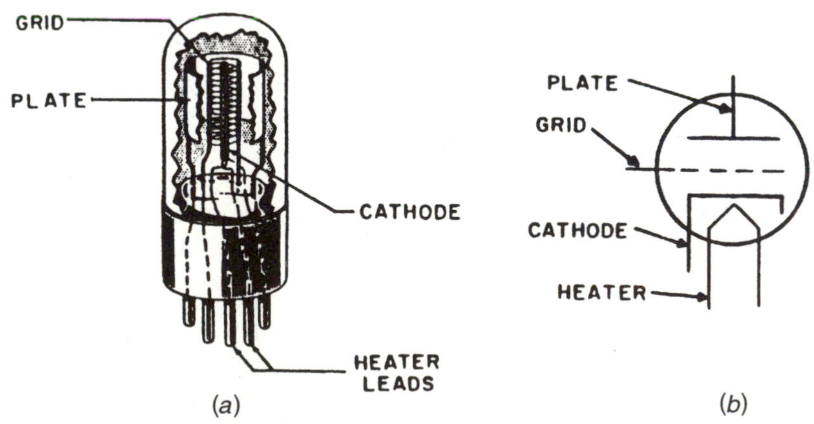

■ **3-6** *(a) Triode construction and (b) schematic symbol.*

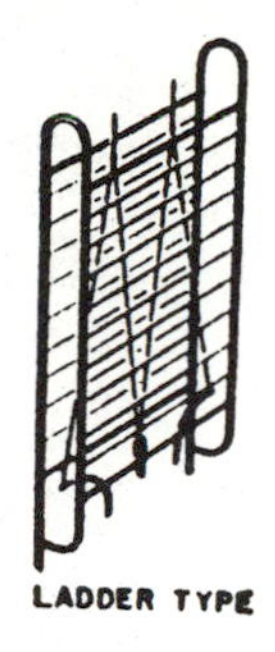

■ **3-7**
Grid construction.

The function of the control grid is to change the electric field in proximity to the cathode, which in turn controls the number of electrons that leave the cathode and reach the plate as plate current. In this manner, the plate current (and ultimately the power output of the tube) can be controlled by a very small grid current. If the grid is made negative with respect to the cathode, many more lines of force are terminated there, rather than the cathode. As the grid becomes more negative with respect to the cathode, it terminates an increasing number of lines of force. If the potential continues to increase in the negative direction, the grid can be made so negative that all lines of force are terminated there, rather than the cathode. In this condition, as the tube has no electron current flow (or plate current), it is said to be cut off.

Grid construction and placement is not haphazard, as it is a very important design consideration. Tube operational characteristics are a function of the internal construction of the device. Electrode spacing, internal location, and shape all contribute to the electrical characteristics of a vacuum tube. Engineers use the factors to predict with a great deal of accuracy the operational characteristics of a tube design with various cathode, control grid, and plate voltages. These characteristics are known as constants. The three main ones are *AC plate resistance*, *amplification factor*, and *transconductance*.

AC plate resistance is the opposition to AC electron current flow that the tube processes. The amplification factor is the effectiveness of the control grid as compared to that of the plate in controlling plate current. As this is an important design consideration, tubes are grouped according to amplification ability. A tube's amplification factor is expressed as μ. The three general classes of amplification are low μ, medium μ, and high μ. A low medium tube

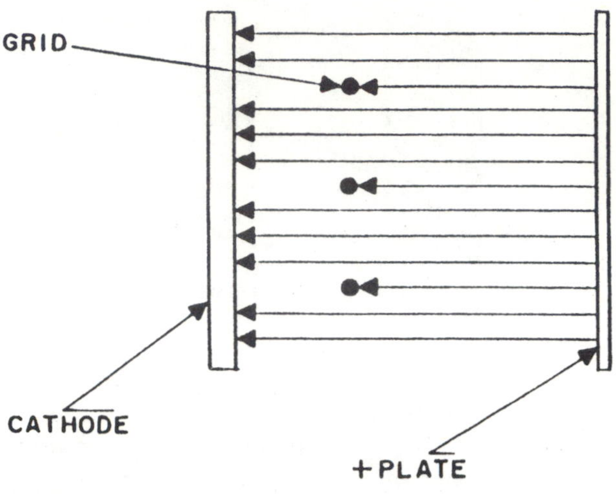

■ **3-8** *Triode electric field.*

has an amplification factor of less than 10, a medium μ from 10 to 50, and a high μ tube is greater than 50.

The final characteristic is transconductance, and it is the ratio of change in plate current with respect to a change in grid voltage. For the figure to be valid, the plate voltage must be a constant value. The resulting value is expressed as *mho*. This is a very important characteristic, as it is a method of comparing similar tubes to ascertain overall performance. The higher the transconductance of a tube, the greater the signal output that it can provide. The amplification factor of a tube is a relatively constant value, as it is determined by the physical geometry of a tube regardless of applied operating voltages on the plate and grid. Plate resistance varies based upon the applied operating voltages. The points where this is most evident are when a high negative bias is applied to the gird, a low plate voltage, or both. The amplifications factor can be increased to a higher level by increasing the separation between the plate and grid. As a drawback, that also increases the AC plate resistance. Transconductance varies inversely with the AC plate resistance. As can be seen from the formulas, the three tube constants are interrelated and are a concern of engineers.

The primary function of a triode is to amplify. For this important action to occur, the tube requires external support components. The best way to visualize how any amplifier functions is to visualize it as a voltage divider network as shown in Figure 3-9. The amplifier, either a tube or a transistor, is represented by a variable resistor connected in series with a load resistor. If the variable resistor has a low resistance, most of the voltage is dropped over the load resistance. Change the variable resistor to a higher resistance, then most of the voltage is dropped over it. The setting of the variable resistor is determined by the grid (or bias) voltage.

The simplified schematic illustrated in Figure 3-10 depicts a simple triode amplifier and how it would function. The high positive plate voltage is supplied by the source E(bb). E(cc) is the negative bias voltage applied to the grid. The function of the load resistor R(L) is to develop the output signal. This happens because 100% of the plate current flows through the load resistor, which results in a voltage drop over the resistor. If the tube is cutoff and not conducting, the voltage felt on the plate will be source voltage. That is because the tube will appear electrically as an open circuit. If further increases in bias cannot increase current flow through the tube, it is conducting as hard as it can and is in the state of saturation. The resulting very high plate current causes a large voltage drop over the load resistor. In this instance plate voltage will be low, as most of the voltage is dropped over the resistor. In this manner, a very small current variation on the grid controls a much greater plate voltage.

Amplifier operation is as follows and is illustrated in Figure 3-10. All amplifiers have a static or quiescent state. That is circuit operation with no signal applied. The external support components set the quiescent state through bias on the tube. In this case, static operation results in 150 volts output from the amplifier. A small AC input signal is applied to the grid. As it swings positive, at some

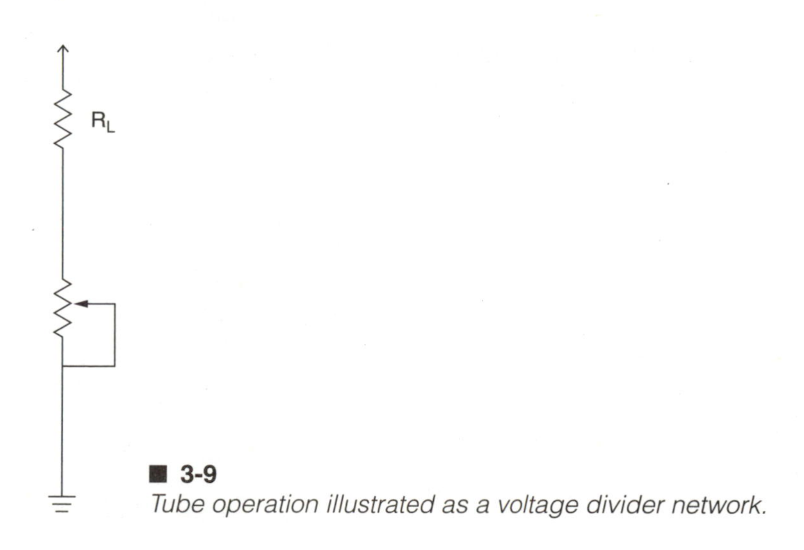

■ **3-9**
Tube operation illustrated as a voltage divider network.

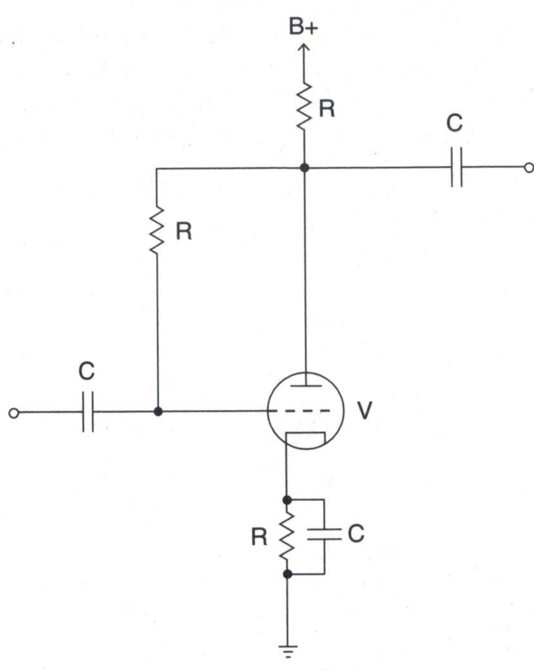

■ **3-10** *Simple vacuum tube amplifier.*

point it is more positive than the static DC bias on the grid. That action causes electron current flow to increase through the tube. As the plate current has increased, the current flow through the load resistor increases. An increased voltage drop over R(L) leaves less voltage to be felt on the plate. Therefore, the output voltage of the triode decreases, or swings negative. As the input signal on the grid decreases, the output of the tube continues to decrease until plate current is maximum and saturation is reached. As the input signal voltage then swings negative, plate current begins to decrease. That is because the lines of force in the field are now beginning to terminate at the grid rather than the cathode. That reduces current flow through the tube. The decreased plate current decreases the voltage drop over the load resistor, leaving more voltage to be felt on the plate. If the input signal continues to swing negative, eventually all current flow through the tube stops, as it has reached cutoff. As can be seen, a small input signal has controlled a much larger output signal, resulting in amplification. Figure 3-11 compares the signal associated with a triode amplifier. This clearly illustrates how a small input voltage on the grid has a direct affect on the plate current. As plate current changes, the voltage drop over the load resistor changes. The more plate current, the greater the voltage drop over the load resistor, the lower the voltage felt on the plate, and hence the lower the output of the amplifier. By comparing the small input signal with the much larger output signal, you can see that first, the signal has been amplified; second, it has been inverted.

As with the diode, the triode also is not linear throughout its entire range of operation. Normally amplifiers are not operated close to the cutoff and saturation points so as to avoid distortion. A desirable characteristic of any amplifier is fidelity—the faithful reproduction of the input signal. A good example of a distorted signal is when a stereo amplifier has the volume set too high. The resulting sound is noisy and difficult to understand. Bias is used to set the static point in the linear region. External component

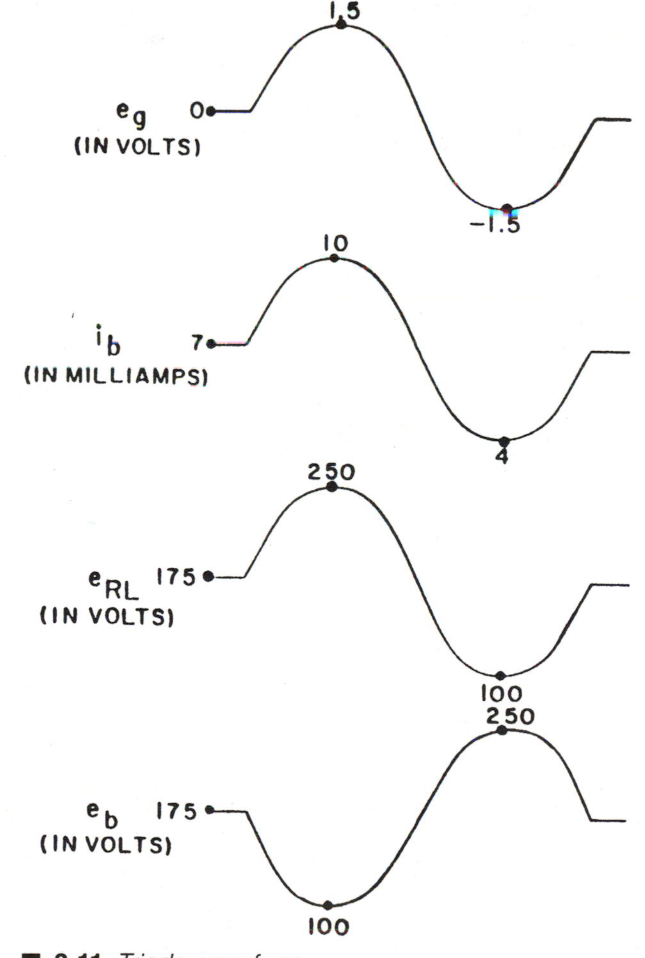

■ **3-11** *Triode waveform.*

values are selected to prevent a normal input signal from driving the amplifier outside of the linear region.

A problem associated with triode vacuum tubes is interelectrode capacitance. In any tube, the plate, cathode, and grid(s) form an electrostatic circuit, as depicted in Figure 3-12. Each electrode acts as the plate of a very small capacitor. The plate and grid act as one capacitor and the grid and cathode as another. The capacitance that exists between the grid and plate is the most critical. In high-frequency RF amplifier circuits, the interelectrode capacitance can produce an unintentional connection between the input and output of the tube. The input circuit is considered to be the grid-cathode and the output circuit the grid-plate. This inadvertent path could cause instability, noise, and distortion.

To provide amplifier tubes suitable for high-frequency applications, the tetrode was developed. The name tetrode came from the fact that the device has four electrodes, as shown in Figure 3-13. By adding a fourth electrode between the grid and plate, interelectrode capacitance can be reduced. Called the screen grid, it functions as an electrostatic shield between the plate and grid. The addition of the screen grid reduces interelectrode capacitance from several picofarads to less than .1 picofarad.

Although the screen grid was initially developed to reduce interelectrode capacitance, it has another useful characteristic in that it makes plate current almost totally independent of plate voltage. For normal operation, the screen grid has a positive potential in reference to the cathode, which causes electrons emitted by the

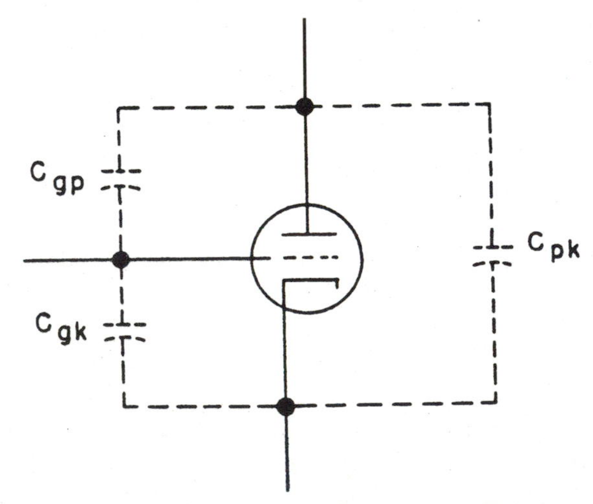

■ **3-12** *Triode interelectrode capacitance.*

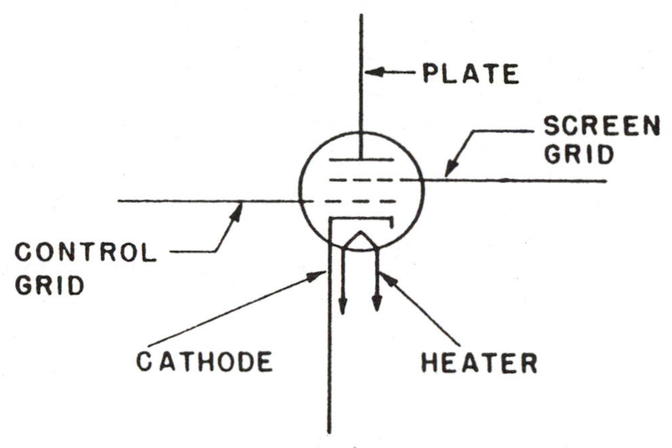

■ **3-13** *Tetrode schematic symbol.*

cathode to be attracted. As the screen grid is constructed from thin wires separated by wide spaces, most of the electrons simply pass through and continue on to the plate. In effect, the screen grid provides an electrostatic force by attracting electrons from the cathode to the plate, while at the same time providing a shield for the cathode. In doing so, the plate exerts comparatively little electrostatic force on electrons in the space cloud near the cathode.

For proper operation, the plate voltage must be more positive than that felt on the screen grid. Because tube conduction is controlled more by the screen voltage than the plate voltage, a tetrode is capable of higher levels of amplification than a comparable tetrode. That, coupled with the lower interelectrode capacitance, makes it a superior RF amplifier. However, the tube does suffer from a drawback—secondary emission. Secondary emission is the action of electrons being broken free from the plate. That phenomenon happens when the plate is bombarded by high-velocity electrons that are emitted from the cathode and are attracted by the plate's positive potential. It is an undesirable effect, as it causes a negative resistance region. That is because unattracted electrons are drifting in the free space between the screen grid and plate. The drifting is caused because both grids have a positive potential. The presence of these electrons repels some of the electrons in the plate current stream. The positive potential of the screen grid provides a strong attraction to the secondary emission electrons. The effect is to reduce plate current and limit the plate voltage swings in tetrodes, which in turn reduces amplification. To compensate for the problem, circuits are designed to use the tubes in the linear portion of their operational range.

Secondary emission effects can be minimized through the addition of a fifth electrode, between the plate and screen grid. Called the suppressor grid, it is connected either internally or externally to the cathode. Figure 3-14 is the schematic symbol for pentodes, with external and internal connections. As it has a negative potential with respect to the plate, it retards the travel of secondary emission electrons and deflects them back to the plate. The main stream of plate current electrons are not affected by the negative potential of the suppressor grid, due to its open construction. Pentodes have found use in RF amplifiers, as they allow higher voltage amplification.

A beam power tube is a tetrode with a directed electron beam. The main feature of this tube is a greater power-handling capability. Figure 3-15 is the internal view of this design. In this type of component, the beam-forming electrodes are positioned so that secondary emission from the plate is repelled back to the plate by the concentrated electron beam. Electrically, the beam-forming electrodes are at the same potential as the cathode to prevent stray electrons from being attracted by the screen grid.

The beam power tube features a high rate of efficiency, high power output, and high power sensitivity because of the concentrated beam of electrons and low screen grid current. These highly desirable characteristics are the result of its internal construction. The screen and control grids are fabricated from spiral wires. The two grids are wound so that each turn of the screen grid is shaded from the cathode by a control grid turn. This alignment of the grids forms the plate current stream into sheets of electrons. As a result, very few electrons are attracted by the screen grid.

The first vacuum tubes were of a general design. As an example, the triode was first used in all applications. Due to the many requirements in communications electronics, it was unsatisfactory

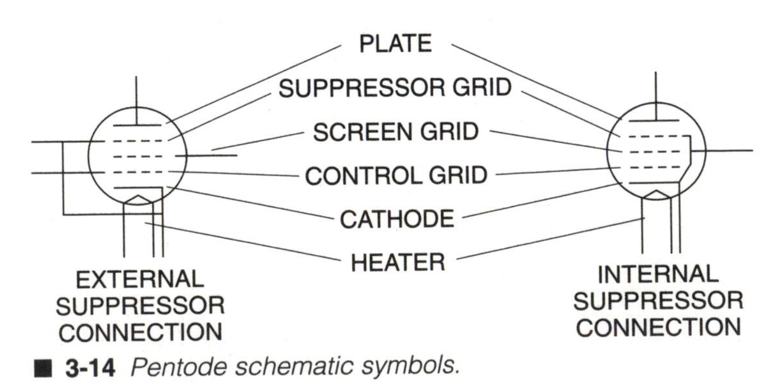

■ **3-14** *Pentode schematic symbols.*

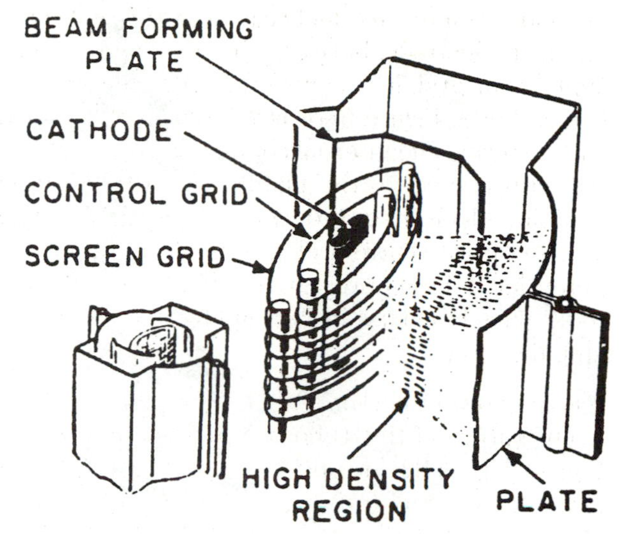

■ **3-15** *Beam tower tube construction.*

in many applications. As the field expanded, specialty tubes were developed. These tubes were optimized for certain applications, or combined multiple functions in one evacuated envelope.

One family of specialty tubes is the multi-electrode tube. This class includes tetrodes and pentodes, which have been presented. Other multi-element tubes have up to seven electrodes. Figure 3-16 is the schematic symbol for the pentagrid tube. This device has five grids to refine operation for a highly specialized application.

Amplifier Classification

There are several different classes that amplifiers, both vacuum tube and semiconductor, can fall under. The first are two broad categories: *voltage amplifiers* and *power amplifiers*. Power amplifier tubes are components that have a low amplification figure and plate resistance. That allows them to utilize high plate voltages and control substantial plate currents. The second broad category is the voltage amplifier. This type of tube is used in applications where the most important consideration is a high voltage gain. This tube has a high amplification factor supplying a high-impedance load.

Another method of classifying amplifiers is by operating frequency range. No doubt you have heard the terms *audio amplifier*, *RF amplifier*, and *video amplifier*. Amplifier tubes and circuits are designed to operate over specific frequency ranges and applications.

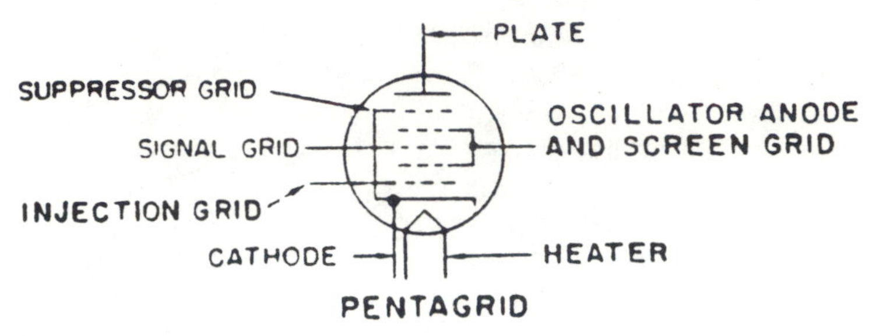

■ **3-16** *Pentagrid schematic symbol.*

Without the proper support components, such as capacitors and inductors, amplifiers would not be possible.

Another way to classify amplifiers is by their class of operation. The values of tube grid bias and voltage determine overall operation. The four classes are A, AB, B, and C, and the corresponding input and output waveforms are illustrated in Figure 3-17a. A class A amplifier is biased such that plate current flows continuously. If a sine wave were the input signal, it would drive into conduction and provide an output for the full 360 degrees. Figure 3-17b is class AB amplifier, which has a grid bias that causes the tube to conduct for greater than 180 degrees, but less than 360 degrees. A class B amplifier, Figure 3-17c, is biased so that with no signal input, no plate current is flowing. When a sine wave is applied to the grid, the tube conducts for only one-half cycle, or 180 degrees. The final amplifier classification is class C, and the input and output waveforms associated with it are illustrated in Figure 3-17d. This type of tube is biased such that there will be no plate current for most of the input signal. This configuration conducts for less than 180 degrees.

As support circuitry determines how a tube amplifier will function, there is another method used to classify vacuum tube amplifiers. Figure 3-18 has the schematic diagrams for the three basic amplifier configurations. The first is the grounded cathode. Notice that the input signal is applied on the grid. The cathode, through the cathode resistor and capacitor, is connected to ground. The plate is tied back to the power supply, B+, through the load resistor. The function of R1, the cathode resistor, is to hold the cathode at a slightly positive level. C1, the capacitor connected in series with R1, has the function of preventing any AC variations from appearing on the cathode. The component has a low impedance to any AC frequencies that might be felt on the input grid. R2, the plate load resistor, is used to develop the output signal and keep B+ off the plate.

This type of configuration is the most common you will encounter. It has good voltage and power gain. Due to the circuit characteristics, it has high input and output impedances. This amplifier features phase reversal. That is when the input signal swings positive, the output swings negative. Due to the limits of the circuit, it has an amplification factor of less than one. That means that the circuit has a lower amplification factor than the tube is capable of delivering.

The next configuration is the grounded grid, depicted in Figure 3-18b. In this configuration, the input signal is applied to the cathode. The grid is grounded, while the cathode is held at some low positive potential. This configuration yields a high voltage gain, but low power gain. It features a low input impedance. Grounded-grid

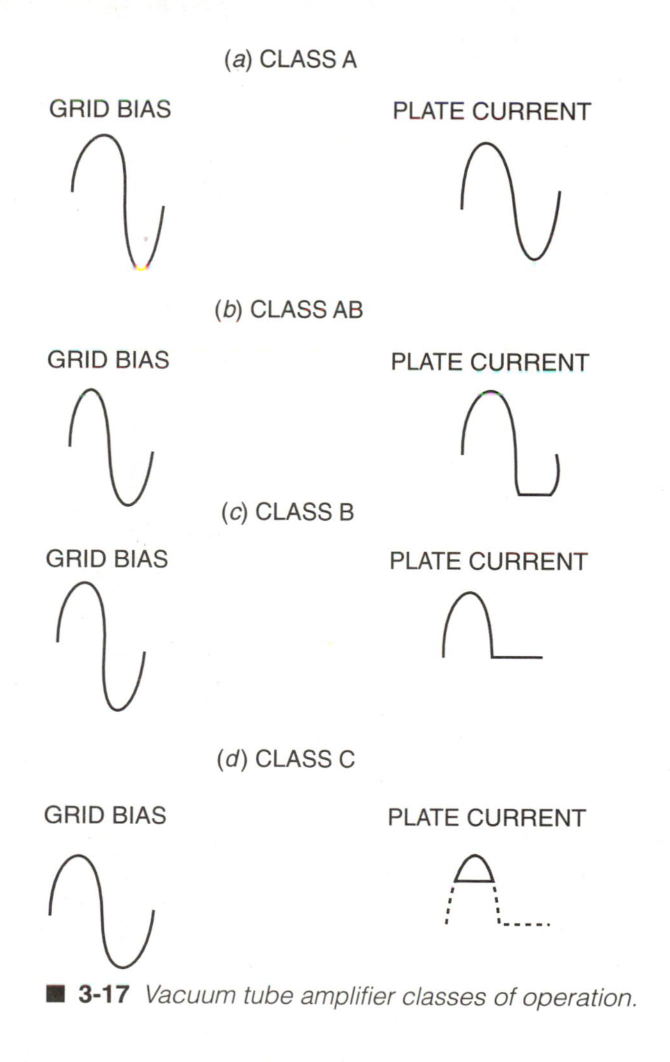

■ **3-17** *Vacuum tube amplifier classes of operation.*

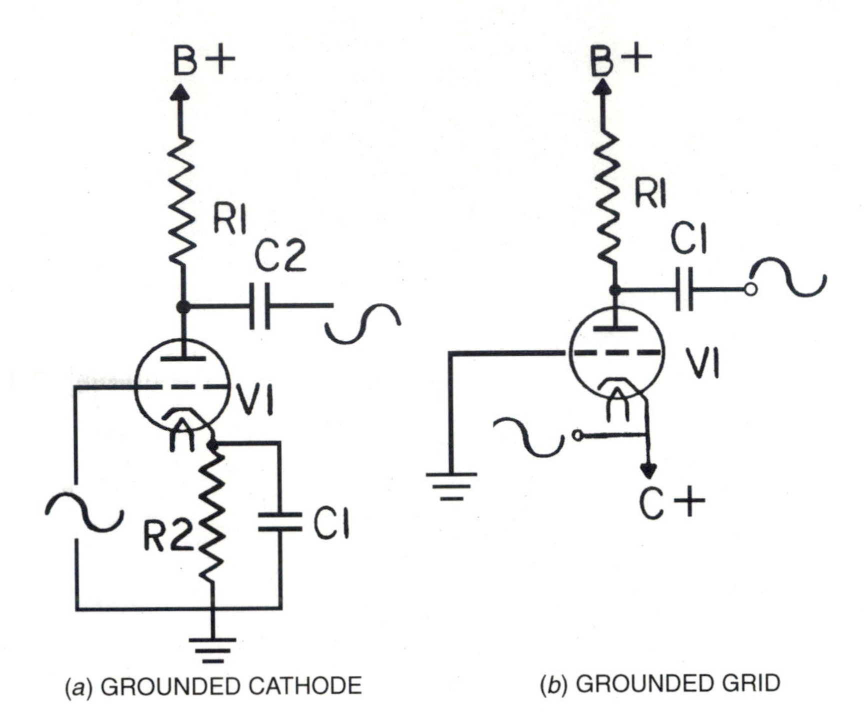

(*a*) GROUNDED CATHODE

(*b*) GROUNDED GRID

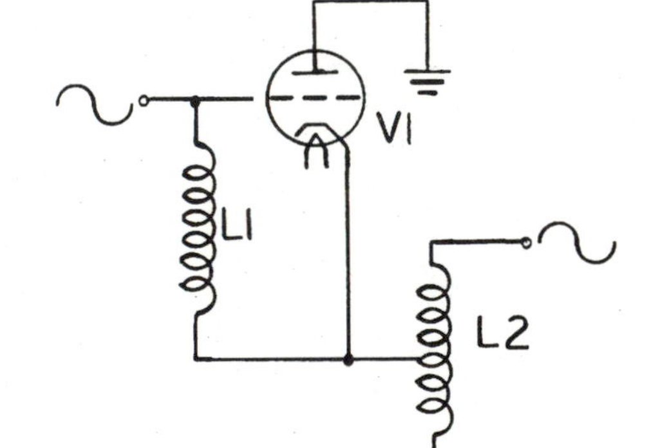

(*c*) GROUNDED PLATE

■ **3-18** *Vacuum tube amplifier configurations.*

configurations are typically found in higher-frequency applications such as UHF and VHF amplifiers, as the grid acts as a shield between the input cathode and output plate. This circuit does not have signal phase inversion, as can be seen in the diagram.

The final configuration is the grounded plate, shown in Figure 3-18c. Notice that plate is tied back to ground and the cathode is held at a negative level by B-. The input is on the grid and the output is taken from the cathode. A common name for this design is the cathode follower. This circuit's characteristics are a high input impedance and a low output impedance. It has a voltage gain of less than one and a low power gain. The most common use for this configuration is for isolation. Isolation is to prevent the circuits that it feeds from reflecting back and producing interaction or interference. This application does not have phase reversal.

Biasing has been mentioned several times and it is another way to classify amplifiers. Self bias is that which is provided automatically by the flow of current electrons through tube. Cathode bias is illustrated in Figure 3-19. The insertion of a resistor between ground and the cathode ensures that the input grid will be negative with respect to the cathode. This is typically found in class A and AB amplifiers. A capacitor must always bypass the cathode resistor. That prevents any AC variations from appearing on the cathode. If the variations were present, then overall gain of the amplifier is noticeably reduced. Any time the gain of an amplifier stage is reduced, it is known as degeneration.

Grid-leak bias is when the grid current provides the bias to set the operating points of the tube and is found in two configurations. Shunt grid-leak bias is shown in Figure 3-20a and series grid-leak bias in Figure 3-20b. Although both configurations appear different, they operate almost identically. As can be seen, the input signal is applied to the grid. The grid is tied back to ground through the grid resistor. A DC voltage is developed that is proportional to the signal input. The coupling capacitor C1 is in series with the input signal.

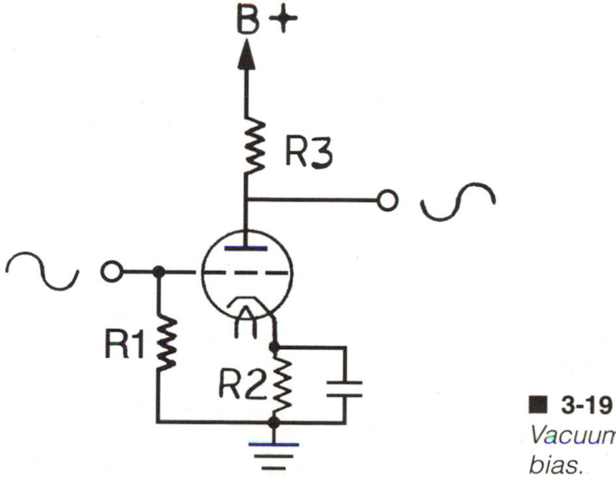

■ 3-19
Vacuum tube cathode bias.

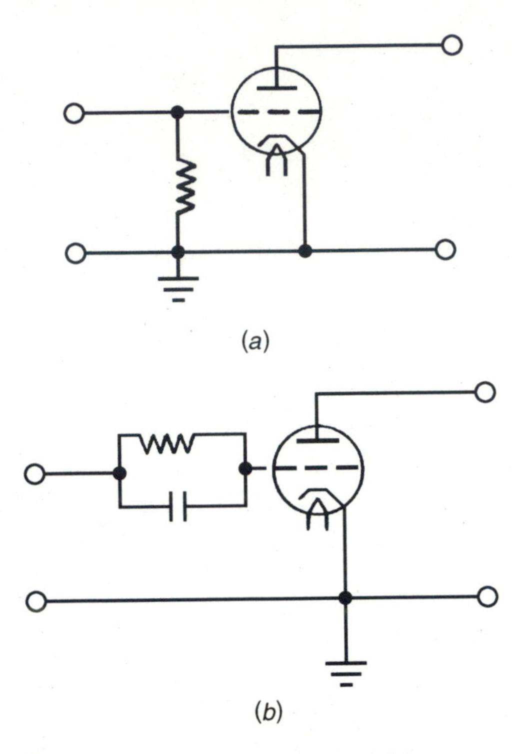

■ **3-20** *Vacuum-tube grid-leak bias.*

During the positive portion of the input signal, the capacitor charges. When the input signal begins to decrease to a point lower than the charge on the capacitor, it begins to discharge back into the circuit. By component selection, the capacitor's charge time is rapid and discharge time very slow. During the next positive portion of the input signal, the capacitor charges again. After only a few input cycles, the grid bias voltage stabilizes, as the capacitor discharges only slightly during each cycle. The level of bias developed depends upon the amplitude and frequency of the applied signal and the charge-discharge time of the capacitor and resistor.

The Semiconductor Diode

Vacuum tubes are excellent amplifiers, however, there are drawbacks to their use: they are very large and bulky, and high voltage is required for proper operation. High voltage means that equipment cooling is a prime consideration. All these factors together result in large, heavy, difficult-to-move equipment. The answer is in the form of the semiconductor diode. Its advantages are small size, low power requirements, reduced equipment size, and much less heat. Figure 3-21 compares the physical sizes of a typical vac-

■ **3-21** *Comparison of a vacuum tube, transistor, and integrated circuit.*

uum tube, diode, transistor, and integrated circuit. All of these components are in use today. As can be seen, the diode is a dramatic size reduction from the tube. Widespread diode and semiconductor adaptation has led to the electronics and computer revolution that we are in the midst of now.

All solid-state devices are based on the electrical properties of semiconductor elements. All elements can be categorized electrically as conductors, insulators, and semiconductors. Basically, under all conditions, conductors allow the flow of electrons. Insulators will block electron movement, unless catastrophic damage results. A semiconductor, however, will conduct under some circumstances and not conduct under others.

The most common semiconductor materials are silicon and germanium. Both elements in pure form have a characteristically high resistance. Through doping, or the addition of impurities under controlled conditions, the electrical properties can be changed. To obtain the desired electrical characteristics, sometimes only a few molecules of an impurity, such as arsenic, might be required.

Figure 3-22 is an electrical representation of a semiconductor diode. It is formed from two blocks of semiconductor material, such as

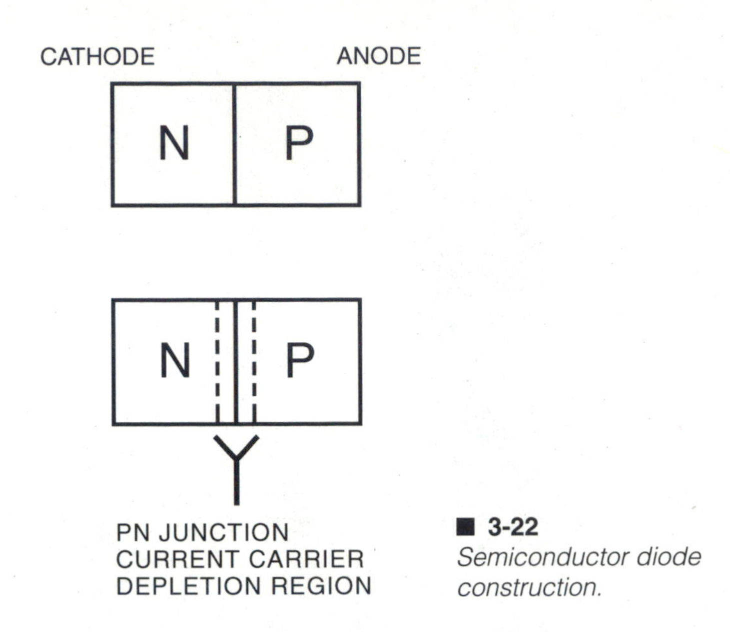

■ **3-22**
Semiconductor diode construction.

silicon or germanium. A diode consists of two elements, the cathode and anode. The cathode is fabricated from N-type material. N material is when semiconductor material is doped with an impurity that adds extra electrons. Ideally, these impurities easily give up electrons from the outer shell when under the influence of a very small voltage. Common impurities to add extra electrons are elements such as phosphorus, arsenic, and antimony. As these impurities add extra electrons to the semiconductor material, they are known as donors. Current in N material is the movement of electrons.

P material is formed when a semiconductor material is doped with an impurity that is lacking electrons in the outer shell. These elements attract electrons to fill the empty outer shells. As a result, current carriers in P material are the lack of electrons, or holes. Common doping materials are aluminum, gallium, and indium.

A block of P material and one of N material are required to fabricate a diode. The point where they come into contact is called the PN junction. The instant the two blocks are brought together, a depletion region forms around the PN junction. What happens is that the excess electrons found in the N material migrate across the junction into the P material, which is lacking in electrons. The excess electrons then fill up all the empty outer shells of the P-material molecules. At the same time, holes, or the lack of electrons found in the P material, appear to migrate across into the N material. The result is that a small region several molecules thick has all the molecules of semiconductor material stable on the subatomic level. All outer shells of the individual atoms are filled with the maximum number of electrons. The result is that this region is depleted of current carriers, either electrons or holes. Current flow is now not possible, unless a bias is placed across the device to electrically control the size of the depletion region.

Diode operation is dependent upon the bias felt across it. With no voltage across it, a diode functions as an open circuit, and the depletion region is relatively large. When voltage of the proper polarity is placed across the device, it functions as a virtual short circuit. That is because the positive potential on the anode and the negative potential on the cathode decrease the size of the depletion region until it virtually disappears. Reverse the polarity and it is an open circuit, as the depletion region has increased in size to the point where electrons cannot cross it. Diode operation is determined by the bias placed across the device. The diode consists of two electrical elements—the anode and the cathode. The anode is the P material and the cathode is the N material. Electron current flow is from the cathode to the anode. Figure 3-23 illustrates circuit action when the anode is positive with respect to the cathode, and the device is forward biased and conducts. Electron current flow is from the cathode to the anode. That is because the negative potential on the cathode repels electrons from the cathode across the depletion region.

When the anode is negative with respect to the cathode as depicted in Figure 3-24, it is reverse-biased and there is no conduction of

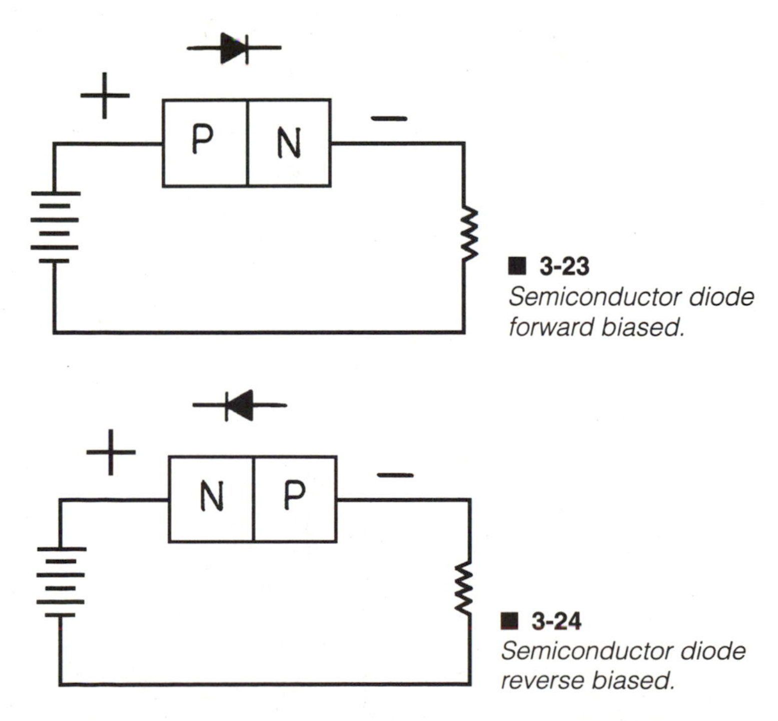

■ 3-23
Semiconductor diode forward biased.

■ 3-24
Semiconductor diode reverse biased.

current through the device. The key to diode operation is the depletion region. The negative potential on the anode attracts electrons, filling more holes and expanding the depletion region in the P material. The positive potential on the cathode attracts electrons in the N material, pulling them away from the PN junction and enlarging the depletion region in the N material.

Diodes are very important and common components in electronics. The most common uses include rectification, demodulation, clippers, clampers, and limiters. Figure 3-25 has the schematic diagrams for the three most common rectifier circuits: the half- wave, full wave, and bridge rectifiers. A half-wave rectifier, shown in Figure 3-25a converts AC voltage into a pulsating DC voltage. This is accomplished through the natural characteristics of the semiconductor diode. On the positive alternation of the input AC, the diode is

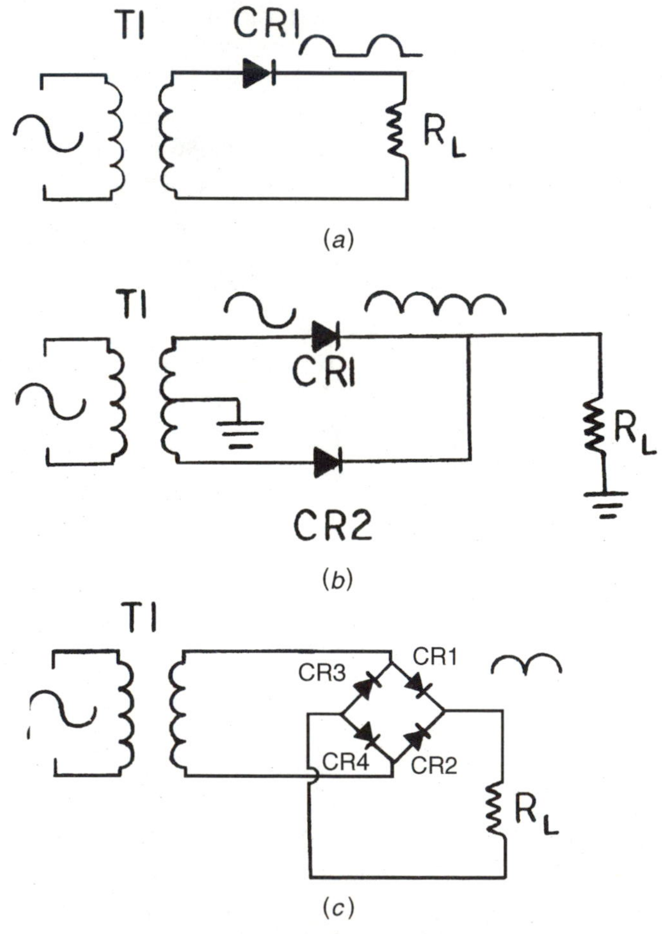

■ **3-25** *Semiconductor diode rectifiers.*

forward-biased. That is because the anode is more positive than the cathode. When forward-biased, the diode acts like a short, passing current. On the negative alternation, the anode is less positive than the cathode, reverse-biasing the device and causing it to act like an open. The output of the half-wave rectifier is a pulsating DC with the frequency of the input AC. Filter circuits are used to smooth the pulsating DC into a smooth, ripple-free DC. If the diode were installed backwards, with the cathode connected to the input transformer, then the output of the rectifier would be pulsating negative DC.

The full-wave rectifier utilizes both the positive and negative alternations of the input AC and is illustrated in Figure 3-25b. The reason why is the center-tapped transformer secondary. During the positive alternation of the input AC, CR1 is forward-biased and conducting, while CR2 is reverse-biased and cut off. During the negative alternation of the input AC, CR1 is reverse-biased and cut off, while CR2 is forward-biased and conducting. As with the half-wave rectifier, the output of the full-wave rectifier is pulsating DC. Because both input AC alternations are converted to DC, the frequency of the pulsating DC is twice that of the input. Filter networks are then used to smooth out the pulsating DC.

A drawback with this type of power supply is that the center-tapped transformer provides only half of the input voltage to each diode. The bridge rectifier, illustrated in Figure 3-25c, eliminates that concern. A network of four diodes is used to convert the input AC into DC. On each alternation, one pair of diodes is forward-biased and conducting while the other pair is reverse-biased and cut off. As both alternations are utilized, the pulsating DC frequency is twice that of the input AC and full power is converted.

Other important uses for diodes include *clippers*, *clampers*, and *limiters*. The function of a clamper is to provide a predetermined DC level to a signal waveform. This type of circuit is also known as DC reinsertion or DC restoration. A clipper is used to remove an unwanted portion of a signal waveform. A good example would be when a square wave is converted into a positive trigger by an integrator circuit. The output of the circuit is a positive and negative trigger that is the leading and trailing edge of the waveform. As only one trigger is desired, a clipper eliminates the unwanted one.

A clamper has the function of referencing, or *clamping*, a signal waveform to a specific DC level. Also known as a DC restorer, it is used to restore the DC component to a video signal after it has suffered losses in capacitive-coupled amplifiers. This is common in interfacing and signal processing circuits.

A limiter is a circuit that is used to limit the amplitude of a signal to some predetermined level. This type of circuit is also called an amplitude limiter, automatic peak limiter, peak clipper, and peak limiter. All of these circuits require the natural characteristics of the diode for proper operation.

Diodes have characteristic limits of voltage and current that cannot be exceeded without damaging the component. Figure 3-26 is the characteristic curve for a representative diode. Maximum reverse voltage is the highest level of reverse-bias voltage that can be placed across a diode without destroying it. This figure is divided into peak, which is the highest instantaneous value, and average, which is the greatest sustainable level over a reference period of time. Average will be somewhat less than peak. Maximum leakage current is the expected minuscule current flow with a given voltage level under the reverse-bias conditions. This is an important figure for rectifiers and wave-shaping circuits. Forward voltage is the value of forward bias that is required to bring a diode into conduction. There are also temperature limits for proper diode operation. If it is exceeded, the diode will draw excessive current and burn open.

The Transistor

As useful as the semiconductor diode is, it lacks one important characteristic. Diodes are electronic devices that cannot amplify.

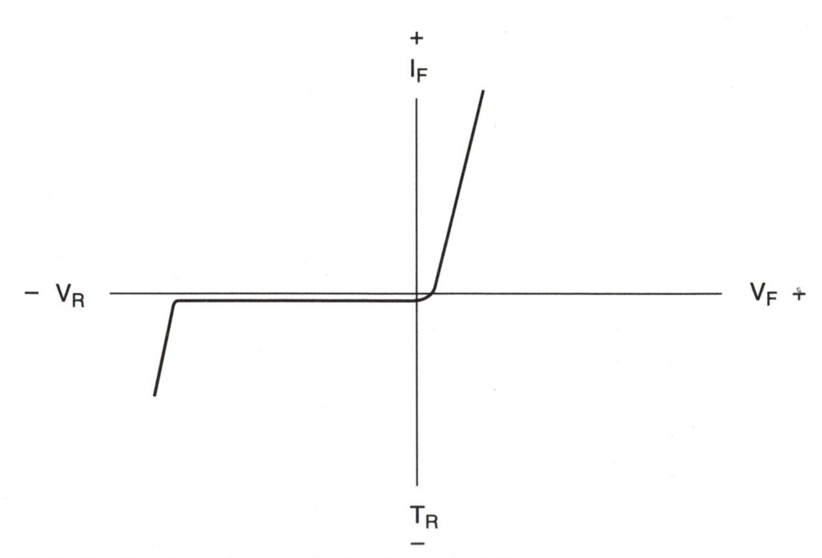

■ **3-26** *Semiconductor diode characteristic curve.*

Amplification is the ability of an electronic device to increase the strength of a signal. When the voltage, current, or power of a signal is increased, then amplification has occurred. In a transistor, amplification is the ability of a semiconductor device to control a relatively large amount of current in the output element with a small amount of current in the input element. As can be seen in Figure 3-27, transistors are constructed from the same materials and techniques that are used to fabricate diodes. Rather than two elements, a transistor consists of three elements—an *emitter*, a *base*, and a *collector*. There are two broad categories of transistors, the *PNP* and *NPN* types.

Electrically, a transistor is nothing more than two PN junction diodes joined together. The device has two depletion regions that must be properly biased for conduction to occur. The key to transistor operation is the physical size of the elements that make up the device. The base is the smallest region. In some types of transistor, it might only be a few molecules thick. The emitter, which produces the current carriers, is the next largest. The collector, as it has to receive the current carriers, is the largest.

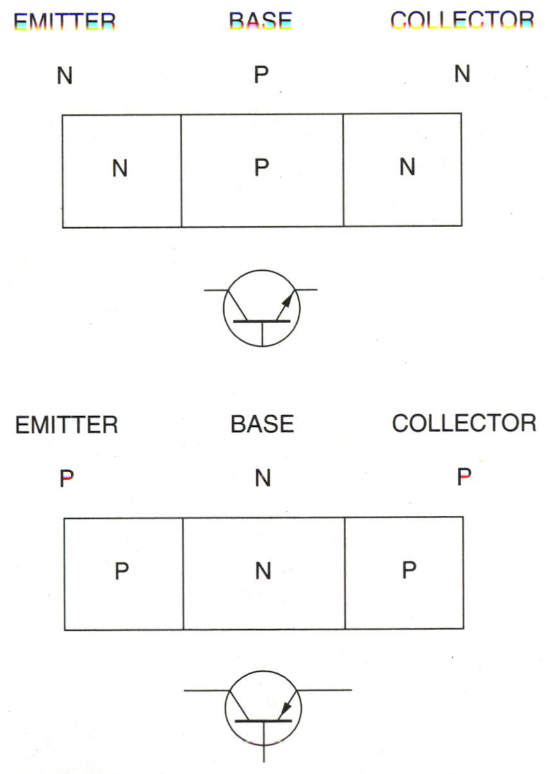

■ **3-27** *Transistor construction and schematic symbols.*

The NPN transistor will be covered first. As three blocks of semiconductor material are required for construction, the device has two depletion regions. Included in Figure 3-27 is the schematic symbol for both the NPN and PNP transistors. In the NPN, the angled line with the arrow on it indicates the emitter element. *The arrow points against the direction of electron current flow.* The base is the vertical line attached to the emitter. The collector is the angled line on the upper part of the device attached to the base. When the transistor is forward-biased, current flows through it. For normal operation, 100% of total current flows through the emitter, 5% through the base and 95% through the collector. The base and emitter current may vary, but must total 100%. In the NPN transistor, the predominant current carriers are electrons. The PNP transistor is opposite, as it has holes, or the lack of electrons as the principal current carriers. The two function very similarly, except the PNP requires negative voltage, rather than positive for proper operation.

Figure 3-28 illustrates a simple transistor amplifier. The emitter is tied to ground, the collector, via the collector load resistor to V_{cc}, the power supply, and the signal input is on the base. Due to bias, the transistor conducts when no current is applied. A bias-free state is called *static*. As V_{cc} is 12 vdc, that places about 6 volts on the collector. A sine wave is applied to the base. As it swings positive, it increases conduction through the device. As conduction increases through the transistor, that increases the conduction through the load resistor. As the increased current flows through the resistor, it increases the voltage drop over it. That leaves less voltage to be felt on the collector. The collector

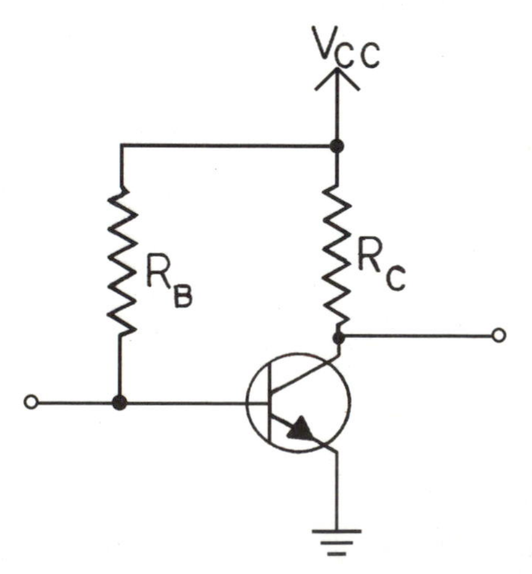

■ 3-28
NPN transistor amplifier with fixed bias.

voltage begins to decrease from 6 vdc toward 0. After reaching maximum positive, the input waveform then begins to swing negative. As the waveform decreases, it decreases conduction through the device. As conduction decreases through the transistor, that decreases the conduction through the load resistor. A decreased current flow through the resistor decreases the voltage drop over it. That leaves more voltage to be felt on the collector. The collector then begins to increase toward V_{cc}. Just like a vacuum tube, a transistor can be driven into cutoff or saturation.

Transistor amplifiers are no different than a vacuum tube amplifier in that proper bias is required for operation. Refer to Figure 3-28, as it is a schematic diagram for a simple transistor amplifier using fixed base bias. As can be seen, the transistor is an NPN. The input waveform to be amplified is applied to the base. The output is taken from the collector. R1 is the collector load resistor, and is required to keep V_{cc} off the collector and to develop the output signal. R_b is the base bias resistor. Its value is such that a small positive voltage will be felt on the base. The result is that a small level of current is flowing through the transistor. When a signal is applied to the amplifier, there will not be a change in circuit conditions until it changes the voltage felt on the base. If the input signal swings positive, it will cause the base voltage to go more positive. A more positive emitter-base bias causes the transistor to conduct harder. Harder conduction causes more current to flow through the transistor and the collector load resistor. An increased current flow through the resistor increases the voltage drop over it. An increased voltage drop over the load resistor leaves less voltage to be felt on the collector. At this point, the output of the amplifier is swinging negative. If the input waveform continues to go positive, that action causes the voltage on the base to go more positive. If it continues until further increases in base voltage do not cause an increase in current flow through the transistor, then the device is saturated.

The opposite happens if the input signal swings negative. It will cause the base voltage to go more negative. A more negative emitter-base bias causes the transistor to conduct less. Decreased conduction causes less current to flow through the transistor and the collector load resistor. A decreased current flow through the resistor decreases the voltage drop over it. A decreased voltage drop over the load resistor leaves more voltage to be felt on the collector. At that point, the output of the amplifier is swinging positive. If the input waveform goes so negative that current flow through the transistor stops, then the amplifier is cut off. This type of bias offers low cost and simplicity, but at a price, as this configuration is subject to thermal instability.

Figure 3-29 is a simple transistor amplifier with emitter-base bias to illustrate one way to overcome problems with temperature instability. An emitter resistor is placed between ground and the transistor. Any temperature change affects the conductivity of all the components. If temperature increases, the current flow through the transistor increases. That increase in turn causes the voltage drop over the emitter resistor to increase, leaving less voltage to be dropped over the transistor and the collector resistor, and returning circuit operation to normal.

Voltage-divider bias is shown in Figure 3-30. The bias on the base is determined by the voltage-divider action of R1 and R2.

The final type of bias to be presented is self bias. Self bias is illustrated in Figure 3-31. As can be seen from the drawing, the base-bias resistor R1 is directly connected to the collector of the transistor and the collector load resistor, R2. The result is that any voltage felt on the collector is fed back to the base via R1. In this configuration, the bias voltage is set by the collector voltage. The other components have the same functions that you have already seen. R2 is the collector load resistor, and R3 is the emitter resistor. C1 and C2 are coupling capacitors. Their function is to pass AC signals while blocking DC voltages and levels. C3, the bypass capacitor in the emitter of Q1, is to prevent any DC variations from appearing on the emitter. If that were to happen, then the gain of the transistor amplifier would be reduced.

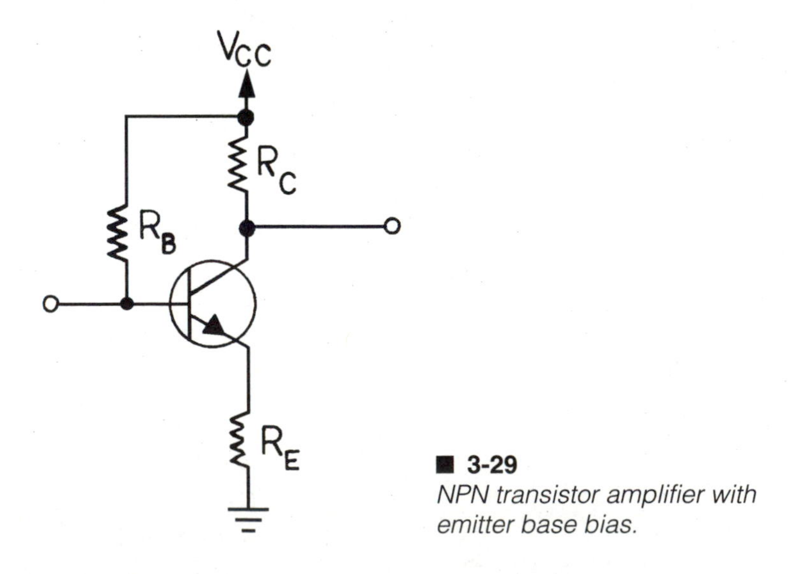

■ **3-29**
NPN transistor amplifier with emitter base bias.

The circuit functions as follows: A positive signal on the base of the transistor increases the positive bias on the amplifier. An increased bias increases the current flow through the collector resistor, the transistor, and the collector load resistor. The increased current flow causes more voltage to be dropped over R3 and R1, leaving less voltage to be felt on the collector of the transistor. The reduced collector voltage is fed back through R1 to the base. That has the effect of reducing the forward bias. A reduced forward bias

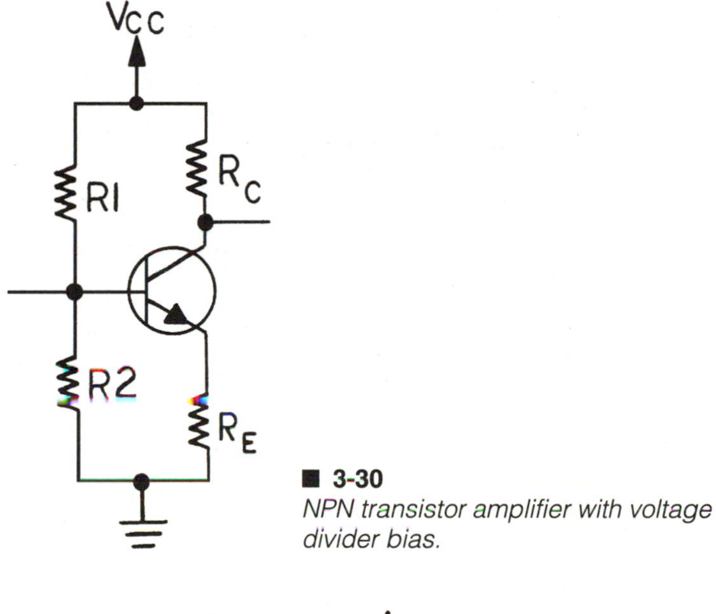

■ **3-30**
NPN transistor amplifier with voltage divider bias.

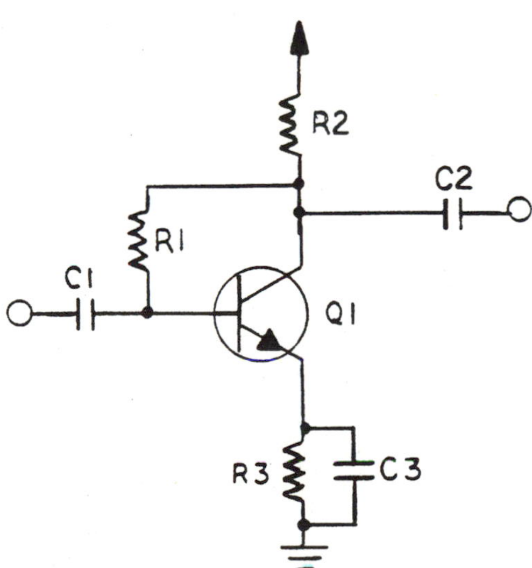

■ **3-31** *NPN transistor amplifier with self bias.*

decreases conduction through the transistor, causing the collector voltage to increase. The result is that a more positive voltage is fed back to the base, reducing bias to normal. This type of bias is common in circuits requiring a stable reference voltage. A very common use for this circuit is in power supplies, to provide a constant output level. If the power-supply output changes due to temperature or load changes, the self-biased amplifier compensates, bringing the output level back to the desired level.

Transistor amplifiers can be operated in Class A, Class AB, Class B, and Class C. Transistor amplifier operation is identical to that of vacuum tube amplifiers. Figure 3-32 illustrates the four classes

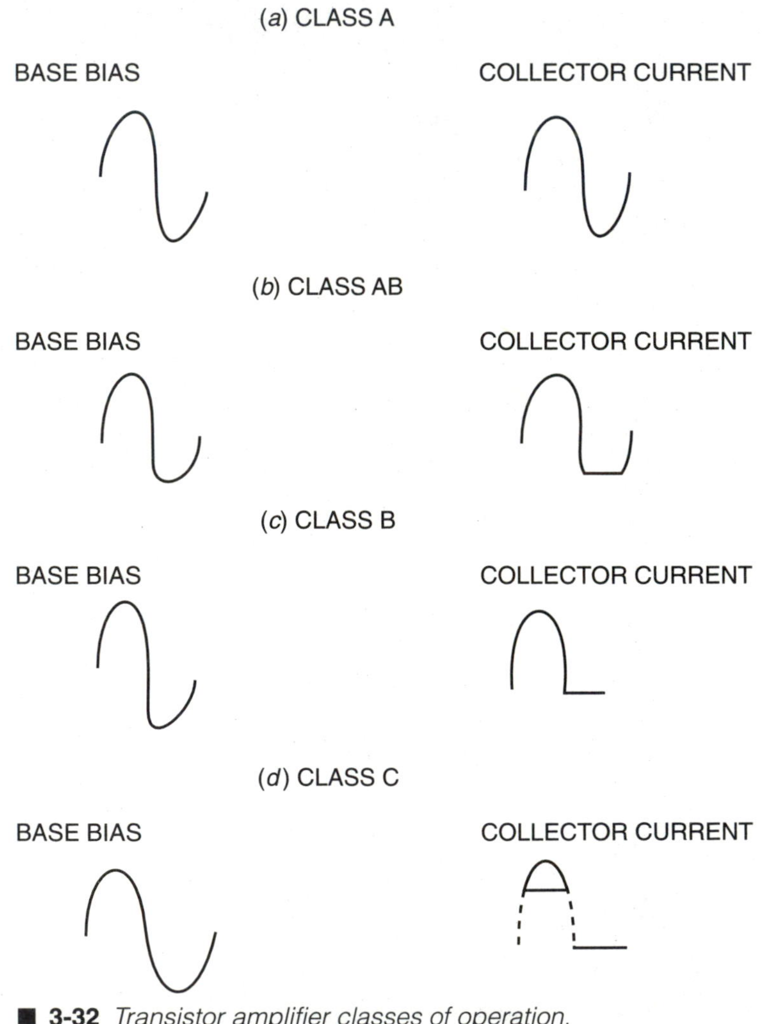

■ **3-32** *Transistor amplifier classes of operation.*

of operation that you will encounter in transistorized equipment. The first is a representative transistorized class A amplifier. A class A amplifier is an amplifier configuration in which collector current flows for 360 degrees of signal input. The wave shape will be identical to that of the input. The class AB configuration is biased such that collector current flows for more than 180 degrees of input signal, but less than a full 360 degrees. The class B amplifier is biased such so that collector current flows for exactly 180 degrees of input signal. A good example of this type of waveform would be that of a diode rectifier. The final configuration is the class C amplifier. In this design, collector current flows for less than 180 degrees.

Transistor amplifiers can be connected in one of three configurations. Figure 3-33 is a chart that lists the major characteristics associated with the three transistor amplifier configurations. It is a good tool for remembering how the amplifier circuits function within equipment. When you examine Figure 3-34, you will notice that there are two separate schematics. One is for an NPN amplifier, the other for a PNP. The only difference between the two circuits is the polarity of V_{cc} and the transistor type. This illustration is the schematic diagram for a simple common-emitter transistor amplifier configuration. Notice that the input and output waveforms are in phase. From the diagram it can be determined that the amplifiers are biased class A. The function of the capacitor on the input of the amplifier is to pass the desired AC signal and block any DC levels from the preceding stages. The signal is

AMPLIFIER TYPE	COMMON BASE	COMMON EMITTER	COMMON COLLECTOR
INPUT/OUTPUT PHASE RELATIONSHIP	0°	180°	0°
VOLTAGE GAIN	HIGH	MEDIUM	LOW
CURRENT GAIN	LOW(α)	MEDIUM(β)	HIGH(γ)
POWER GAIN	LOW	HIGH	MEDIUM
INPUT RESISTANCE	LOW	MEDIUM	HIGH
OUTPUT RESISTANCE	HIGH	MEDIUM	LOW

■ **3-33** *Transistor amplifier configuration characteristics chart.*

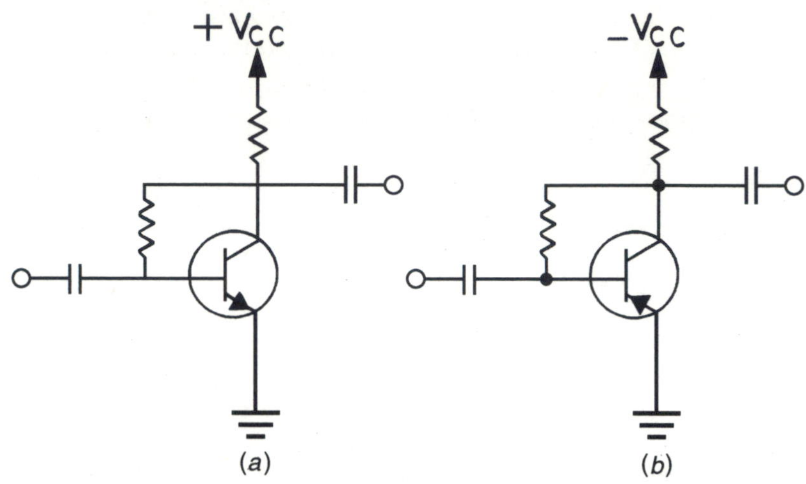

■ **3-34** *(a) NPN and (b) PNP transistor amplifier common emitter configuration.*

applied to the base and taken from the collector. In this type, there are 180 degrees of phase inversion between the input and output signals. Input and output resistance of this amplifier is medium. The input resistance is determined by the emitter-base (input) and the base-collector (output) junctions. Voltage gain and current gain are both medium. That results in a high power gain.

The common-base configuration is illustrated in Figure 3-35. The input is on the emitter and the output is taken from the collector. As can be seen, the input and output signals are in phase. This circuit has a high voltage gain. However, current gain and power gain are both low. Input resistance (emitter-base) is low, and output resistance (base-collector) is high.

The final configuration, depicted in Figure 3-36, is the common-collector arrangement. The input is on the base and the output is taken from the emitter. This design is also known as an emitter follower. In this amplifier circuit, there is no phase inversion. As the input is on the base-collector junction, it has a high input resistance, and as the output is taken from the collector, it has a low output resistance. Voltage gain is low and current gain is high. That gives it a high power gain. This configuration is ideal for impedance-matching applications. It is very common as the output stage for function generators and other test equipment.

Transistors also are affected by interelectrode capacitance, as shown in Figure 3-37. The emitter-base and base-collector pairs exhibit capacitance in high-frequency applications. The leads connecting the three elements to the rest of the circuit and the junctions function as very small capacitors. Frequency compensation is a critical design criterion in many communications designs.

Oscillators

Oscillators are a vital function within modern communications systems. Failure-free communications requires the use of many stable and accurate oscillators to provide error-free frequencies. Another use for oscillators is in the area of system timing. Not long

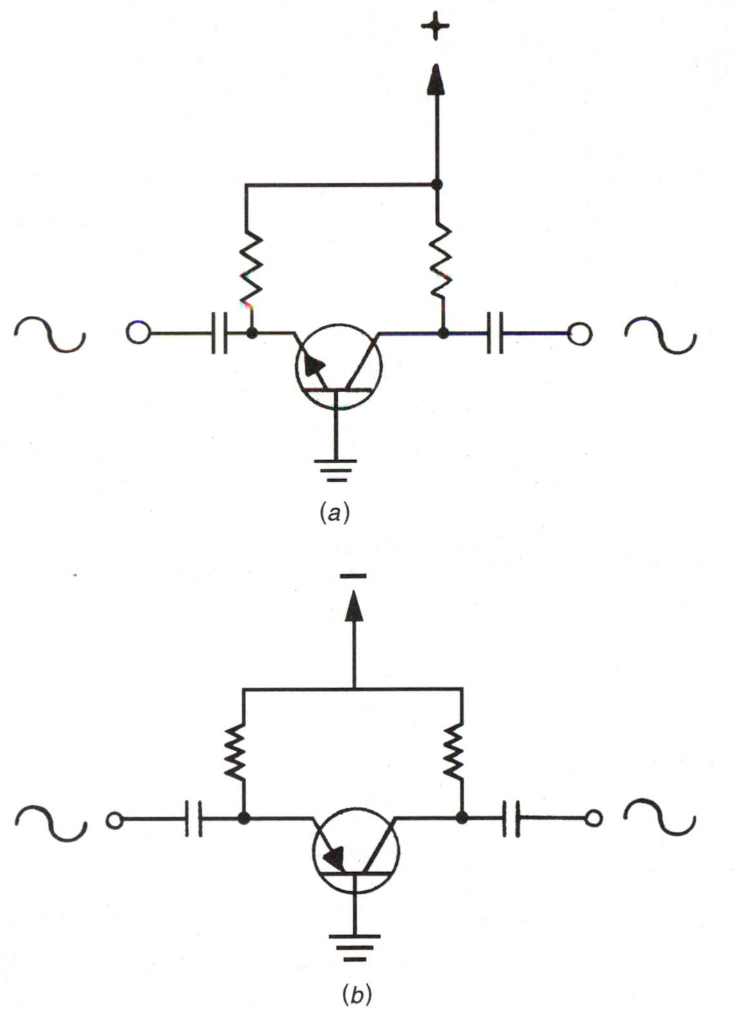

■ **3-35** *(a) NPN and (b) PNP transistor amplifier common base configuration.*

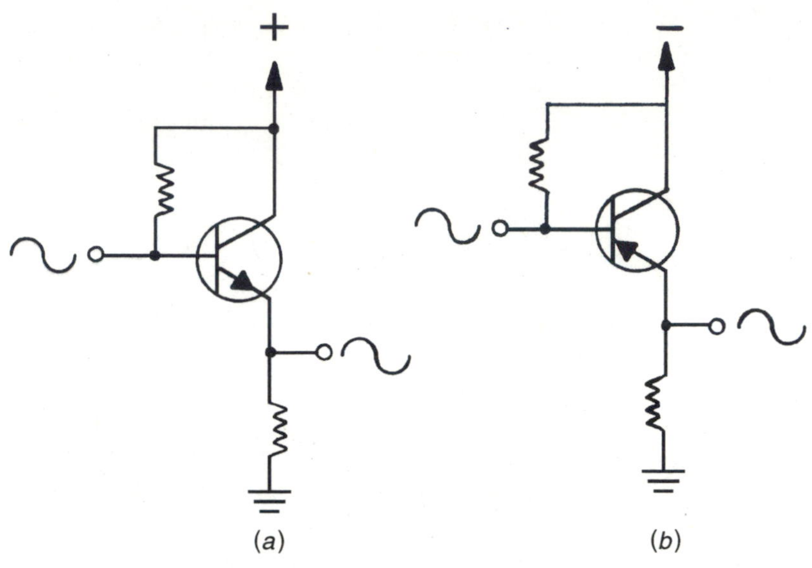

■ **3-36** *(a) NPN and (b) PNP transistor amplifier common collector configuration.*

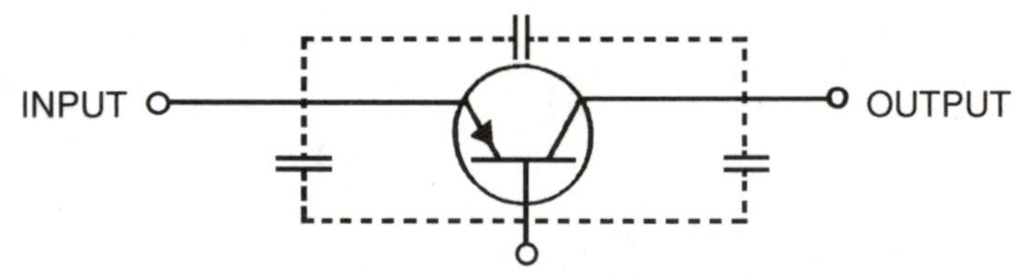

■ **3-37** *Transistor interelectrode capacitance.*

ago, timing was possibly the most important consideration in radar and sonar systems. With the widespread adaptation of computer circuits and signal processing, timing is found in all areas of electronics. All system timing signals and resulting triggers are based on the operation of one type of circuit—the oscillator. By definition, an oscillator is an electronic circuit that generates a periodic repeating waveform of known amplitude, frequency, and duration. As the output must be constant in amplitude and frequency, a high degree of circuit stability is required. Stated simply, oscillations are the result of the controlled periodic storage and release of electrical energy.

One of the earliest types of oscillators, and in many respects still the best, is the crystal oscillator. Its operation is based on a natural phenomenon associated with quartz crystals. It has many advantages, including stable operation, a high degree of accuracy, inherent ruggedness, and longevity. Crystals have been known to produce accurate oscillations without failure for many years. However, severe mechanical shock or excessive voltage can damage them.

To provide a usable output, the low-level oscillations that a crystal naturally produces must by amplified by external active components. A crystal oscillator can be fabricated with only a few external components, such as a transistor, resistors, and capacitors. As the resonant frequency of a crystal is determined during the manufacturing process by its physical dimensions, it is very difficult to alter its base frequency. A major disadvantage associated with crystals is that they are somewhat temperature-sensitive. If used in a major electronics installation such as a communications or radar system, an oven can be used to maintain a constant operating temperature, thus ensuring the required degree of stability.

Figure 3-38 illustrates a finished crystal's internal construction. Two metallic plates sandwich a very thin slice of quartz crystal. To ensure good mechanical contact, small springs are installed to provide tension. The crystal and metal support parts are hermetically sealed to keep out moisture, dust, and dirt. Electrical connection to the crystal is accomplished by extending the plates through the bottom of the case. The plate leads are connected to external support components by either soldering in place or by plugging into a socket. If the equipment has several operating frequencies, a socket mount provides a means to rapidly change crystals. This used to be a common practice in communications equipment, both for the military and for consumers.

Crystal operation is based on the curiosity known as the piezoelectric effect. Figure 3-39 depicts the concept. A crystal has two axes, X and Y, that pass through it. The X axis is located through

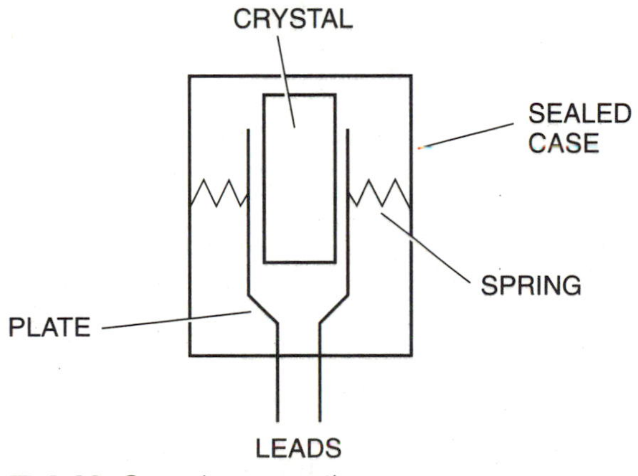

■ **3-38** *Crystal construction.*

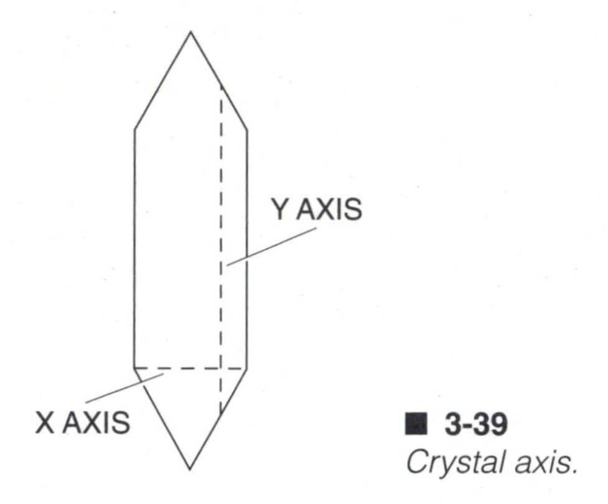

■ **3-39**
Crystal axis.

the crystal's corners. Lying perpendicular to the X axis and on the same plane is the Y axis. The action of placing a mechanical stress along the Y axis results in the development of a voltage along the X axis. The process can be observed in reverse by applying a voltage along the X axis, which induces a mechanical stress along the Y axis. The mechanical stress is in the form of naturally occurring vibrations, or oscillations. Through amplification, these oscillations can provide an accurate basis for triggering circuits.

The operation of a typical crystal oscillator can be described using a simple transistor-based crystal Pierce oscillator as drawn in Figure 3-40. By placing a voltage across the crystal X-1, it begins to vibrate, or resonate. As the oscillations are very small, transistor Q-1 must amplify them to a usable level. If more than one frequency is required by the equipment, several crystals can be installed in parallel, selectable with a switch. By turning the switch, a different active crystal can be selected, changing the resonant frequency of the oscillator rapidly and accurately. The basic frequency of a crystal oscillator can be increased through the use of external circuitry, called frequency multipliers, but this often leads to stability and reliability problems.

The Multivibrator

The multivibrator is a very common circuit found in many equipments for timing purposes. Classified as a relaxation oscillator, the circuit has been in use for timing applications since the dawn of electronics technology. The first examples were constructed using vacuum tubes as the active components. As technology advanced, vacuum tubes were replaced by transistors, followed by integrated circuits. Circuit operation is based on the charge-discharge time of a capacitive-resistive network interconnecting the two active components.

Figure 3-41 is a schematic of a basic transistorized multivibrator circuit. This example is classified as a free-running multivibrator because, with the application of power, the circuit begins to function automatically, producing a square-wave output of a constant frequency and amplitude. Notice that the two halves of the circuit are a mirror image of one another. Opposite components will have the same value, such as C1 and C2, R2 and R7, and R1 and R6. With the initial application of power, one capacitor will charge faster. That is due to the slight variations that exist in any number of components with the same nominal value. The charging capacitor causes the base current of one of the transistors to rapidly increase. That charging action drives the transistor into saturation, decreasing its output voltage to minimum. Notice that the transistor collectors are tied back to the base of the opposite transistor. Due to that interconnection, the other transistor is cut off by the action of the saturated transistor, increasing its output to maximum, or V_{CC}. This condition lasts until the charged capacitor discharges through its discharge resistor back to the base of the opposite transistor. That drives the saturated transistor into cutoff and the cutoff transistor into saturation, reversing the output voltage levels. The cycle then repeats itself as long as power is

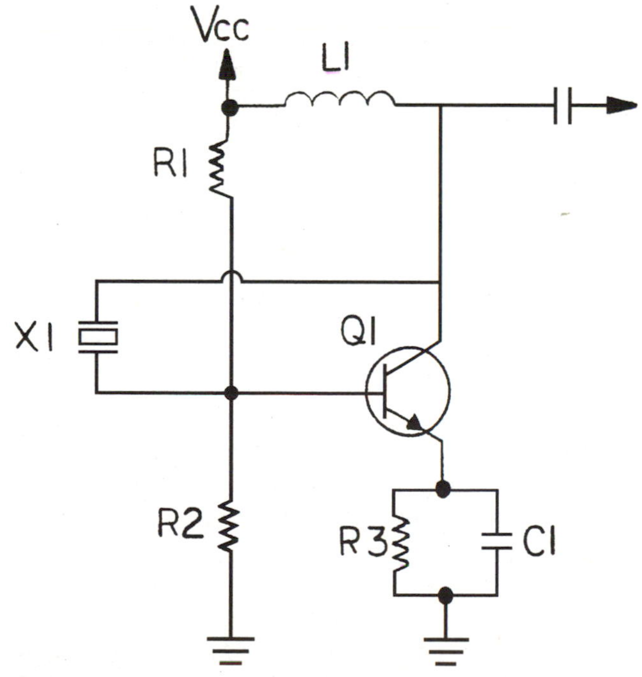

■ **3-40** *Pierce crystal oscillator.*

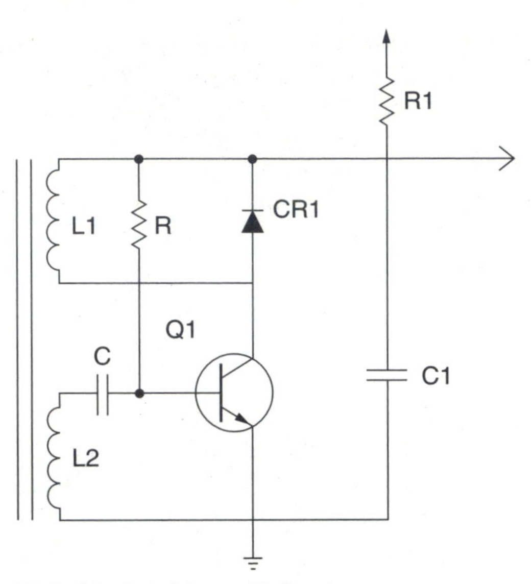

■ **3-41** *Astable multivibrator.*

applied to the circuit. The above operation illustrates a basic multivibrator characteristic: The circuit has only two possible output conditions—conduction and saturation. During operation, the two outputs will always be opposite; any other combination indicates a failed circuit. When one is high or cut off, the other one has to be low or saturated. These conditions are referred to as states, and the outputs can be called "HIGH" and "LOW," "ON," and "OFF," or "1" and "0." This type of circuit is called an *astable multivibrator* because it constantly changes states, or free runs, as long as power is applied.

Other common multivibrator circuits include the *monostable* and *bistable* types. These types are different in that both require an external trigger to initiate circuit operation. Figure 3-42 is the schematic drawing and timing diagram for the transistorized monostable multivibrator. The major difference between this and the preceding astable circuit is the addition of an input on the lower left side of the schematic. Triggered multivibrators are used to modify a basic trigger for the control or initiation of various functions throughout a major electronics system. When quiescent, or static, the output from the circuit is Q HIGH, Q(NOT) LOW. That is because the circuit is designed for Q1 to be cut off and Q2 conducting. The application of an input trigger brings Q1 into conduction, driving Q2 into cutoff. Due the value of the RC components in Q2's base circuit, this state lasts for a predetermined length of time. At the conclusion of the timing period, the capacitor discharges, returning the circuit to its stable condition. It will remain this way until the application of another external trigger. If a trigger is received during the timing process, the circuit is unaffected. This type of circuit is also known as a one-shot multivibrator.

The bistable multivibrator, illustrated in Figure 3-43, is a relaxation oscillator that has two stable states. In other words, when power is applied, one output goes high, while the other low. It will remain in that condition until the application of an external trigger causes it to change states. Once more it will stay in that stable condition until the application of a second trigger, which will cause it to change states again, returning to the original condition. Whatever type of multivibrator is used, they all share several very important characteristics: easy fabrication, low cost, and stable operation.

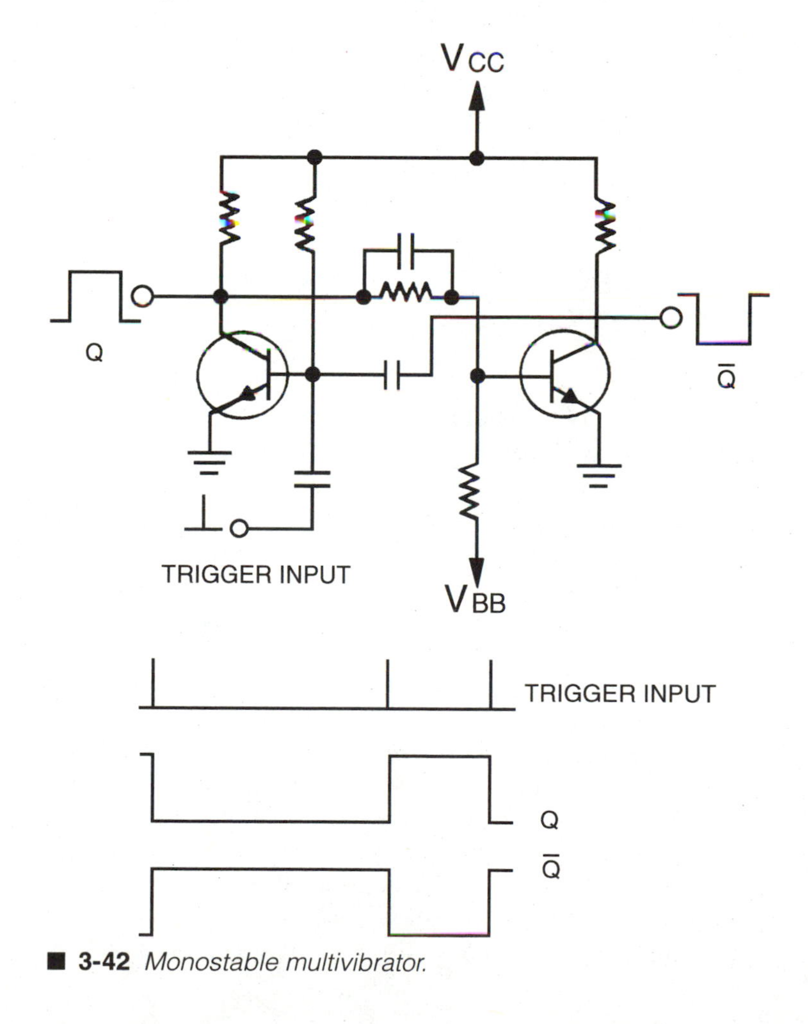

■ **3-42** *Monostable multivibrator.*

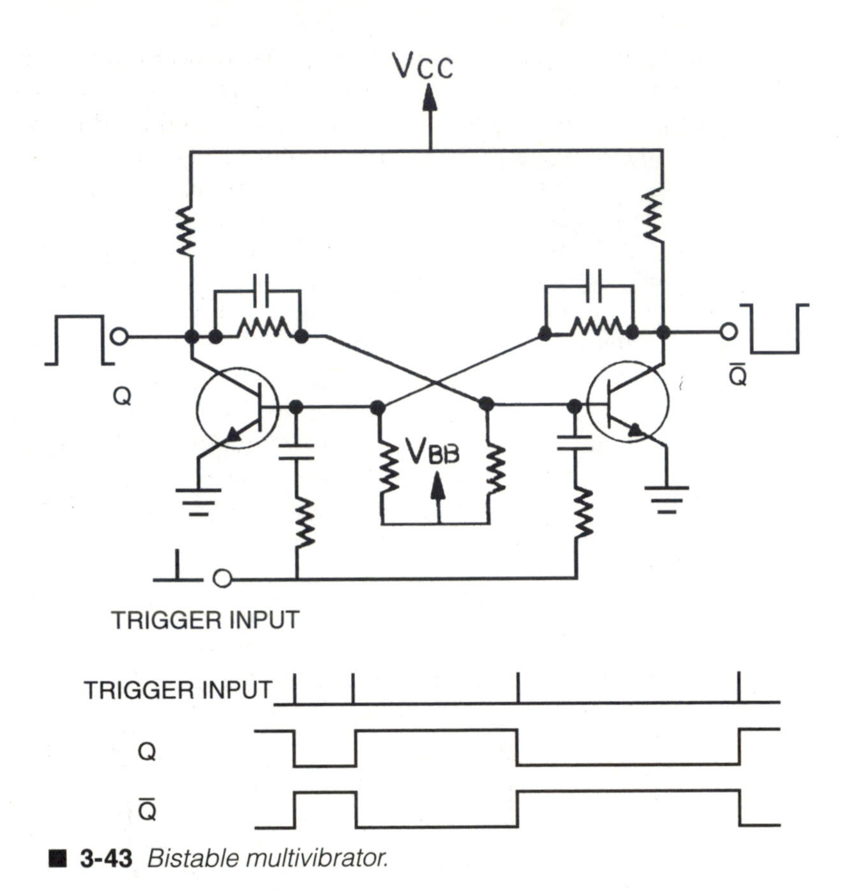

■ **3-43** *Bistable multivibrator.*

Increasingly, modern communications systems are using computers as their basic technology for equipment circuitry, design, processing, and control. Found in system timing, signal processing, and control circuits, their application has revolutionized the field. Now it is common to encounter such computer functions as *shift registers*, *random access memory* (RAM), and *central processing units* (CPUs) within communications equipments. Control has reached the point that on the latest U.S. Navy communications systems, frequency selection, antenna tuning, transmitter power out, and individual equipment selection are all capable of being controlled by computer. In civilian applications, the ever-more-popular cellular phones are computer-controlled. For proper operation these types of circuits require a synchronization signal called a *clock* to operate. A clock signal must have an exceptionally stable operating frequency due to the tolerances involved. The waveforms can be symmetrical, which is when both positive and negative alternations are equal in time duration. An asymmetrical clock is when one alternation has a greater time duration than the other. The more complex an electronics system, the more clocks that are required to ensure synchronization of the entire system. It is not uncommon for advanced state-of-the-art communications systems to require literally hundreds of clocks and synchronizing signals to ensure accuracy. As you progress in the field, you will find that a clock circuit is nothing more than a digital timer or multivibrator.

The vacuum tube and transistor multivibrator circuits are being superseded by digital gates connected as stable, two-output-producing circuits, such as the 555 integrated circuit. The reasons for the changes are that as the digital devices decreased in size, cost, and power consumption, reliability exponentially increased. To further help, circuit design, fabrication, and packaging have became easier. The result is that equipment sizes and cost have declined in relative terms while capabilities have dramatically improved. Just on the size basis alone, digital circuitry is far superior to transistors and vacuum tubes. It is now common for a microprocessor IC such as the Pentium to contain more that five million individual transistors.

Possibly the most common integrated circuit, the 555 chip is an eight-pin dual in-line package (DIP). Figure 3-44 is the drawing of a basic 555 timer circuit. Notice that for operation it requires only four external support components. Capable of operation with power supply voltages that range from 5 to 18 vdc, it is a very versatile circuit. External components determine the frequency of the timer's output waveform. With just DC voltages, a 555 timer can produce a stable square wave output. By adding an adjustable

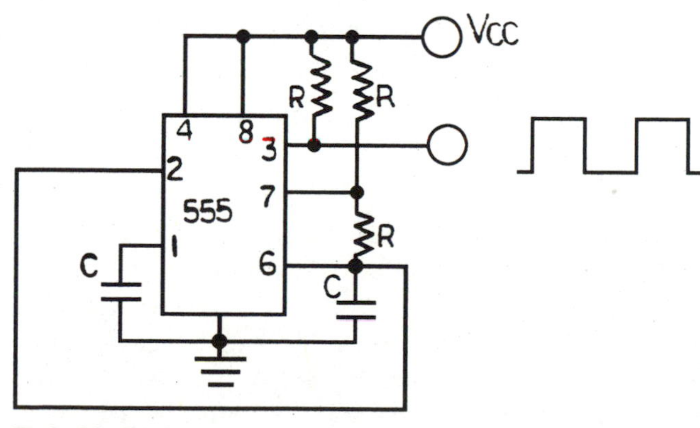

■ **3-44** *Digital 555 timer.*

resistor, the timer is capable of operation over a range of frequencies instead of one fixed frequency.

Circuit operation is straightforward. C2 is a small value component installed to increase circuit stability. R1, R2, and C1 determine the output frequency of the circuit. When power is first applied, C1 begins to charge. As charging begins, the input from C1 is applied to pin 2, the circuit's trigger input. Because C1 is just beginning to charge, the input on pin 2 is low. In this chip design, a low is an enabling input, so on pin 3, the output is active, going high. As C1 controls the 555, when its charge voltage reaches two-thirds of the power supply, the internal circuitry turns on, disabling the chip and driving pin 3 low. C1 then begins to discharge through pin 2 and the 555 timer's internal circuitry. When the charge on C1 reaches one-third that of the DC power supply, the timer is once again enabled, forcing the output on pin 3 to go low. The cycle repeats itself as long as power is applied with a nearly symmetrical, stable square wave as the resulting output.

Digital gates have revolutionized nearly all circuits, including multivibrators. Digital multivibrators have been engineered using basic gates, such as the AND, OR, NAND, and NOR gates. Flip flops have become very common in modern radars and other electronic equipments as they provide an inexpensive way to obtain accurate timing, processing, and storage circuits.

The final oscillator to be presented is one that is relatively new, and yet has become very common. Modern communications equipments and consumer radios feature advanced electronic tuning techniques such as scanning, search, memories for several hundred frequencies, and rock-stable performance. While these features are not new, they used to be performed by mechanical gear trains. Electronically, these features have been available for years on military equipments. Instead of crystals and mechanical drivetrains, simple, low-cost electronic circuits are used.

The voltage-controlled oscillator (VCO) provides for the automatic electronic control of output frequency. A VCO is an oscillator that provides an output frequency that is controlled by an input voltage. Figure 3-45 is a simplified schematic of a VCO. This circuit is built around a unijunction transistor (UJT). Constructed from only four components—UJT, two resistors, and a capacitor—it requires no power supply. The input control voltage provides all necessary power. The output frequency is inversely proportional

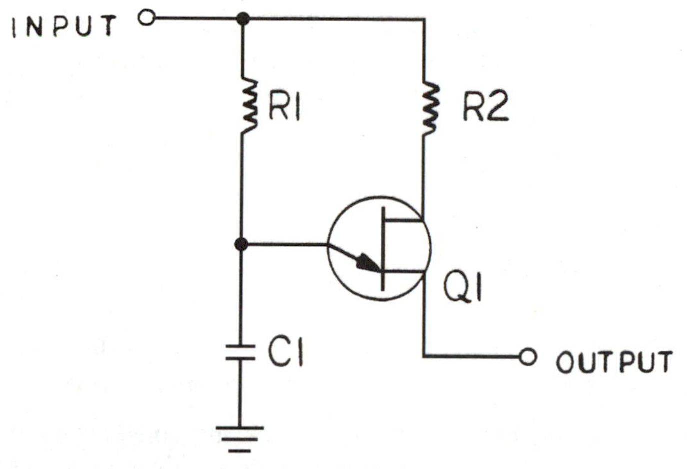

■ **3-45** *Simple voltage-controlled oscillator.*

to the input control voltage. R1 and C1 are the components that control the frequency range of the circuit.

Figure 3-46 illustrates a more complex VCO, the phase-locked loop (PLL). This type of circuit can be used as a phase detector, low pass filter, or VCO. A PLL consists of three functions: phase detector, filter, and VCO. The phase detector has two inputs, the signal input to the PLL and a feedback signal from the output. As long as the input signal and the feedback signal are in phase, which indicates the same frequency, the phase detector does not generate an error signal. If there is a difference in frequency, then the phase detector generates an error signal, changing the VCO frequency.

PLL operation is actually very straightforward. Use the schematic and timing signal in Figure 3-46. As an example, this circuit generates a square-wave output. The center frequency of this PLL is 10 kHz. The input control voltage can vary it +/– 10 kHz. With no input, the output is 10 KHz. If the input signal increases to 15 KHz, then the phase detector determines that there is a phase difference between the feedback loop and the input frequency. It then generates an error signal, which is applied to the VCO. The increased voltage applied to the VCO drives its frequency up. It continues to increase until the feedback signal and the input frequency are equal, and therefore in phase. If the input signal decreases to 5 KHz, then the phase detector determines that there is a phase difference between the feedback loop and the input frequency. It then generates a negative error signal which is applied to the VCO, driving its frequency

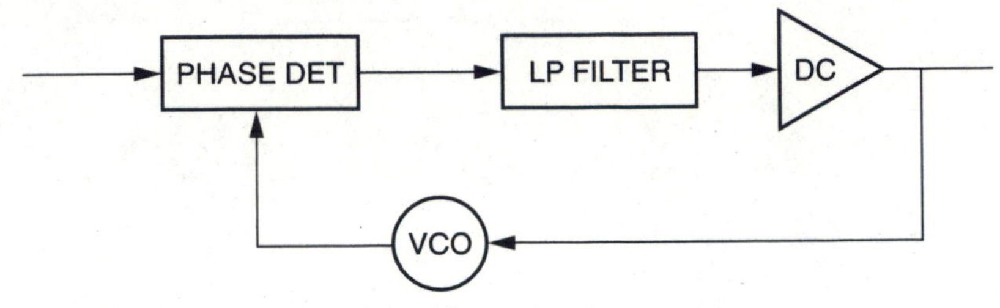

■ **3-46** *Simple phase locked loop circuit.*

down. It continues to decrease until the feedback signal and the input frequency are equal, and therefore in phase.

This chapter has introduced many concepts and basic building blocks that are required to form a complete communications equipment. Future chapters will build upon this information to discuss transmitter, antenna, and receiver theory.

4

Communications Transmitter Fundamentals

ANY COMMUNICATIONS LINK, REGARDLESS OF THE complexity, must consist of several basic subassemblies—a transmitter and antenna, separated by some distance from an antenna with a receiver.

Due to technological advances, it is now common to combine the transmitter and receiver functions in one unit, called a *transceiver*. Such advances have allowed advanced communications techniques to be packed into small packages so that users can be in contact regardless of the location. No matter what type of system we are talking about, the best point at which to begin any study of electronic communications is with the transmitter.

There are many theories and facts that are characteristic of any transmitter application. Electronic transmitters are used in many electronics fields, including electronic communications, radar, and sonar. Any effective transmitter must perform two basic functions. First, it must generate on the correct frequency a signal of sufficient power to reach the area of interest. Secondly, it must provide a means of changing the intelligence frequency to one suitable for transmission so it has the capability of covering great distances. *Intelligence* can be speech, code, music, or video. To perform these fundamental functions, it must have power supplies, a frequency determining circuit or oscillator, a modulator, and RF amplifiers. There are several methods used to classify a transmitter, with operational frequency and type of modulation the most common.

The output frequency of a particular transmitter is determined by several factors, including the type of communications link, required range, time of day, season, physical location, and type of intelligence to be transmitted. Because of all the variables that go into a successful communications system, it is a complex series of often-conflicting considerations and compromises. As the function of a communications system is the ability to exchange intelligence over long distances, the best starting point is with modulation techniques.

Modulation Defined

A communications system provides the ability to disseminate information of distances greater than can be traveled by the human voice. To do so, the information, be it voice, video, or some type of code, must be raised in frequency so that it is suitable to be transmitted through free space as electromagnetic radiation. Free space is considered to be the atmosphere, or outer space. By definition it is a region high enough so that the radio frequency energy radiated by a transmitting antenna is not affected by terrestrial objects such as topographical features, trees, and buildings. To convert intelligence to a form suitable for transmission through free space, an electronic technique called *modulation* is used.

The definition of modulation is the action of varying a characteristic associated with one wave in accordance with a characteristic associated with another wave. The base, or carrier wave, can be varied in frequency, amplitude, bandwidth, or duration of transmitter "ON" time. Whatever type of modulation is used, the receiver in the communications link must be capable of demodulating it to extract the intelligence component.

Communications systems are classified according to the type of modulation employed. The simplest and oldest type of modulation is *continuous wave* (CW). In this form of modulation, the transmitter produces an RF output that is constant in power and frequency. Intelligence can be transmitted only by using an encoding system, such as Morse code. The transmitter ON and OFF time is controlled to produce an RF output consisting of long and short pulses of RF energy separated by silence. The pulses form dots and dashes that represent letters, punctuation, and numbers in Morse code. It is still common on the HF (3 MHz to 30 MHz) communications band among radio amateurs, professionals, and commercial applications. It has several advantages, including simple equipment design, long-range capability, narrow bandwidth, and a high degree of intelligibility under even the most severe atmospheric noise conditions.

Figure 4-1 is the block diagram of a basic CW transmitter. As can be seen from the drawing, it consists of four major functions: oscillator, power amplifier, a means to turn the RF oscillations on and off, and power supplies. The function of the oscillator is to

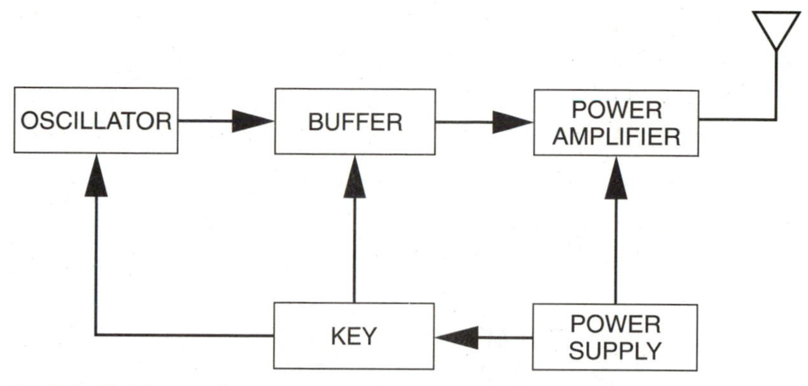

■ **4-1** *Basic continuous wave transmitter block diagram.*

generate the RF carrier used to convey the intelligence. It is imperative that the RF carrier be of a stable frequency and amplitude. The primary function of the buffer amplifier is to provide isolation of the RF oscillator from the power amplifier to improve stability. In more advanced designs it can provide additional amplification and frequency multiplication to the RF carrier. As the final stage, the power amplifier provides final amplification of the signal prior to transmission. Just below the buffer is a block labeled Key. The function of the key is to turn the buffer amplifier on and off, modulating the carrier. It can be a manual key with a slow rate of speed or an electronically controlled key for remote operation. In CW modulation, the intelligence is impressed upon the carrier by simply turning it on and off. As an example, Figure 4-2a is the resulting output waveform from the CW transmitter when the letter C is radiated. The other waveforms in Figure 4-2 are representative waveforms for other types of modulation, which will be covered shortly.

With CW modulation, the transmission speed of intelligence is slow and restricted to some type of code that requires a skilled operator. Today's complex civilization requires the rapid dissemination of audio and video information. *Amplitude modulation* (AM) allows for the more complex types of intelligence to be transmitted by a communications system. With no modulating signal, the transmitter produces a continuous RF carrier. When modulated, the normal amplitude of the carrier frequency is varied by the modulating intelligence. If you refer to Figure 4-2c, you can see how a typical AM waveform would appear.

The block diagram for a representative AM transmitter is illustrated in Figure 4-3. Notice that the only difference between the CW and

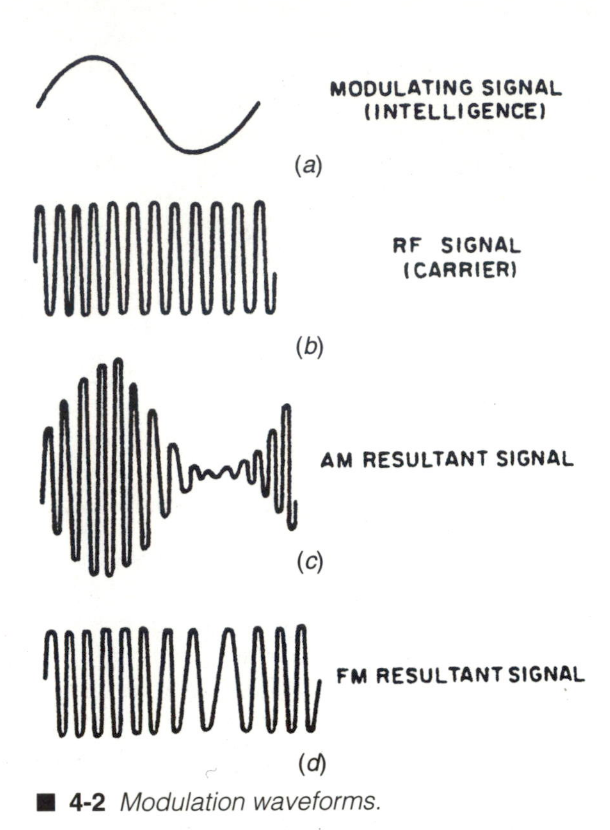

■ **4-2** *Modulation waveforms.*

AM transmitters is in the modulator. As this is a radiotelephone, or voice system, the microphone provides the modulating signal input. The microphone converts the audio input into electrical energy. The function of the modulator is to amplify the audio signal to a level sufficient to modulate the RF carrier. The oscillator generates the RF carrier at the required frequency. As with the CW transmitter, the buffer provides isolation between the power amplifier and the oscillator stages for stability. The final stage, or power amplifier, combines the carrier and modulating signal to produce the amplitude-modulated RF carrier suitable for transmission. In AM modulation, the much lower frequency of the intelligence is impressed on the higher frequency carrier. As an example, a broadcast AM station has a carrier frequency of 1200 KHZ—the carrier. The intelligence, or speech of the announcer, has a frequency of 5000 Hz. When the two are combined, the result is that the 1200 KHZ carrier is varied at the 5000 Hz rate of the voice frequency. The resulting 5000 Hz shape of the carrier is known as the envelope.

AM is suitable for the long-range transmission of audio, but it does have some drawbacks. *It lacks sufficient fidelity, or sound qual-*

ity, for the transmission of quality stereo music. Also, it suffers from susceptibility to atmospheric noise and interference. To reduce the noise problem, *single sideband* (SSB) was developed. In AM, the transmitter produces a modulated output waveform consisting of the carrier and two sidebands, an upper and a lower. The modulating signal varies the carrier frequency simultaneously above and below the carrier center frequency. Therefore, the intelligence appears above and below the carrier center frequency. As the intelligence is contained within both sidebands, only one needs to be transmitted. The result is SSB, an improvement over AM. In this type of modulation, the carrier frequency and one sideband are suppressed, or eliminated. Only one sideband is transmitted through free space. The result is that the SSB signal requires less power for a given range, takes up less valuable band space, and offers superior operational characteristics in a noisy environment. However, it is not suitable for the transmission of music.

Figure 4-4 is the block diagram for a typical SSB transmitter. The heart of this unit is the carrier frequency generator. It provides a stable, accurate frequency for the SSB generator and frequency multiplier. The audio amplifier boosts the low-level input audio signal to an amplitude suitable for modulation purposes. The SSB generator mixes the audio input and carrier frequency to produce a typical AM waveform consisting of a carrier and two sidebands. Through the use of filters and active electronic circuits it then removes the carrier, leaving just the two sidebands. An output filter assembly is then used to suppress one sideband and pass the other for further processing. The sideband is then applied to a mixer assembly. The other input to the mixer stage comes from the frequency multiplier stage. A frequency multiplier is very important, as most SSB carrier frequency generators operate at a relatively low frequency to ensure stability. The multiplier is then used to multiply the base carrier frequency to the one suitable for transmission. The output of the frequency multiplier is then routed to

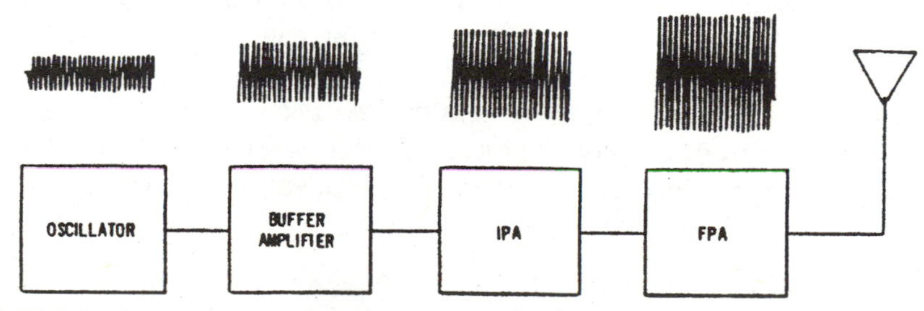

■ **4-3** *Amplitude modulation transmitter block diagram.*

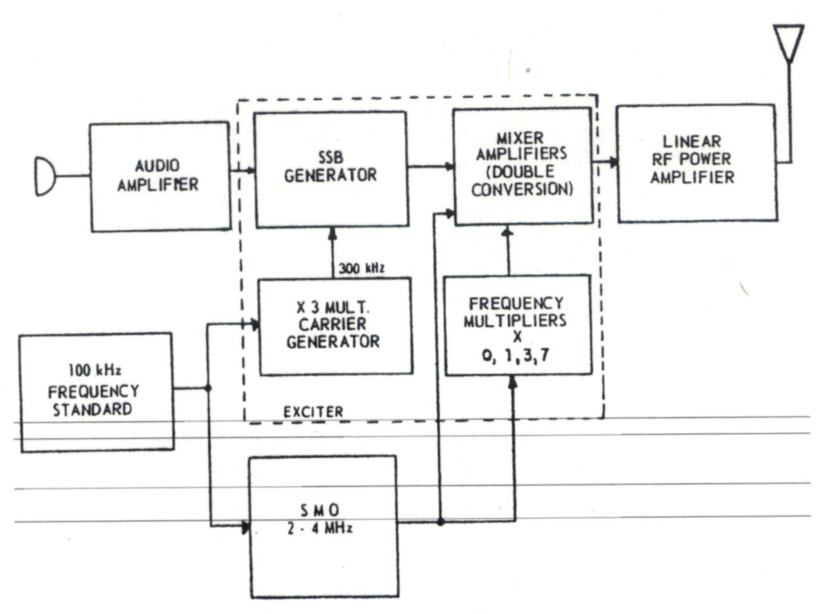

■ **4-4** *Single sideband transmitter block diagram.*

the mixer stage, where it is mixed with the one sideband. The mixer stage increases, or *translates*, the sideband frequency to a much higher one that is to be transmitted. The final stage is the power amplifier, which boosts output power to a level required for transmission for the desired range.

CW, AM, and SSB all lack the ability to provide for high-fidelity audio signals. To remedy the problem, frequency modulation (FM) is used. In this type of modulation, the frequency of the carrier is varied in respect to the amplitude of the modulating signal. Figure 4-5 is the block diagram for a simple FM transmitter. The modulating signal is applied to the unit via a *reactance tube*. This device has its reactance vary according to the input modulating signal. With no input signal, the output of the reactance tube causes the oscillator to produce a steady, stable center frequency. The output of the oscillator is then applied to the frequency multiplier stage, where it is increased to one that can be transmitted. The final stage is the power amplifier, which has the function of increasing the signal to the level required for transmission. The prime advantage of FM modulation is superior fidelity—faithful reproduction of the modulating signal. Its disadvantage is the amount of frequency space required for a signal.

One criterion for the selection of one type of modulation over another is a factor known as noise. Noise can be classified as atmospheric, galactic, and man-made. Atmospheric noise is caused mainly by lightning. Its affect to the communications signal is inversely proportional to the frequency of the communications link. Galactic noise is external to the atmosphere, and a good example of this natural occurrence would be solar flares. The final group of noise is man-made, and examples would include heavy electrical equipment and high-voltage power lines. In amplitude modulation and continuous wave modulation, noise superimposes a random component on the modulation envelope. Any lightning strike would appear as a simple spike. Frequency modulation is virtually unaffected by all but the most severe interference. Single-sideband modulation techniques, due to the narrower bandwidth requirements and stricter filter design criteria, is much less affected than AM by noise. CW is capable of penetrating heavy noise due to the simplicity of the modulation.

An AM Transmitter Block Diagram

Figure 4-6 is an expanded block diagram of a representative AM transmitter. As can be readily seen, it is more complex than the one illustrated in Figure 4-3. This is a normal progression encountered in technical repair manuals. A simplified block is used to isolate a failure to a function. An expanded block is used to locate a failed stage within the function. At that point, an in-depth schematic diagram is used to find the failed component. That is because even physically small electronic equipment is often too complex to use a component-level schematic diagram for troubleshooting without confusion.

One difference with this diagram is the addition of a microphone and required amplifiers and drivers. With a CW transmitter, all that was required was a telegraph key to turn the transmitter on and off. With AM, SSB, and FM, a microphone or other device is needed to couple the audio intelligence into the transmitter to

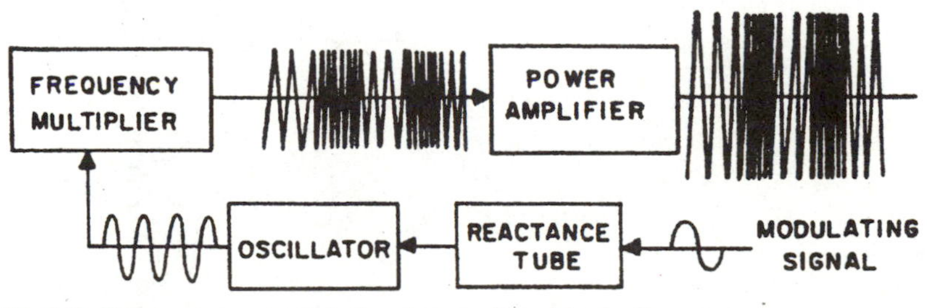

■ **4-5** *Frequency modulation transmitter block diagram.*

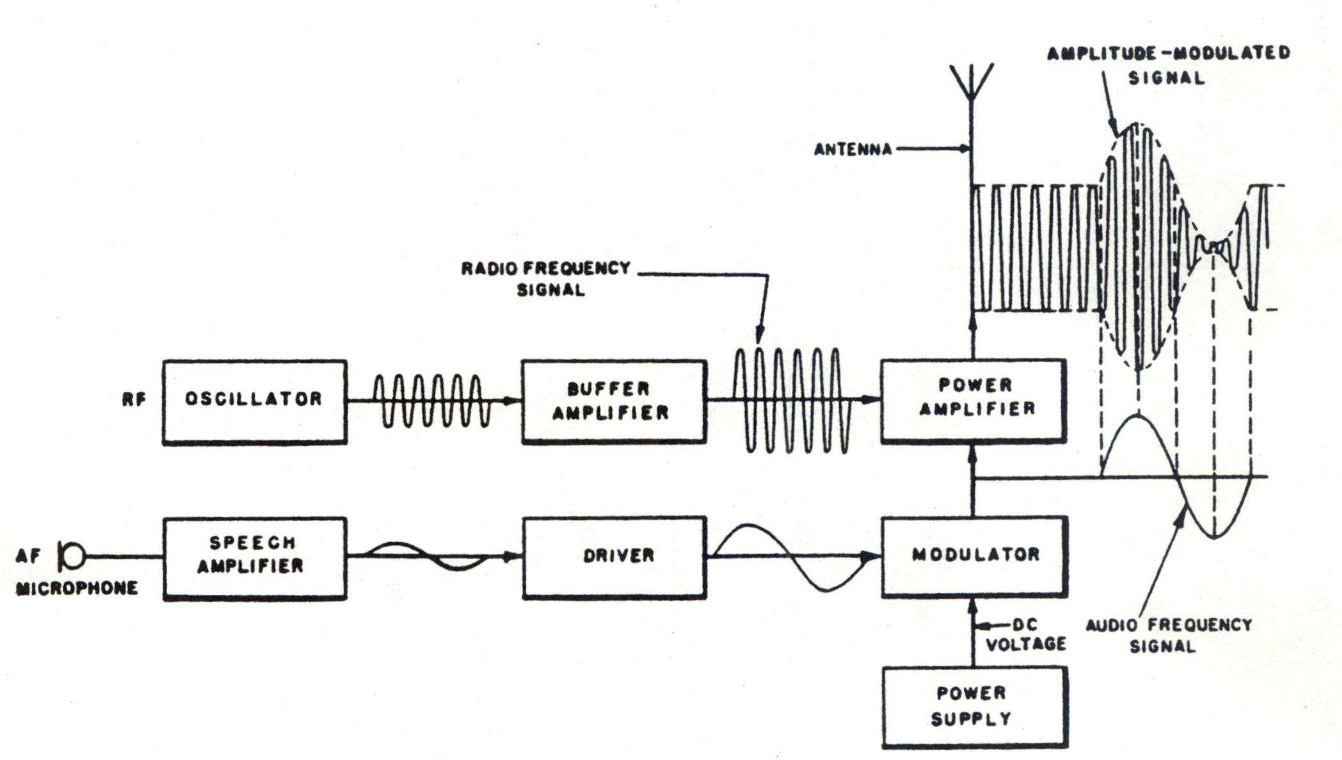

■ **4-6** *Exanded amplitude modulation transmitter block diagram.*

provide the modulating signal. Figure 4-7 is a drawing of a dynamic microphone.

The function of a microphone is to convert sound energy into electrical energy. The outer covering of the microphone is a diaphragm, which is connected mechanically to the voice coil through the use of a piece of small-gauge nonconducting wire. Sound energy strikes the diaphragm, causing it and the voice coil to move. That is because of the differential air pressure that exists on the opposite sides of the diaphragm when sound energy strikes it. The sound striking the diaphragm causes increased pressure, pushing it inward. Notice that the voice coil is located inside a stationary magnetic field. The motion of the diaphragm causes the voice coil to move through the magnetic field. That action induces an EMF in the voice coil that is in proportion to the pressure exerted on the diaphragm by the sound waves. A disadvantage with this type of microphone is its susceptibility to external electromagnetic fields, so a balanced (twisted pair) cable must be used to minimize interference.

Other types of microphones include the crystal, carbon, and condensor designs. Most microphones produce a very low-level signal output that must be amplified to a useable amplitude. To do so, a preamplifier is required to boost the signal. A preamplifier is desirable because it can provide impedance-matching and low-noise amplification to a low-level signal input. Impedance matching is important, as a good match results in the maximum transfer of electrical energy. As an example, if a microphone has an internal impedance of 50 ohms, then the input amplifier circuit should have an impedance of 50 ohms. A mismatch results in a decreased value of electrical energy for modulation purposes. A high-quality

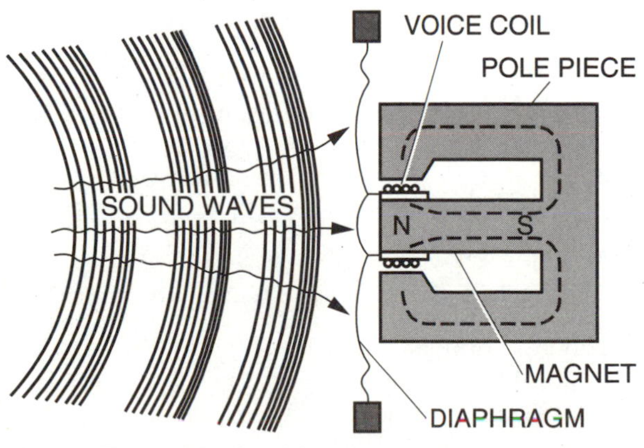

■ **4-7** *Dynamic microphone.*

preamplifier will boost signal level without losing any signal quality or inducing noise. The output of the preamplifier is further amplified and conditioned by the driver. The function of a driver is to provide a high-current signal to the modulator stage of the transmitter. The modulator stage is used to vary the amplitude of the carrier frequency in accordance with the frequency of the intelligence signal.

There are several points within the equipment where the modulation action can be accomplished. The carrier can be amplitude modulated at different component points in the RF section of the transmitter. In equipment design, modulation techniques are classified as to the element of the amplifier to which the modulation signal is applied. Commonly, low-powered applications use semiconductors and high-powered systems use vacuum tubes. As technology advances, the vacuum tube will eventually be completely replaced. The most common type of modulation is collector modulation in semiconductor equipment and plate modulation in vacuum tube designs. In this classification, the modulating signal is imprinted on the DC power supply for the collector or plate of one of the RF amplifiers. In this manner, the frequency of the RF carrier output of the modulated stage is varied by the frequency of the intelligence as the power supply for the stage is varied by the modulation signal.

A less common modulation technique is the application of the modulation signal to the base or grid of an amplifier. In this method, the modulation signal varies the bias of the amplifier. That in turn controls conduction, modulating the RF carrier. A modulation signal can also be applied to the emitter or cathode of an amplifier. It is similar to the base/grid modulation, as it varies the amplifier bias. In both types, all DC power supply voltages are held constant.

Modulation can also be classified by level. High-level modulation is found in collector/plate modulation. In this class, all amplifiers are operated class C. Although it is very efficient, a high level of voltage amplification is required for proper operation. Low-level modulation is typically applied to the grid/base or cathode/emitter of the amplifier. This type is less common, as it is difficult to obtain high degrees of modulation with these methods.

Collector modulation is more common than plate modulation, as the amplifier can operate at both low and high levels of modulation. Figure 4-8 is the schematic diagram for a representative collector-modulated transistor amplifier. As this is an RF circuit, transformer input and output coupling is used to provide for the maximum transfer of RF energy at the center frequency of the carrier. The amplifier stage consists of Q1 and support circuitry. T1 is used to couple the unmodulated carrier waveform into the stage. The transistor has a fixed bias applied to the base due to the action of R1, R2, and Vcc. The function of C1 is to bypass R1, preventing RF from feeding back into the power supply. If that were to happen, then RF would be felt throughout the transmitter, degrading operation. R3, tied to the emitter, is used to provide thermal stabilization. It is also bypassed by capacitor C2, which has the function of preventing AC (RF) variations from appearing on the emitter. In the event that were to happen, then the amplification factor of the amplifier would be reduced, reducing the output of the stage. That would occur because as a signal on the base caused the transistor to conduct harder, the increased current flow through R3 would increase the voltage drop felt over it. That in turn would leave less voltage to be felt over the transistor and collector load resistor, actually decreasing the output of the stage. As can be seen, the audio modulation signal is applied via transformer T3. Notice that T3 is also bypassed by a capacitor, C6. It has the function of decoupling the RF from the power supply and audio circuits. T3 is connected in series with the collector of the amplifier. T2 primary and C3 form a tuned tank. The function of the resonant circuit is to pass the desired frequency band of the modulated carrier, while blocking any undesired frequencies.

Due to the fixed bias, the transistor amplifier will operate at Class A, B, or C, depending on the amplitude of the audio signal input. The RF carrier signal is coupled into the stage via T1. The modulating

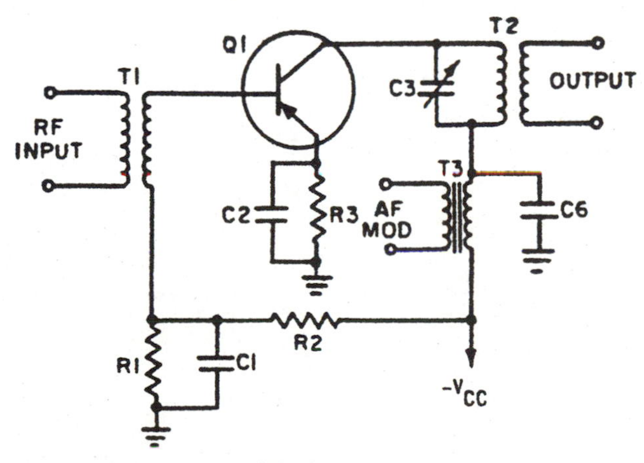

■ **4-8** *Collector modulation.*

AF signal is applied to the modulator stage via T3, which is tied back to the collector of the amplifier. The effect is that the transistor collector is in series with the output tank circuit (T2, C3), modulation input transformer, and Vcc. As the AF modulation signal changes, it either aids or opposes Vcc. The net effect is that the modulation signal varies the transistor gain, impressing the lower audio frequency on the stable carrier signal and forming the modulation envelope. During Class A operation, the transistor conducts for a full 360 degrees of carrier signal input. In Class B or C operation, the transistor is cut off for a portion of the time. Any time it is cut off, the output tank circuit (consisting of T2 and C3) restores any missing portion of the signal cycle to maintain the carrier signal.

Plate modulation is the vacuum tube version of output element modulation and it is illustrated in Figure 4-9. If you compare this circuit to the one in Figure 4-8, you will notice many similarities. The AF modulating signal and the modulated output are both coupled via transformers. The components have the same function in both circuits.

In this circuit, vacuum tube V1 is a driver that amplifies the low-amplitude audio signal. It is transformer-coupled to the grid of modulator tube V2. V2 is an AF power amplifier that provides the modulation signal input to the plate of RF power amplifier V3. In this configuration, the AF circuits are operated Class A, while the RF power amplifier is Class C. That arrangement obtains a high degree of efficiency and provides a nonlinear amplifier stage for desirable modulation. In this circuit, R1 (the cathode resistor for V2) is bypassed by C1. Its function is to prevent AC variations

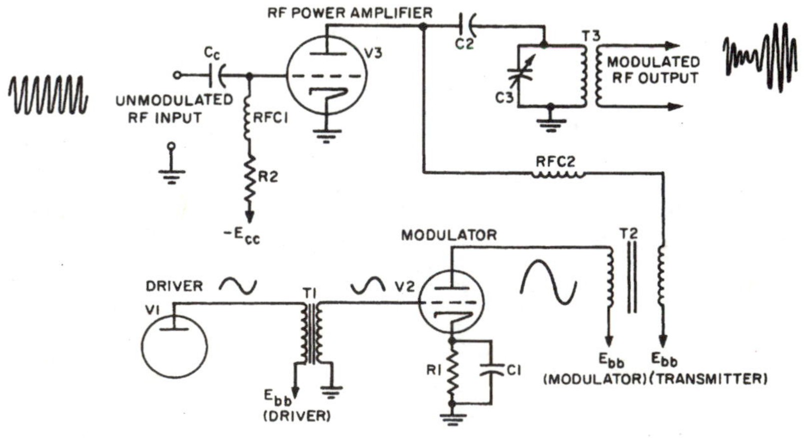

■ **4-9** *Plate modulation.*

from reducing the gain of the stage. Tank T3-C3 is tuned to the center of the carrier frequency band to pass maximum RF energy out of the stage. V3, the power amplifier, has its grid bias set by R2 and RFC1. An RFC is a radio frequency choke that has a high impedance to current at RF frequencies and low impedance to low-frequency or DC current. Its sole function is to prevent RF from feeding back into the power supplies. If that were to happen, then RF would be felt through the transmitter, degrading operation and affecting frequency stability.

With no audio signal applied, the output of the stage is an amplified version of the carrier input signal frequency. With a signal applied to the grid of driver V1, operation begins. V1 amplifies the signal and applies it to the plate transformer T1. T1 provides phase reversal and couples the signal to the grid of modulator tube V2. When the input AF waveform swings negative, driving the bias negative, V2 decreases conduction, increasing the voltage felt across T2. That in turn is phase inverted by T2, decreasing the plate voltage on V3. As R2 provides the total voltage to V3, the output of the power amplifier swings negative. As the unmodulated RF is applied to the grid of V3, the changing audio voltage on its plate causes the RF carrier to vary its amplitude at the audio frequency rate. The output from the stage is the high-frequency carrier varied at the lower audio frequency.

It is common for high-power modulation applications to use tetrode and pentode amplifier tubes with plate and screen grid modulation techniques. Figure 4-10 illustrates a representative plate and screen grid pentode modulator. In this design, both the grid and plate voltages are varied to produce a modulated signal. In this configuration, efficiency is very high and produces maximum power. This circuit is capable of producing signals of 100% modulation. Notice that in this design only one tube is used, instead of the two in the plate-modulated version. Notice that this circuit uses capacitors and radio frequency chokes (RFC) for power supply decoupling to prevent RF variations from being fed to other circuits. Factors such as cost, size, weight, power, and complexity determine what combination of RFCs, capacitors, and resistors are used to perform the vital task of RF decoupling. Once more, transformers are used to couple the AF modulating signal into the stage and the modulated signal out.

In plate and screen modulation, the unmodulated RF signal is applied to the screen grid. Decoupling is accomplished by RFC1 and C1. Due to combination bias determined by the grid resistor

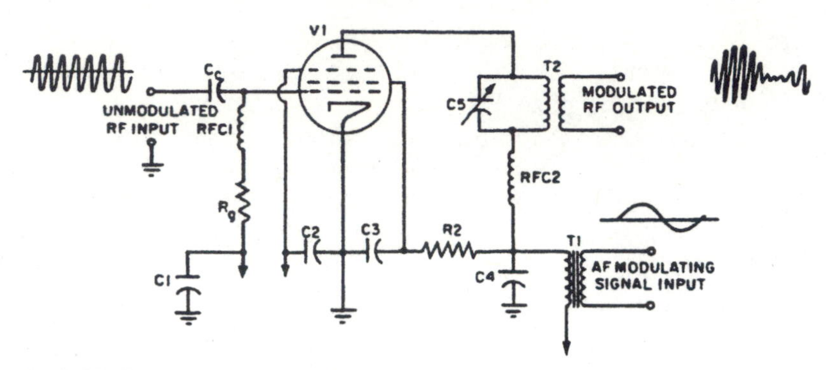

■ **4-10** *Plate and screen grid modulation.*

Rg and grid leak bias, the stage operates Class C. The screen grid is used as an input for the AF modulating signal. As the screen grid requires a lower voltage than the plate, R2 is either a high value to drop the power supply voltage to the proper level, or a separate low-voltage power supply is employed. As so often is the case, the choice is driven by cost and complexity. The AF modulating signal is brought into the stage via T1, and it is then routed to the screen grid via R2 and the plate circuit by the output tank circuit and RFC2.

With no signal input to the stage, static DC voltages are applied to both the plate and grid of the amplifier. As the carrier is constantly fed to the stage, the output is an amplified version of the unmodulated carrier. As the tube is biased Class C, it conducts for only one-half of the input oscillations. However, the positive alternations that are passed by the tube are sufficient to bring the tank circuit into oscillations, restoring the missing signals. That is because the tank is tuned to the center of the carrier frequency.

The modulation process begins with the application of the audio modulation signal. The modulation signal, routed via T1, is applied to the control grid through R2 and the plate via the tank circuit. When the AF signal is swinging positive, the induced voltage in the secondary is additive with the plate supply. That action increases the voltage felt on the screen grid and plate. A negative AF signal is subtractive with the plate power supply, decreasing voltage on both the plate and grid. If only the plate or grid voltage were varied with the AF signal, modulation would result, but it would be limited. By applying the AF signal to both the grid and plate, enough plate current variation can be obtained for 100% modulation.

In simple, low-powered transmitters, the output of the modulator would be routed directly to the antenna for radiation. More complex and high-powered transmitters would use further stages of power amplification. If you refer to Figure 4-6, notice that the representative transmitter uses a power amplifier.

The heart of any transmitter is the oscillator. The function of the oscillator is to provide the stable, accurate oscillations to be the basis for the carrier frequency. An oscillator must produce an output waveform of constant amplitude and frequency under all conditions. As was mentioned in Chapter 2, an oscillator can be any one of several types. The crystal oscillator, due to its stability and years of trouble-free operation, is in many ways the best. The output of the oscillator is applied directly to a buffer amplifier. The function of the buffer is to isolate an oscillator from the effects of impedance load variations caused by subsequent electronic stages. The output of the buffer amplifier is applied to the power amplifier. Depending upon the complexity and design requirement for the transmitter, this can consist of the modulator in low power applications or multiple stages in high-powered equipments. Multiple stages of amplifiers are often used when a single stage cannot provide sufficient power amplification without severe distortion. Figure 4-11 is the simplified block diagram for a high-powered equipment that uses multiple stages of amplification. Notice that the output of the buffer stage is 1 watt. The first intermediate amplifier (IFA) increases the power by a factor of 10 to ten watts. The second IFA generates an increase to 50 watts, an amplification factor of 5. The final stage, the RF power amplifier boosts the output to 500 watts, an amplification factor of 10. So, from a small signal of 1 watt, a high powered output signal of 500 watts was obtained. As can be seen from this, power amplifiers are very important circuits.

Although CW communications systems were the first to be developed, the communications bands around the world are filled with the sounds of Morse code, as it is still an important method of ensuring a communications link with low-technology equipment in severe interference situations. AM is used for commercial broadcast, HF communications systems, and international medium-wave stations. In many applications it has been replaced by single-sideband and frequency modulation systems, but will be in use for many years to come.

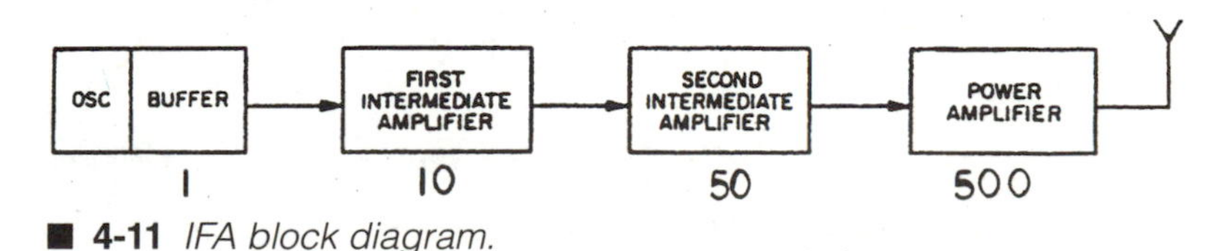

■ **4-11** *IFA block diagram.*

SSB Techniques and Transmitter Block Diagram

A single-sideband (SSB) communications system is one in which the RF carrier and one sideband are eliminated, or suppressed. Such systems have the advantages of requiring less power and a narrower frequency band. The result is the capacity for more communications links in the same frequency space. As the signal bandwidth is decreased, it is less susceptible to interference, both man-made and natural. Figure 4-12 compares a normal AM signal with an SSB signal. Notice that the AM signal is composed of a carrier frequency and two sidebands. The carrier frequency is the center frequency of the signal. Each sideband is equal in frequency width. That is because the modulation process causes an equal frequency change on either side of the center frequency. In the case of the signal illustrated in Figure 4-12, the carrier frequency is 20 kHz. Each sideband has a width of 2.8 kHz. The total bandwidth for the AM signal is 6 kHz.

In SSB modulation, the carrier and one of the sidebands are suppressed prior to modulation. If one were to examine the RF radiated by an SSB transmitter, the resulting waveform would be only half the width of a comparable AM signal. Suppression of the undesired carrier and one sideband occurs after the intelligence is used to modulate the carrier, leaving only one sideband to be applied to the final RF amplifiers for final amplification and transmission. The result is that all the RF energy radiated by an SSB transmitter is used to convey the intelligence in the form of one sideband, and not "wasted" in the carrier and second sideband.

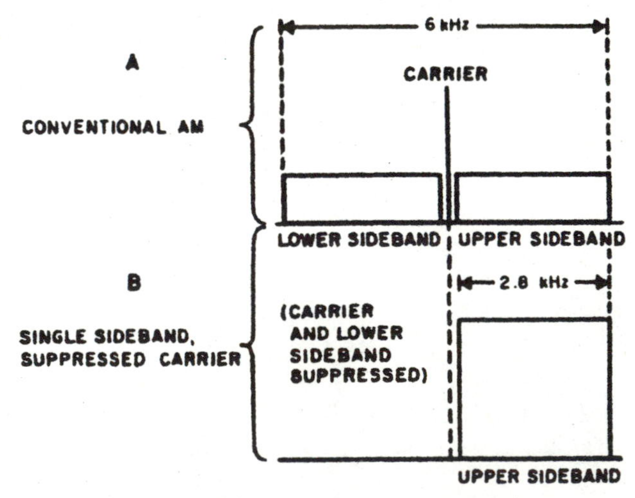

■ **4-12** *AM and SSB waveforms compared.*

Typically, after the carrier has been suppressed, only the two sidebands—the upper sideband and lower sideband—are left. That is possible, as the primary function of the carrier frequency is to raise the low frequency of the intelligence to a much higher frequency suitable for transmission through free space.

An SSB communications system does have several disadvantages over a comparable AM system. Most importantly, SSB equipment is more complex than equivalent AM equipment. Frequency stability is critical in the proper operation of all SSB equipment—both transmitters and receivers. Also, as an SSB signal is very narrow, it is possible for a weak signal to be masked completely by more powerful ones. With the current advances in electronic circuitry, these problems are not insurmountable and are far outweighed by the advantages.

Figure 4-13 is the block diagram for a simple, basic SSB transmitter known as the *filter type*. Notice in the diagram that the unit is divided into two sections: the SSB exciter and the RF section. The basis for the transmitter is the RF carrier oscillator, located in the SSB exciter. As frequency stability and accuracy are primary concerns in SSB communications, a crystal oscillator is usually employed. Typically, the RF carrier oscillator operates in the 100- to 500-kHz range. The single output of the RF carrier oscillator is routed to the balanced modulator. The function of the modulator circuit is to combine the RF carrier oscillator frequency and the input audio modulating signal. The audio signal is impressed on the RF carrier, and is amplitude modulated. After modulation, the circuit suppresses the carrier frequency. As a result, the output of the stage is the upper and lower sidebands. Any audio component left after the modulation process is blocked from the rest of the transmitter by the narrow bandpass of the modulator's final output circuit. The output of the balanced modulator is applied to the sideband filter, which blocks one sideband and passes the other. This type of filter is very critical, as it must have a very sharp cutoff frequency to allow passage of the desired sideband and elimination of the undesired one without signal loss or distortion. The filter can be a mechanical, crystal, or reactive component design. The output of the exciter is a stable, low-power, low-frequency single-sideband signal.

The stages following the exciter in the RF section are very similar in design and operation to a standard AM transmitter. The balanced mixer and HF variable oscillator in the RF section serve as a frequency multiplier, which has the function of boosting the

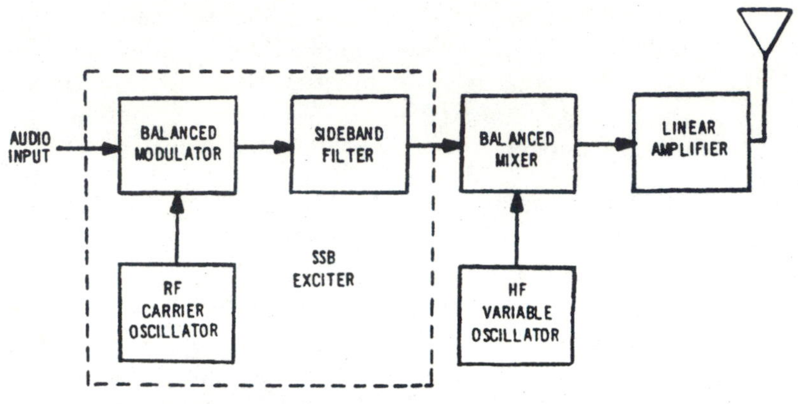

■ **4-13** *Basic SSB transmitter block diagram.*

frequency of the single sideband signal to a frequency suitable for transmission. Although only one step of frequency multiplication is shown, any number of multiplying stages could be used to achieve the desired frequency and stability. The final stage of the transmitter is a linear amplifier, to ensure the frequency stability of the SSB signal.

As was stated previously, SSB operations require a high degree of frequency accuracy. Not only can the equipment (receivers and transmitters) induce error, but so can natural phenomena. The movement (or *propagation*) of the radiated RF through the ionosphere can induce a frequency error due to the Doppler effect. The Doppler effect is when the relative motion between a source (transmitter) and an observer (receiver) causes a change in frequency. The result of the distance between the two decreasing is an increase in the received frequency. An increasing distance would cause the receiver frequency to decrease. One cause of the problem is that the atmosphere is not a stable medium, as it is composed of individual gas molecules that are under constant motion. Radiated RF from an antenna produces a wavefront, known as the sky wave, that reflects off the ionosphere. The resulting frequency shift is small—only a few hertz—but can induce some distortion. A greater problem exists when the relative movement of the receiver and transmitter are explored.

The Doppler effect is most evident when the transmitter and receiver are in relative motion, such as communications between two aircraft, or between an aircraft and a ground station. The amount of frequency shift is also dependent upon the frequency used for the communications links. A frequency of 20 MHz is a good example, as many military communications links are found around that frequency. U.S. Navy fleet communications, Air Force strategic communications, and tactical communications links are found between 15 MHz and 25 MHz. When in that frequency range it has been calculated that every 670 miles per hour of velocity difference between the two communicating stations results in a Doppler shift of about 20 Hz. As stated earlier, the maximum allowable error for SSB communications is 50 Hz, leaving only a maximum error of 30 Hz in the transmitter and receiver circuits. To ensure compliance, most military, space, and commercial applications require an absolute accuracy standard of ±.1 per 1,000,000 cycles. As it is not uncommon for high frequency (3 MHz to 30 MHz) SSB communications systems to have several thousand channels, the design requirements are intimidating. The advent of digital and computer technology has made the design of this equipment easier, but proper equipment operation and maintenance is mandatory.

A technological answer to accuracy requirements was in the development of the frequency synthesizer. By definition, a frequency synthesizer is an electronic circuit that produces a signal frequency by the action of heterodyning and combining frequencies that are not harmonics of a fundamental frequency. Frequency synthesis has several very important and desirable characteristics. It features the frequency stability of a crystal oscillator. All-electronic frequency control can be used, allowing for the application of computer technology. Finally, it is very reliable and reduces equipment cost, as it eliminates many mechanical and electromechanical components. As this type of circuit has been in use for many years, there are numerous designs in use. Figure 4-14 is a simplified frequency synthesizer block diagram that illustrates the concept. In this example, two crystal oscillators are used: 4 MHz and 6 MHz. The mixer stage combines the two frequencies and produces four outputs—the two original frequencies, the sum of the two frequencies, and the difference between the two frequencies. Filters are used to eliminate unwanted frequencies and pass the desired ones. In this case, the sum and difference frequencies are the ones to be utilized. Filter one is the sum leg and passes the frequency of 10 MHz, shunting the others to ground. Filter two is the difference path, passing the frequency of 2 MHz.

Figure 4-15 is the expanded functional block diagram of a typical SSB transmitter that includes a frequency synthesizer. This is considered to be a *phase lock loop (PLL) synthesizer*, as a phase detector is the basis for the function. As this unit is expected to operate over a frequency range extending from 1.7 MHz to over 30

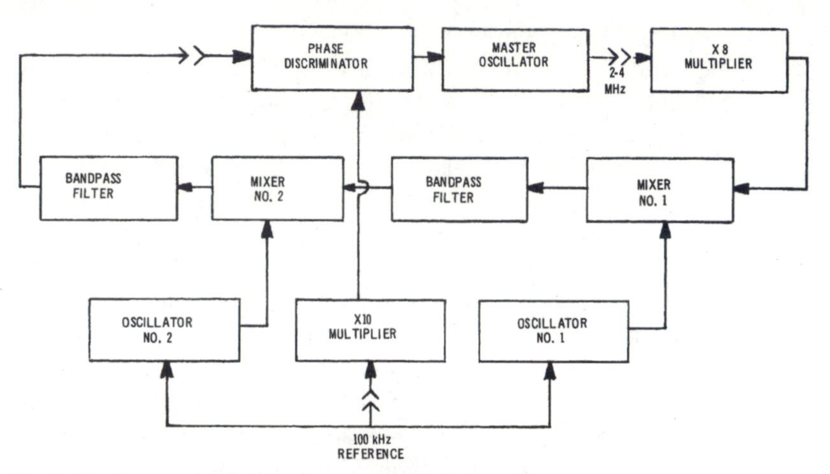

■ **4-14** *Basic frequency synthesizer block diagram.*

MHz, it is considerably more complex than the simple one illustrated in Figure 4-13. The frequency synthesizers and mixers are broken down into a greater detail.

The first points to consider are the frequency multiplication functions. The main function of the frequency synthesizer is to provide an accurate and stable reference frequency for a *variable-frequency master oscillator* (VMO), or as it is more commonly known, a *stabilized master oscillator* (SMO). The frequency multiplier stage consists of two multipliers, a 3X and 7X. Through the use of two frequency multipliers, a more flexible and stable frequency range is obtained. The function of the buffer amplifier is to provide isolation, shielding the SMO from any load variations. In a transmitter, that would be caused by the high-powered amplifier stages.

The operation of the transmitter is as follows: Front panel switches are used to select transmitter operation parameters such as power output and frequency. To go through frequency selection and how the frequency multipliers function, an arbitrary frequency of 300 kHz for the SSB carrier generator will be used. To select a transmitter frequency of 3 MHz, the equipment action is as follows: To obtain an output frequency of 1.7 MHz, the FREQUENCY SELECT control on the front panel in conjunction with inner switches sets up the frequency synthesizers for proper operation. In this frequency example, the SMO is used as the prime determining function. The function produces a base signal, a 2 MHz SMO signal frequency that is applied through a buffer amplifier to the subtractive mixer stage. A second input to the subtractive mixer stage is

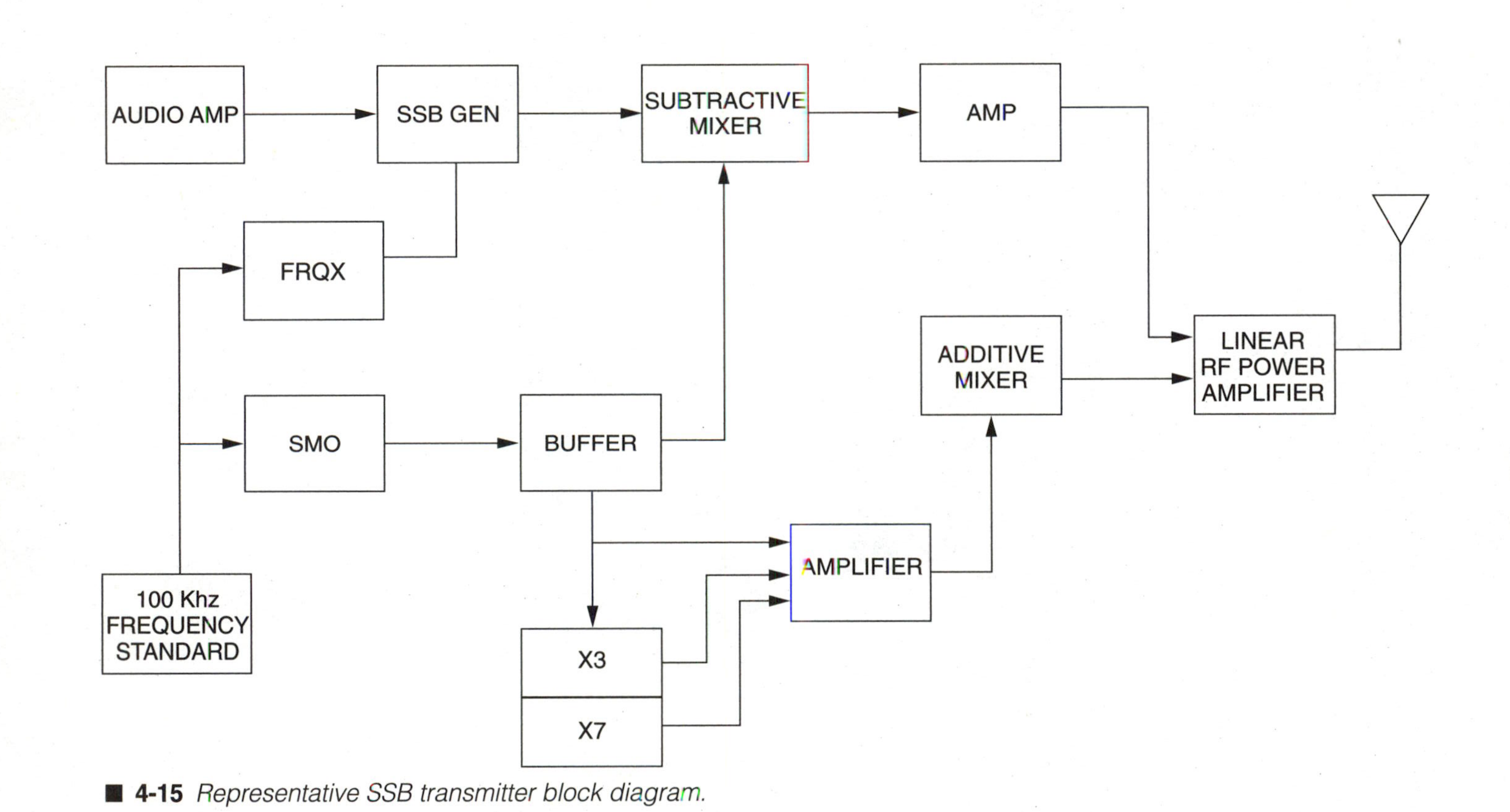

■ **4-15** *Representative SSB transmitter block diagram.*

the 300 kHz SSB generator carrier frequency. The subtractive mixer combines the two input frequencies, with a 1.7 MHz output signal as the result. The 1.7 MHz SSB modulated signal is then routed through an amplifier stage for application to the linear RF power amplifier. The RF amplifier provides for final power amplification and transmission of the signal.

Through use of front panel controls and internal switches, the outputs of the multiplier stages can be selected to provide stable higher-frequency signals suitable for transmission. The three times multiplier stage produces a signal with an output range of from 6 MHz to 12 MHz. The seven times multiplier stage is a function capable of producing a much higher output signal, and has a range from 14 MHz to 28 MHz. In conjunction with the SMO, an overall transmitter frequency range from 1.7 MHz to 31.7 MHz can be obtained.

To set the transmitter to operate on a frequency of 7.7 MHz, the following steps are required. The front panel frequency switch must be placed in the 7.7 MHz position. As the switch does not control the SMO, it still provides an input to the subtractive mixer, a frequency of 1.7 MHz. The output of the three times frequency multiplier is used. It has a single input, the 2 MHz signal from the SMO and buffer amplifier. The function of the three times multiplier stage is to increase the base frequency to 6 MHz, which is routed to the additive mixer via an amplifier stage. The same switch that selects the output of the frequency multiplier also selects the additive mixer. The additive mixer has two inputs, the 6 MHz from the multiplier and the 1.7 MHz output of the subtractive mixer. The two frequencies are mixed, with a resulting output of 7.7 MHz. Therefore, through the selection of mixers, additive circuits, and subtractive circuits, any frequency between 1.7 MHz and 31.7 MHz can be obtained.

Specialized SSB Circuits

At this point, a good question would be "How are frequencies mixed and combined to produce the synthesized frequencies and still have system stability maintained?" The answer is in the SMO function. The technique is known as *frequency translation*, which is the ability to shift a modulated RF carrier to a different frequency without any impact on the carrier frequency or sidebands.

As a point of reference, remember this was illustrated in Figure 4-13 as a single block. The single input is on the left and is a 100-

kHz reference from the frequency standard. The single output is the 2- to 4-MHz waveform from the master oscillator. The output frequency is electronically adjustable from 2 MHz to 4 MHz in 500-hertz increments. Frequency accuracy is maintained because of the feedback loops providing error signals. Oscillator number 1 on the upper left-hand side of the diagram is the coarse adjustment, with a range from 19.5 MHz to 35.5 MHz. Its range is adjustable in 160 steps of 100 kHz. Each 100 kHz change causes the master oscillator to change its frequency 12.5 kHz. The fine frequency adjustment is performed by oscillator number 2, with a range of 2.4 MHz to 2.5 MHz in twenty-five 4-kHz steps.

The key to frequency translation and frequency stability is the use of feedback, mixers, and filters. The 100 kHz reference is applied to a ten times multiplier, which is used to provide a 1-MHz reference signal to the phase discriminator to stabilize the master oscillator.

Function operation is straightforward and is as follows: For this explanation, the SMO is operating with an output of 2 MHz. For this to occur, oscillator number 1 has an output of 19.5 MHz and oscillator number 2 one of 2.5 MHz. Notice that the output of the master oscillator is also applied to the eight times multiplier, which boosts the frequency up to 8 MHz and routes it to mixer number 1. The second input to the mixer is the 19.5 MHz output of oscillator number 1. As will be presented later, a signal mixer receives two inputs and produces four outputs—the two original frequencies, the sum of the two frequencies, and the difference of the two frequencies. Bandpass filters are used to select the desired output and block the other three. As a result, the desired output of the first mixer is 3.5 MHz, which is routed to mixer number 2 via the bandpass filter network. Oscillator number 2 has an output of 2.5 MHz, which is also applied to mixer number 2. With two inputs, one 2.5 MHz and the other 3.5 MHz, the desired output of mixer number 2 is 1 MHz. Bandpass filter number 2 blocks the original frequencies and the sum frequency, passing only the difference frequency. The 1-MHz difference frequency is applied to the phase discriminator. As there is no change in the phase of the reference frequency or the feedback frequency, the oscillator is stable.

With an increase of 500 Hz in transmitter output frequency, the following changes occur in the frequency translation system: The frequency shift begins with oscillator number 2, with its output decreasing 4 kHz to 2.496 MHz. Routed to mixer number 2, it is combined with the 3.5-MHz output of mixer number 1. The bandpass filter network passes the difference frequency of 1.004 MHz while blocking all other frequencies. The 1.004 MHz frequency is routed to the phase discriminator, where it is compared with the 1-MHz reference signal. The phase discriminator, a phase detector, senses a feedback signal that is higher in frequency than the reference signal. The result is that the phase discriminator develops a negative error signal to drive the frequency of the master oscillator to increase its output 500 kHz. The now-2000.5 MHz master oscillator signal is routed externally to the mixer and multiplier stages of the transmitter to increase the frequency of the radiated signal. Internally it is applied to the eight times multiplier stage, which increases the signal frequency to 16.004 MHz. From the multiplier it is applied to the mixer, where it is mixed with the 19.5 MHz output of oscillator number 1. As the desired frequency output from mixer number 1 is the difference frequency, you would observe a 3.496 MHz signal at this point. After the unwanted frequencies (the two original and the sum) are eliminated by the bandpass filter, the 3.496 MHz signal is applied to mixer number 2. Mixer number 2 combines the 3.496 MHz feedback signal with the output of oscillator number 2, which is 2.496 MHz. The difference frequency is 1 MHz, which is routed to the phase discriminator. As the 1-MHz reference signal and the feedback signal are the same frequency, an error signal is not produced by the phase discriminator. That indicates that the transmitter is operating on the proper frequency.

Single-Sideband Circuits

By examining the block diagrams of various types of SSB transmitters, it should be apparent that for proper operation, high-quality filters and precise frequencies are required. Figure 4-16 is a drawing of a common SSB filter, the mechanical filter. Typical bandpass filters found in SSB applications are highly selective, with a common bandpass of 100 kHz to about 600 kHz. The signals passed through the filters must have steep rise and fall characteristics with a flat bandpass. Included in Figure 4-16 is the ideal response curve for a mechanical filter. Notice that the waveform is approximately 6 kHz wide. The rise and fall for the waveform are sharp, with a pulse width at the half-power points of about 4 kHz.

The mechanical filter has several advantages over crystal and reactive component filters. The devices are high Q (quality), physically small, very rugged, and exhibit excellent rejection of frequencies outside of the designed bandpass. As can be seen in Figure 4-16, the filter consists of moveable metal discs, rods, and

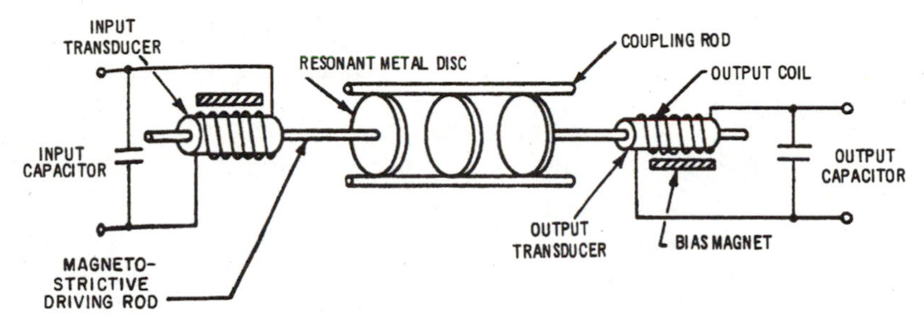

■ **4-16** *Mechanical filter diagram.*

a transducer. An input transducer is used to convert electrical energy into mechanical energy through the use of a physical phenomenon known as the magnetostrictive effect.

The magnetostriction effect is similar to the piezoelectric effect associated with crystals. It functions through the application of a changing magnetic field to an iron rod. The force of a magnetic field causes a change in rod length due to strain. In effect, the magnetic field squeezes the rod, making it smaller in diameter and longer. Conversely, when an iron rod is placed within a magnetic field, the field changes strength based on the effect that it has on the rod. The changes are dependent upon factors such as the direction of the magnetic field, polarization of the rod, and its material content. Just like resonant electronic circuits, metal bars have a resonant mechanical frequency. If a changing magnetic field with a frequency close to that of a metal bar's mechanical resonant frequency are brought together, oscillations can be set up. The induced change from the magnetic field must be properly phased to enhance the mechanical strain oscillations. If the metal bar is clamped in the center, flexural motion vibrations similar to a tuning fork are produced.

As stated previously, the function of the transducer is to convert the electrical energy of the signal frequency into mechanical energy. Mechanical oscillations are transmitted from the input transducer to coupling rods and metallic disks. Although only three disks are shown in the illustration, any number may be used to obtain the proper bandpass waveform shape characteristics. Each disk functions as a mechanical series resonant circuit, as shown in Figure 4-16. The disks are manufactured to resonate at the center frequency of the bandpass filter's bandwidth. The last (or *terminal*) disk vibrates an output transducer rod, which induces a current into the output transducer coil. To couple the electrical energy into and out of the transducers, capacitors are used as parallel resonant circuits. The resonant frequency of the input and output capacitors must match that of the mechanical filter.

The range of frequencies passed by a particular filter is determined by the physical area of the coupling rods. The frequency is increased by using additional rods or increasing rod size. Mechanical filters are available with bandwidths ranging from .5 kHz to 35 kHz, with center frequencies from 100 kHz to 600 kHz.

Another common SSB filter is the crystal lattice filter, as illustrated in Figure 4-17. The main advantages of this type of filter would be low cost and design simplicity. Although the drawing shows six crystals, more can be used in the construction to obtain the desired results. In effect, the filter is a balanced bridge network.

SSB Stabilized Master Oscillator

As has been stated, frequency accuracy and stability are mandatory requirements for SSB operation. Oscillators were presented

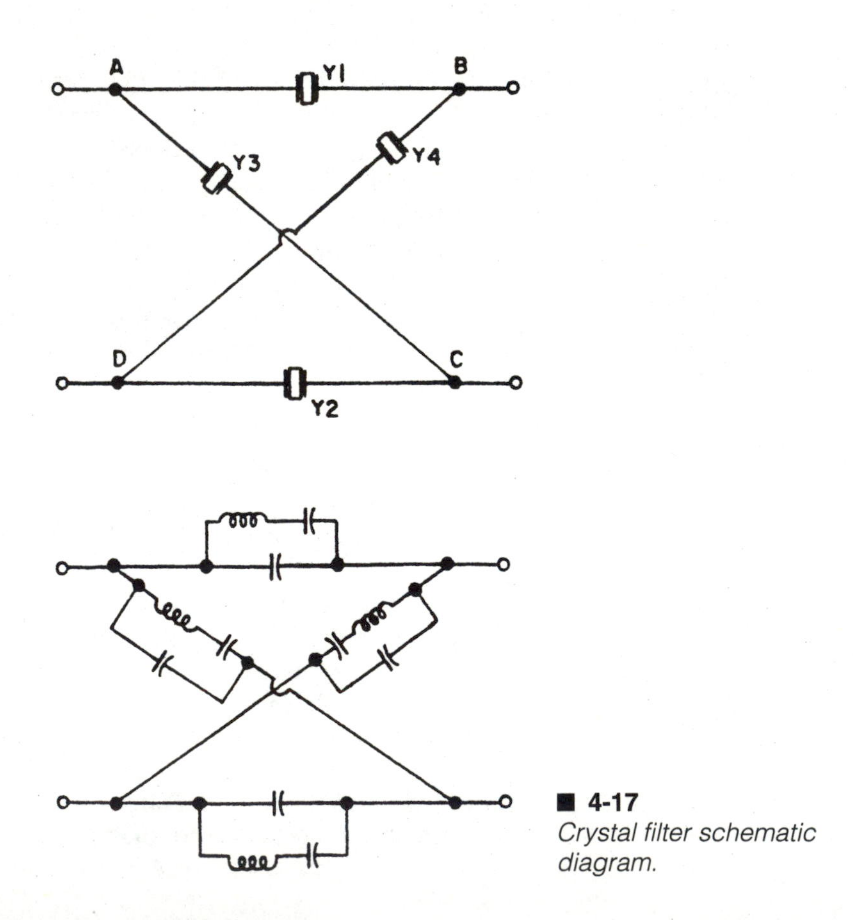

■ 4-17
Crystal filter schematic diagram.

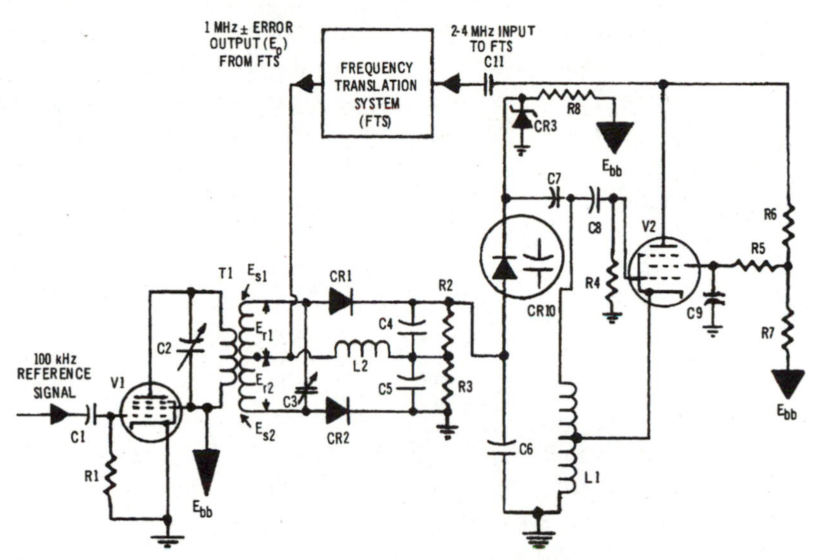

■ **4-18** *Stabilized master oscillator schematic diagram.*

in Chapter 2, but not oscillators of the complexity and precision requirements of an SSB stabilized master oscillator. Figure 4-18 is one example of an SMO. Although it is difficult to see initially, this is a conventional series-fed Hartley oscillator that is fed by a phase detector. The phase detector consists of V1, T1, CR1, CR2, L2, C1, C2, C3, C4, C5, R1, R2, and R3. The Hartley oscillator is composed of V2, L1, CR3, CR10, C6, C7, C8, C9, R4, R5, R6, R7, and R8, all located to the left of the dotted line. Between the two circuits is a box labeled *Frequency Translation System*, consisting of mixers, buffers, and multipliers. The only component that you might not be familiar with is CR10, a varicap. A varicap, or varactor, is a semiconductor diode with an internal capacitive component that is controllable by the value of reverse bias placed across it.

Circuit operation is conventional. The 100-kHz reference signal is applied to the control grid of V1 via coupling capacitor C1. The function of V1, T1, and associated circuitry is to multiply the 100-kHz reference signal to 1 MHz and compare it to the feedback signal from the SMO. If a phase difference exists between the two frequencies, then the SMO Hartley oscillator changes frequency to bring them back into phase. The heart of the circuit is varactor CR10. Notice that the cathode is tied back to the positive power supply. The function of CR3 is to hold the cathode positive to ensure that CR10 will never go into conduction, under normal circuit voltages. If the phase discriminator senses a phase difference between the reference frequency and the feedback from the frequency translator caused by a decrease in frequency, the reverse bias of CR10 decreases, increasing its value of capacitance. The result of the increased capacitance is a decrease in the operating frequency of the Hartley oscillator's tank circuit, which is comprised of C6, C7, and L1. The decreased oscillator frequency causes a decrease in the FTS frequency, which brings the error signal back into phase with the reference signal. That, in turn, causes an increase in the bias across CR10, bringing the frequency of the Hartley oscillator back to normal. After this circuit action, the SMO is locked onto the selected frequency.

If a higher frequency were selected on the front panel, it would cause the phase discriminator to sense a phase difference between the reference frequency and the feedback frequency. That, in turn, would cause the reverse bias of CR10 to increase, decreasing its value of capacitance. As it decreased, the result would be a decrease in the frequency of the Hartley oscillator's tank circuit. The advantages of the SMO are its rock-stable accuracy and ability to rapidly control transmitter frequency, whether by front panel selection or internal components changing value due to environmental conditions or input power fluctuations. These changes would be faster than could be observed by an operator.

Another type of oscillator found in SSB equipment is the magnetostriction oscillator. The magnetostriction effect was first discussed under SSB filters. This type of circuit, illustrated in Figure 4-19, is typically found in use at audio or very low RF frequencies. This oscillator produces a very stable sine-wave output. It has two advantages over crystal oscillators in low-frequency applications—a very simple design and low cost.

This circuit's operation is quite ingenious. The tube is configured with conventional grid-leak bias. The plate is in series with the coil. The basis for circuit operation is that the coils are wound on the metallic bar in such a manner as to generate the same value of magnetic flux for either an increasing plate current or an increasing grid current. This arrangement produces degenerative (negative) feedback to the oscillator. Degenerative feedback is when the feedback signal is 180 degrees out of phase with the input or initial signal. It is used to decrease amplification, to stabilize frequency, and to reduce signal distortion. Also, the two coils are situated in such a manner as to have little or no coupling. Because of these factors, under normal conditions the oscillator is not oscillating. Circuit frequency is controlled by tuning capacitor C1.

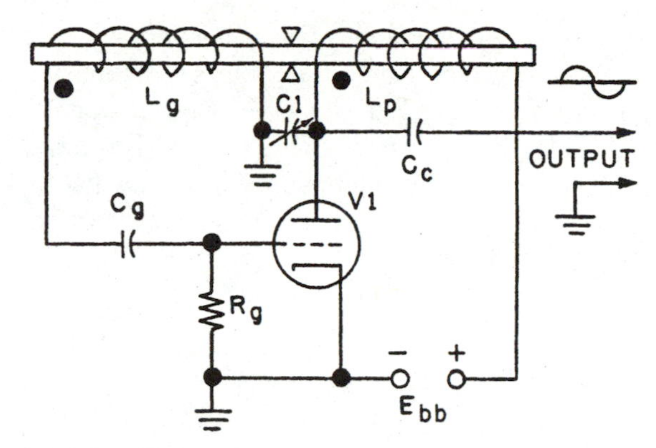

■ **4-19** *Magnetostriction oscillator.*

Circuit operation begins with the application of plate voltage. Current flow through the tube causes a strain on the metallic bar. At that point, an external noise pulse would produce an increase in the flux in the plate coil. That, in turn, causes a compressional wave to start at the plate end of the bar and travel to the grid end.

The phenomenon of magnetostriction causes the compressional wave to move through the conducting bar in a manner similar to the movement of sound waves through a rod. When the compressional wave has reached the grid coil, a lengthening of the bar has occurred due to the magnetic effects. That, in turn, induces a positive voltage in the grid coil. As the grid is fed by the grid coil, there is an increase in plate current. An increased plate current induces a stronger magnetic field in the plate coil, which induces another compressional wave.

The compressional wave in the bar is reflected from the grid back toward the plate, where they are reflected back toward the grid. This action is similar to that of the currents flowing in a tank circuit as the reactive components charge and discharge. The action of the compressional wave motion, changing magnetic field, and induced voltages combine to sustain oscillations when the induced and reflected compressional waves are in phase.

Frequency Multiplications Circuits

Frequency multiplication is based upon an electronic phenomenon known as harmonics. A harmonic is a multiple of a fundamental frequency. As an example, if 5 kHz were the fundamental frequency, then the second harmonic would be 10 kHz, the third 15 kHz, and so on. In a class C amplifier, a pulse of collector or plate current that is not sinusoidal in shape contains a significant harmonic content. Because of this fact, class C amplifiers can be used to generate an output signal that is a harmonic of the input waveform. Often, to obtain the correct output frequency, several frequency multipliers can be used in series. A frequency multiplier is a useful circuit because, as the operating frequency of an oscillator increases, its stability decreases. As has been previously stated, SSB operation is dependent upon stable, accurate operating frequencies.

For frequency multiplication to occur, three conditions must be satisfied. The output tank must be tuned to the desired harmonic frequency, the amplifier must have a high negative bias, and the amplifier must have a high input driving voltage. Frequency multipliers can be constructed from vacuum tubes or transistors.

To illustrate the operation of a frequency multiplier circuit, Figure 4-20 depicts vacuum tube and transistorized frequency multipliers. The first condition for multiplication to occur is the proper tuning of the output tank circuit. If the tank is tuned to the same frequency as the input signal, then no multiplication takes place. By tuning to the second harmonic, a 500-kHz input signal frequency is increased to a 1-MHz output signal frequency. In this case, the circuit would be a doubler. By selecting a different harmonic, higher frequencies can be obtained. If the tank were tuned to the third harmonic, then the same 5-kHz input would result in a 1.5-MHz output. Multipliers can typically be found with the fifth harmonic as the highest one utilized. If the fifth harmonic was insufficient to obtain the correct frequency, then additional multiplication stages would be used. A high value of negative bias is required to drive the amplifier at a more efficient level. Although a reduced gain factor is a small problem, it is more than compensated for by the ability to accurately produce desired high operating frequencies.

SSB Modulators

Although an SSB transmitter shares many circuits and characteristics with a conventional AM transmitter, there are marked and important differences. Frequency stability and accuracy are obvious design concerns that have already been addressed with SSB equipment. Another, less obvious difference is in the modulator design. A typical SSB modulator design usually found in a basic filter-design SSB transmitter is the balanced modulator, as illustrated in Figure 4-21.

In this design, the RF carrier is suppressed and an output consisting of a double sideband is produced. The intelligence modulating

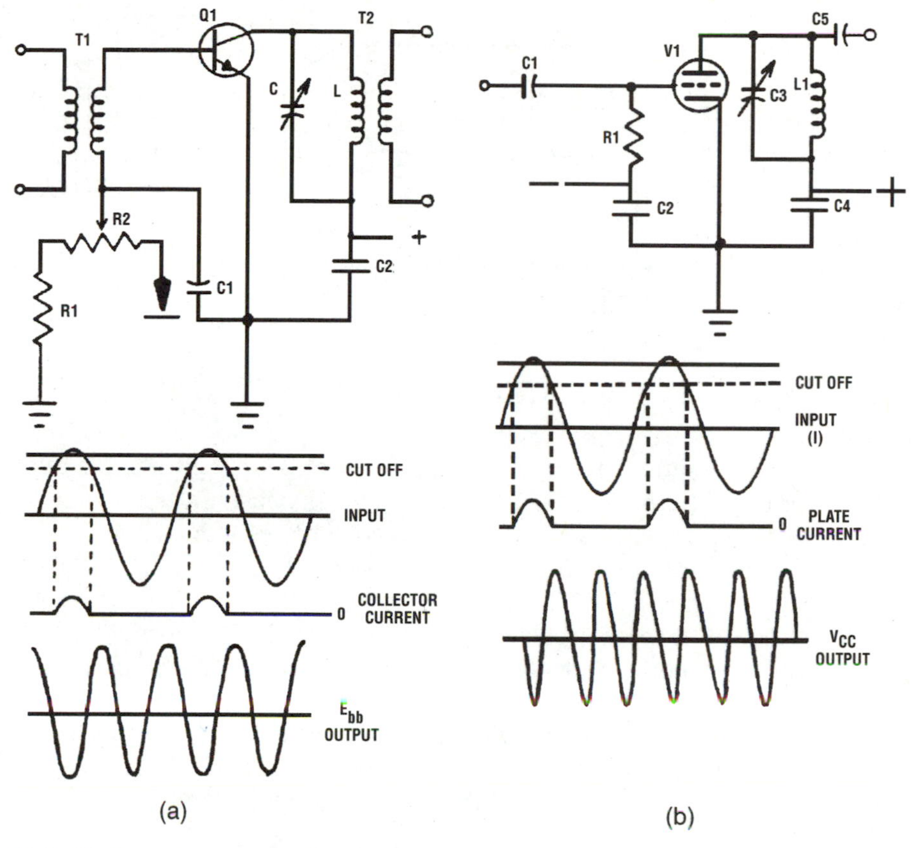

■ **4-20** *Frequency multipliers.*

signal is applied to the input audio transformer T1. A transformer is used to provide isolation from the oscillators to prevent loading affecting the stability and accuracy. *For proper operation, the frequency of the RF carrier must be significantly greater than the frequency of the intelligence (in the range of eight to ten times greater).* The RF carrier is applied to the modulator via C2. Transistors Q1 and Q2 are connected in a push-pull configuration. The simultaneous application of the modulations and RF carrier signals results in the development of the sum and difference sidebands across the transistors. As the RF signal is split into two paths and applied in phase to both transistor bases simultaneously, the RF carrier component of the modulated waveform is canceled across T2. The center tap of T2 is grounded, and due to component value presents a very low impedance to the intelligence frequency, ensuring that it is not passed to the output. The sidebands are out of phase on the transistor collectors because the modulating signals present on the bases are out of phase with each other.

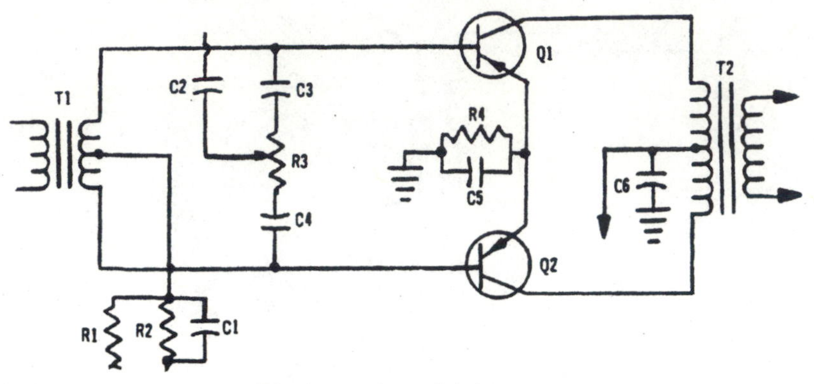

■ **4-21** *Transistorized balanced modulator.*

Each component has a specific function to ensure circuit operation. T1, an audio transformer, provides phase inversion and isolation. Proper amplifier bias is provided by resistors R1 and R2. C1 decouples the audio signal and places the center tap of T1 at ground potential. C2 couples the carrier RF from the carrier oscillator to the carrier balance potentiometer, R3. R3 ensures that the input to both transistors is balanced, canceling the RF carrier. C3 and C4 are coupling capacitors that block DC and pass the signal. The transistors heterodyne the carrier and intelligence signals. R4 provides bias, C5 stabilization, C6 decoupling for the power supply, and the center tap of T2 provides an RF ground.

Circuit operation is as follows: The RF carrier is routed into the balanced modulator via C2 to the potentiometer R3. For proper operation, R3 must be adjusted so that the RF signal is equally applied to both transistor bases. As the input signal swings negative, the forward bias is increased on both transistors, causing both to conduct harder. As the voltage drop across both sides of T2 is positive, equal-but-opposite voltages are developed. With equal currents flowing in opposite directions, they cancel and an output signal is not applied to T2's secondary. On the positive alternations of the input RF carrier signal, the bias on the transistors decreases, decreasing conduction and causing a negative voltage drop across T2. As the currents are still flowing in opposite directions, cancellation occurs. With R3 properly adjusted, the carrier signal is not coupled to the secondary of T2.

With a modulation signal applied, circuit action is as follows: The RF carrier is applied to the transistor bases in phase. The modulating signal is out-of-phase; that is due to T1 acting as a phase inverter to the audio signal. C3 and C4 are chosen to present a high impedance to audio signals, to isolate the transistors. With the RF and AF applied to the bases, modulation is performed.

The out-of-phase modulation signals cause the sideband signals on T2 to be out-of-phase as well. As the transistors are connected in push-pull configuration, the sidebands are coupled to the secondary of T2. Audio is not passed out of the modulator because T2 has a low reactance to audio frequencies. As the RF carrier is suppressed, only the upper and lower sidebands are passed out of the modulator section.

Figure 4-22 is the vacuum tube version of the balanced modulator. If you closely compare this schematic with the one in Figure 4-21, you will see that they are very similar. Once more, the amplifiers are connected in push-pull configuration. Due to the push-pull action, the RF carrier is suppressed, with only the upper and lower sidebands being routed from the balanced modulator.

As depicted in Figure 4-23, a balanced modulator can be fabricated using diodes. Known as a balanced ring modulator, it also suppresses the RF carrier and develops the upper and lower sidebands. Its operation is very uncomplicated. On the positive alternations of the RF carrier, CR1 and CR3 are forward-biased and conducting, while CR2 and CR4 are reverse-biased and cut off. The key to circuit operation is the center-tapped input and output transformers with the modulation signal applied across the center taps. With an RF carrier modulated by an audio signal input, the output is the upper and lower sidebands with the RF carrier suppressed.

Frequency Modulation

As important as amplitude modulation is to the communications field, it does have a major disadvantage: susceptibility to interference. The technique used to combat interference was the development of a new

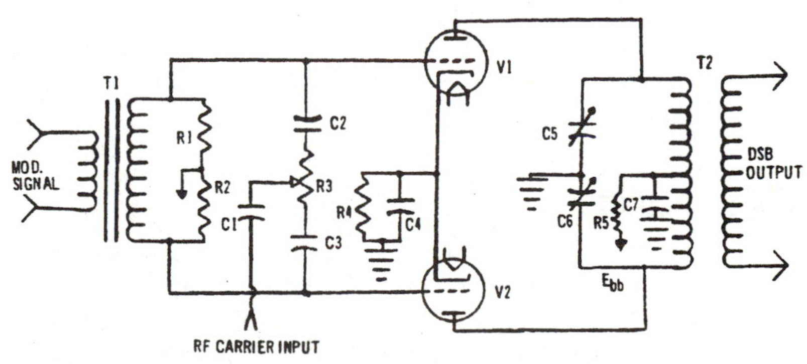

■ **4-22** *Vacuum tube balanced modulator.*

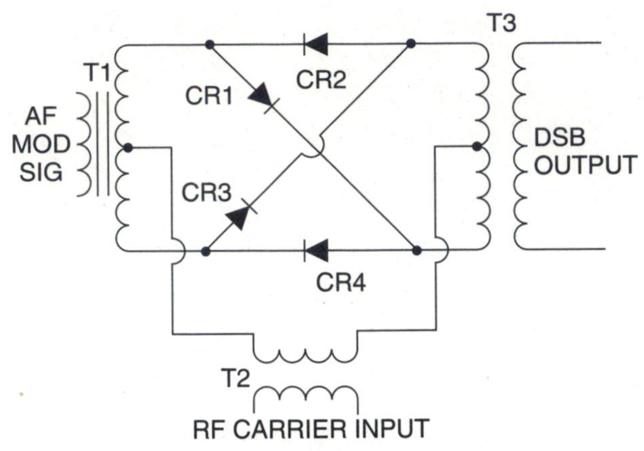

■ **4-23** *Diode ring modulator.*

method of modulation. Use of the intelligence frequency to vary the carrier frequency eliminated atmospheric noise. Widespread adaptation of frequency modulation revolutionized the broadcast industry. In the 1950s, with the popularity of television increasing, it seemed as though it would replace radio as the only form of home entertainment. FM, with its static-free conditions, gave broadcast radio a new lease on life. Communications researchers have used it as a way to provide portable, interference-resistant communications links. In concept, it is quite simple. The RF carrier is modulated by varying its frequency, rather than its amplitude (as in AM or SSB).

There are several terms that are used to describe and define FM operation. The first is *deviation*, which is the amount of frequency shift that results when the carrier is modulated. It is measured in kilohertz and is directly proportional to the amplitude of the modulating signal. As an example, if the transmitter had a carrier frequency of 2000 kHz and when modulated, the frequency varied from 1990 kHz to 2010 kHz, then the frequency deviation would be 10 kHz. Therefore the amount of variation in one direction from the center carrier frequency is the deviation. The total movement (or *swing*) would be 20 kHz.

FM Transmitter

In addition to amplification, intermediate stages of power amplification can also provide frequency multiplication. That can be very important to equipment operation, because as the operating frequency increases, the stability of an oscillator decreases. By using frequency multiplication, the desired operating frequency can be obtained while maintaining equipment stability.

In FM equipment, amplifiers, frequency multipliers, and oscillators are similar to those found in systems that use other forms of modulation. One major difference in FM equipment is in the modulator itself. In an FM transmitter, the frequency modulator changes the frequency of an oscillator in accordance with the modulating intelligence signal. Ideally, for a signal without distortion, the frequency change is proportional to the modulating voltage. Common designs utilize LC tuned circuits to set the frequency of the oscillator. To provide for FM modulation, one of the reactive components must be controllable. In that manner, as the amplitude of the intelligence signal changes, the frequency of the oscillator changes. The simplest and oldest example of this technique is the capacitive microphone. The sound waves that strike the diaphragm of the microphone change its capacitance, which in turn changes the oscillator frequency.

An example of a common frequency modulator found in current FM equipment is the varactor modulator. Figure 4-24 is a simplified varactor diode modulator. The resistors are used to provide proper bias for the modulator. The varactor is the device that actually provides the modulation, as it is part of the tank circuit which also consists of C2 and L1. The function of the transistor is to control the bias felt across the varactor. The varactor exhibits a capacitive component that is inversely proportional to the applied voltage.

Circuit action is as follows: Statically the modulator conducts at a predetermined level that provides enough current to the oscillator to sustain oscillations at the oscillator's center frequency. The modulation signal is applied to the base of Q1. If the input signal voltage swings negative, the conduction through Q1 decreases, increasing the collector voltage. That places a more positive voltage across CR1, which decreases its capacitance, which in turn increases the frequency of the oscillator.

Power Amplifier Theory of Operation

An important circuit that is common to all high-powered transmitters is the power amplifier. A high-powered RF power amplifier is different from conventional amplifiers in several very important ways. First, there are the high power levels and frequencies at which the circuits are designed to operate. To provide stability and

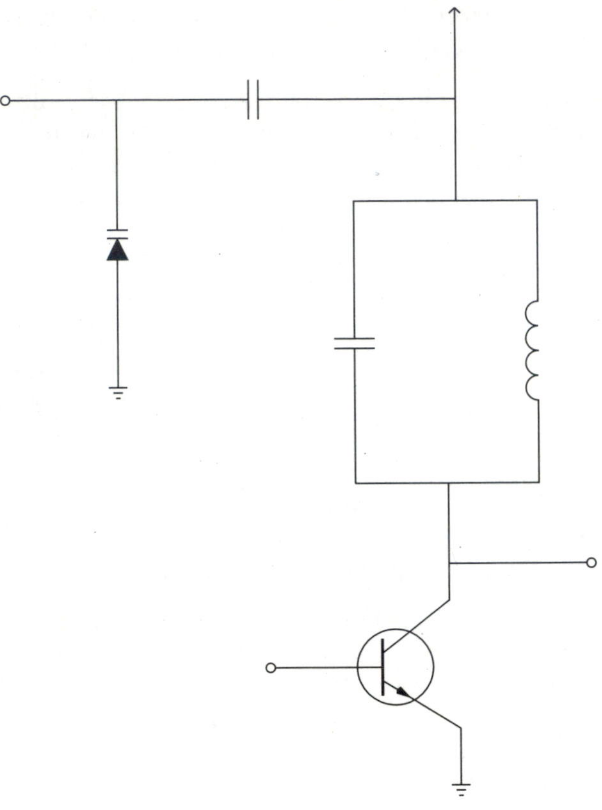

■ **4-24** *Varactor diode modulator.*

accurate frequencies, LC circuits are used for tuning. Finally, the circuits are used to amplify a narrow band of frequencies.

Figure 4-25 is a representative high-powered RF amplifier with tuned input and output tanks. Notice that the capacitors in the tanks circuits are connected with a dotted line. That indicates that the two adjustable components are connected or ganged together mechanically. In that way one control changes both at the same time, keeping both tanks on the same resonant frequency. The function of C2 is to decouple the power supply, preventing any AC variations from appearing in the power supply.

There is an effect in electronics known as *regeneration* that is encountered from time to time. Depending upon the circumstances, it can be beneficial or detrimental. Regeneration is the action when the output signal is fed back to the input. Under the proper conditions it is used to increase amplifier gain and selectivity. It is also used in oscillator circuits to aid in the sustaining of oscillations. In high-frequency power amplifiers regeneration can be a problem in that it can cause undesirable oscillations, leading to signal loss and distortion. To prevent the problem, neutralization is utilized.

In Figure 4-26 is the schematic diagram for an RF power amplifier with neutralization. In this example, notice that only the output circuit has a tuneable tank on the output to set the frequency range of the amplifier. In the emitter circuit of Q1, notice that the emitter resistor is bypassed by a capacitor. That is to ensure that any unwanted RF variations are shunted to ground and do not appear elsewhere in the equipment. The power supply return is also shunted by a capacitor for the same reason.

Class C amplifiers are often used in high-powered applications, as the class C amplifier is a very efficient design, simple and suitable for CW, AM, and FM transmitters. You would not find this class in SSB operation, as it is nonlinear and does not hold phase relationships. The output of this amplifier is a carrier frequency with two sidebands. Figure 4-27 depicts a class C RF amplifier. In this example the amplifying element is a *field effect transistor* (FET). Electronically, a FET is very similar in operation to a vacuum tube.

RF power amplifiers can be fabricated from transistors, FETs, other semiconductor devices, and vacuum tubes. Figure 4-28 is a

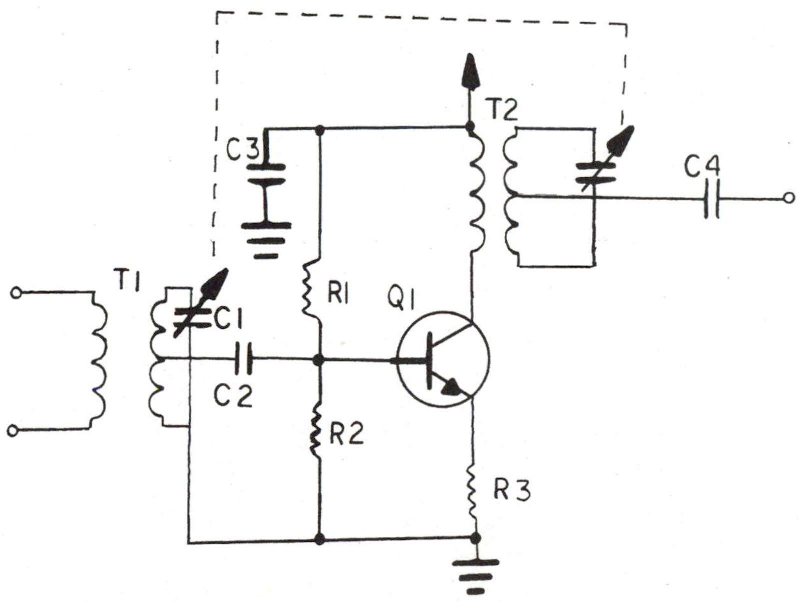

■ **4-25** *RF power amplifier with capacitive tuning.*

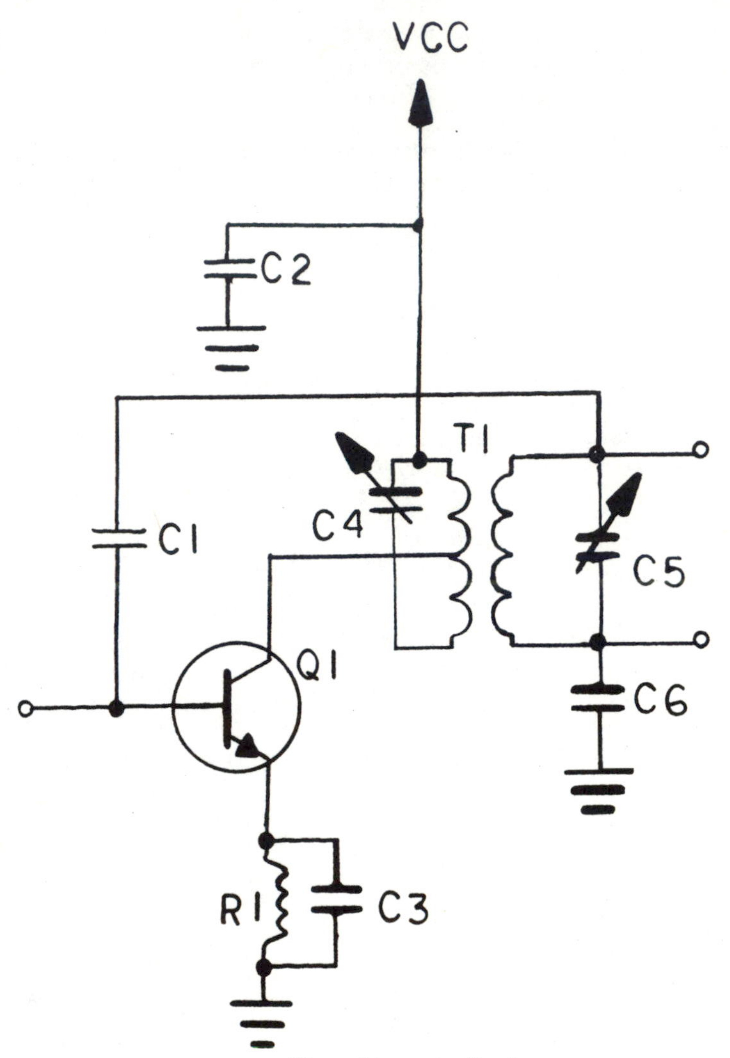

■ **4-26** *RF power amplifier with neutralization.*

vacuum-tube RF power amplifier operated class C. The output tank circuit, with a tunable capacitor, is adjusted so that the resonant frequency matches the desired carrier frequency. To neutralize the amplifier, the power supply and cathode resistors are bypassed with capacitors. The bias level of the amplifier is set by potentiometer R2.

The final RF power amplifier design is the push-pull amplifier. Push-pull amplifiers are typically operated class AB. The advantages are that the circuit provides a linear output with adequate efficiency as compared to other classes of operation. A push-pull amplifier consists of two amplifiers connected so that the input and output signals are 180 degrees out of phase. The amplifiers should be as closely matched as possible to ensure proper amplification.

A final concept to be presented is the method used to express and calculate power in electronics. Power can be expressed in *watts* or *decibels*. Power expressed as watts is simply calculated by multiplying the circuit voltage and current together.

$$P = E \times I$$

This formula is fine as long as you are dealing with power in a simple circuit or source. When calculating power in a complex electronic device or system, a different approach is used due to several factors that might have to be considered. A decibel (dB) is a power or voltage measurement. It can be used to represent signal strength or a signal level. In communications it is used as a measurement of transmitter gain or receiver sensitivity.

Figure 4-29 is a dB chart that compares dB, microvolts, and watts. Direct your attention to the 1000 microvolt position on the chart.

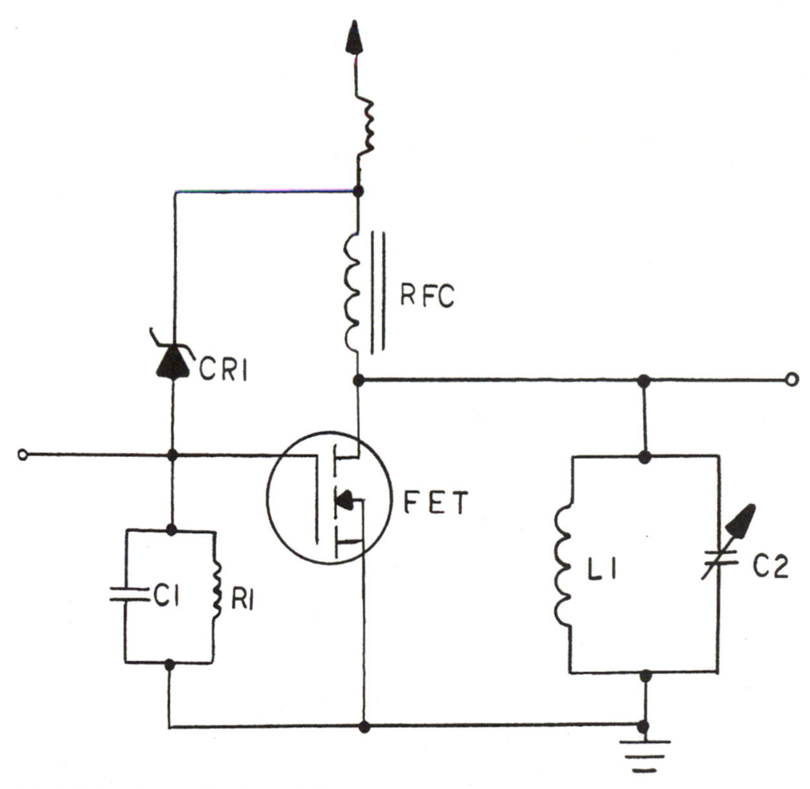

■ **4-27** *Class C RF amplifier.*

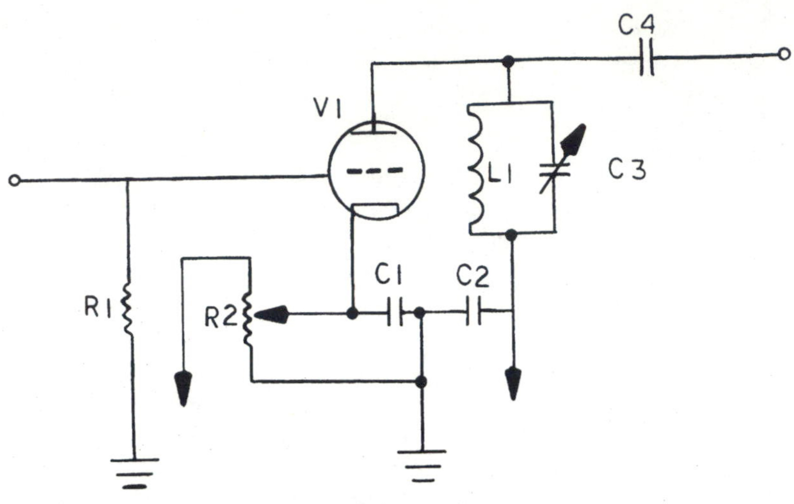

■ **4-28** *Vacuum tube class A RF amplifier.*

Notice that it is equivalent to 0 dBmv. This is a very important figure, as it is an industry standard. In the early years of television development, it was determined that a TV receiver should produce a snow-free picture with an input signal of 1000 microvolts. As it was then considered to be a reference point, it was decided to call 1000 microvolts 0 dBmV.

Microvolts is an excellent method to refer to receiver sensitivity, but is not suitable for power. The power or gain of a stage or equipment can be easily calculated using decibels. In the case of power it is a ratio of the output power to the input power.

$$dB = 10 \log \frac{p2}{P1}$$

In this formula, the output power is divided by the input power, and the result is expressed as a logarithm and then multiplied by 10. As an example, a power amplifier has 10 watts input with a 1000 watt output. The power gain would be expressed in decibels as follows.

$$dB = 10 \log \frac{1000}{10}$$

$$dB = 10 \log 100$$

$$dB = 10 \times 2$$

$$dB = 20$$

As can be seen from the formula, the calculations are fast and accurate. There are also several reference levels that are expressed in dB. When dealing with power, dBW is a value of decibels referenced to 1 watt. If the power is small, then you would use dBm, which is decibels referenced to 1 milliwatt. In voltage calculations, dBV is decibels referenced to 1 volt.

Decibels are an easy way to take several different stages of amplification or losses into consideration. A loss is expressed as negative dB (-dB). This is used when maintenance is being performed on a piece of equipment. In reality, when you are testing a circuit, losses can be inserted with interconnection cables. Often, to observe power and voltages in a stage, cables are required to connect the test equipment and the system under test. Each cable has losses, as perfect impedance matches are often not possible. As an example, you could be measuring the power of an amplifier stage. The bolometer, or input device on a power meter, could insert a loss of -2 dB. The cable to connect the power meter to the stage could have a loss of -1.5 dB. By taking these losses into account, an accurate measurement of the actual circuit power can be obtained.

In this chapter, the concept of the four common types of modulation have been presented. Simple transmitters for each type of

Microvolts	dBmV	Microvolts	dBmV
10	- 40	1995	6
16	- 36	2000	6
32	- 30	3200	10
63	- 24	4000	12
100	- 20	8000	18
125	- 18	10000	20
250	- 12	16000	24
316	- 10	32000	30
500	- 6	63000	36
560	- 5	64000	36
630	- 4	100000	40
700	- 3	127000	42
800	- 2	250000	48
891	- 1	320000	50
1000	0	500000	54
1100	1	1000000	60
1300	2	2000000	66
1400	3	4000000	72
1600	4	8000000	78
1800	5	10000000	80

■ **4-29** *DB chart.*

modulation were shown on the block diagram level. Representative circuits found in many of today's transmitters were presented to show how basic components are interconnected to construct a useful electronic device. Now that the transmitter has been completed, it is time to continue on to antennas and signal propagation. Without an effective antenna, a communications system would be unable to function.

5

RF Propagation and Antennas

Antenna Fundamentals

THE FUNCTION OF A TRANSMITTING ANTENNA IS TO impedance-match the output stage of the transmitter to the impedance of the atmosphere. If the match is good, then maximum RF energy is radiated into free space. On the receiving end, an antenna is also used, but this time the function is to match the impedance of the input stage of the receiver to the impedance of the atmosphere. As with the transmitter, a good impedance match is required to ensure maximum transfer of received RF energy into the receiver for processing. Antenna design is interesting in that in the transmitting application, kilowatts are often propagated, while with the receiving antenna, only microwatts are intercepted. Although many people might claim otherwise, antenna theory is more of a science than an art. Through understanding resonance, currents, and voltages in reactive circuits, and the effect that the atmosphere and sun have on RF energy, excellent communications links are a reality, not a haphazard event.

Antennas, when compared to other assemblies that make up a communications system, are very simple in concept and are often overlooked. The best communications system that can be bought is only as good as its weakest link, and that link is all too often the antennas. The world's finest transmitter and receiver cannot compensate for a poor antenna, either for reception or transmission. Improvements in communications system performance can be obtained for less cost and effort in the antenna system than in either the receiver or transmitter.

Two important criteria in a communications link are range and signal quality at the receiver. As was presented in the previous chapter, power is measured in decibels (dB). Every time there is a 3 dB increase in transmitter power, that is a doubling in effective power. So, from 100 watts to 200 watts is a 3 dB increase, from 100 to 400 watts is 6 dB, and so on. In communications systems a transmitter is often used to drive, or provide an input to, a linear power amplifier. For a radio amateur to purchase a linear amplifier to boost transmitter output power from 100 watts to 1500 watts would cost in the range of $3000. This figure would be higher for a commercial unit. As power has been increased, a more expensive antenna system is needed to handle the additional energy, and there will be far more power consumption and more maintenance. For the investment and increased problems, a 12-dB gain in system performance has been achieved. An alternative is in the area of antenna improvements. For far less money, a directional beam antenna mounted on a small mast would result in almost an 18-dB power gain. It is for this reason that antenna theory is a valuable commodity. With the end of the cold war, military technology will find its way into civilian applications with a greater frequency. Antennas, along with other areas of electronics, will gain from this infusion.

Radio Frequency Wave Propagation

As with many words in the English language, propagation is derived from Latin. The original Latin root is *propagare,* which means *to travel*. Antenna design, installation, placement, and system operating frequency determine how the RF energy is radiated into the atmosphere or free space. *Free space* is defined as the region high enough so that the radiated RF energy is not affected by topography and surrounding man-made objects. Figure 5-1 demonstrates how RF radiates from a representative antenna. RF energy is radiated from the transmitting antenna in a pattern that is similar to that of a doughnut laying on its side. A portion of the radiated energy is reflected off of the ground and is called the *ground wave*. Some RF energy, known as the *space wave* or *direct wave,* is radiated directly to the receiving antenna. The remaining RF energy is directed up into the atmosphere and is called the *sky wave*.

Each wave, whether ground, sky, or space, has range limits. Figure 5-2 illustrates how the ground and sky waves reach a receiving antenna. The ground wave moves out and down from the antenna, coming into contact with the surface of the earth. Because of this it is affected by the conductivity of the earth, from the surface extending to a depth of up to 100 meters. The greater the conductivity, the greater the range that can be achieved by the ground

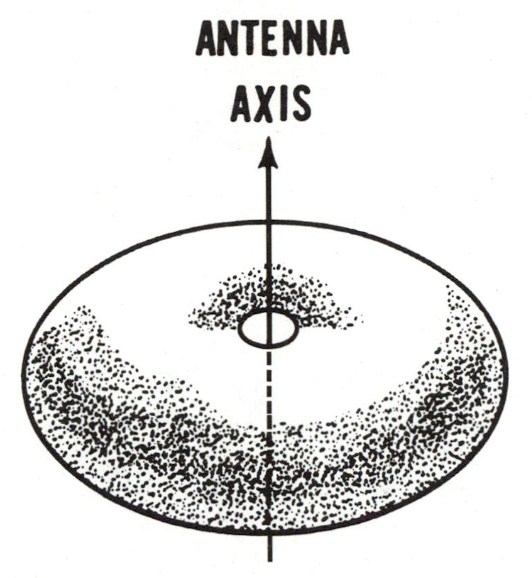

■ **5-1** *RF radiation pattern.*

wave. As an example, the conductivity of salt water is 5,000 times greater than that of dry earth. For this reason, high-powered low-, medium-, and high-frequency transmitters are located as close to the ocean as possible. The ground wave is responsible for most daytime RF signal reception. Higher up on the antenna, the radiated RF is formed into the space wave and is unaffected by the earth's surface. This wave travels in a line-of-sight direction from the transmitting antenna to the receiving antenna. Near the top of the antenna, the RF energy is radiated out and up toward the upper reaches of the atmosphere. Present in the ionosphere are several layers of ionized gases that refract the RF energy back to the earth. This is the sky wave, or indirect propagation.

Figure 5-3 illustrates how the three propagation waves move through the atmosphere. The most common uses for ground-wave propagation are for long-range high-powered transmitters operated at medium and low frequencies and low-powered transmitters at high frequencies and short ranges. Good examples would include the AM and FM broadcast stations in the United States. Space-wave propagation in usually used for very-high-frequency communication links for both long and short ranges. Air-to-ground communications is a good example. Another would be commercial TV and FM broadcasts, and satellite communications. The sky wave, refracting off of the layers of the ionosphere, is the type of propagation used for long-range, high-frequency daylight communications. At night it is used by international

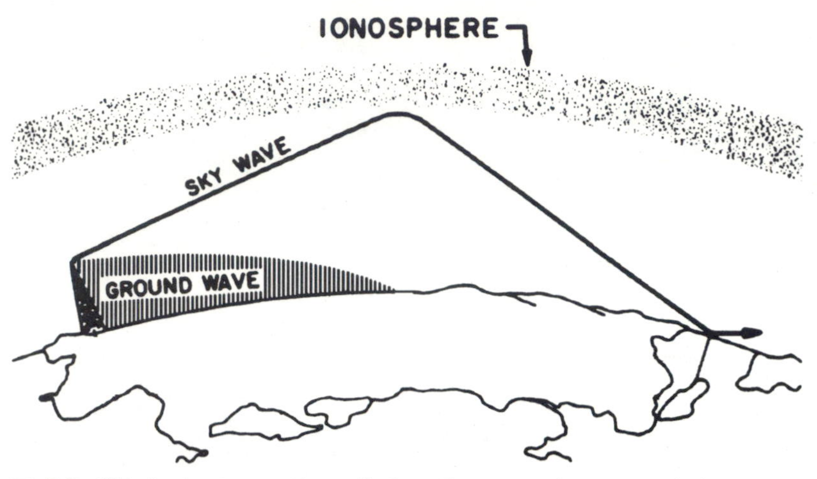

■ **5-2** *RF electromagnetic radiation: the ground wave and sky wave.*

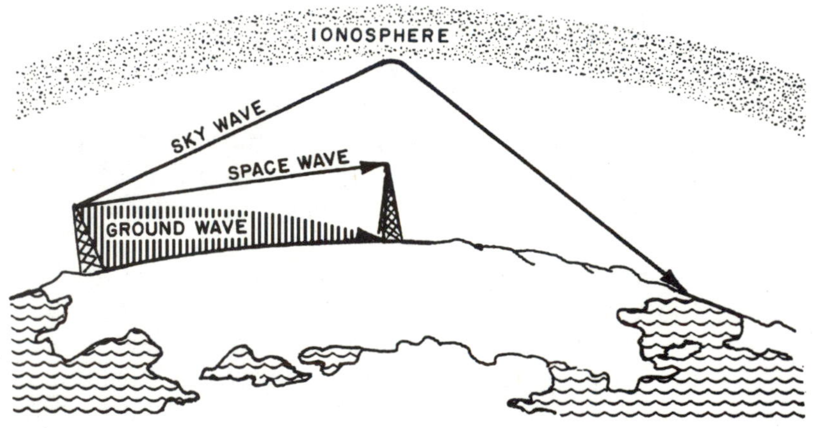

■ **5-3** *Propagation of the ground wave, space wave, and sky wave.*

shortwave broadcasters for intercontinental long-range, high-powered shortwave (3 to 30 MHz) broadcasts.

The RF energy radiated by an antenna is formed into a wavefront. The wavefront is composed of two components—an electric field (E) and a magnetic field (H). The term *polarization* refers to the plane that the E field is in. If the E field is perpendicular to the surface of the earth, the wavefront is vertically polarized. To produce a vertically polarized wavefront, the antenna must be in the vertical plane. A horizontally polarized wavefront would have the E field parallel to the surface of the earth. To form this wavefront, the antenna must be in the horizontal plane. Figure 5-4 compares a vertically polarized and horizontally polarized wavefront.

TV and FM broadcasts are made with horizontally polarized antennas. Below the frequency of 2 MHz, most transmissions are vertically polarized. The frequencies above 2 MHz and extending up to 30 MHz are can be either vertically or horizontally polarized. That is because under some propagation conditions, one will provide superior range. A phenomenon associated with the propagated RF waves is that they can twist due to atmospheric conditions.

The ground wave can only be used with RF signals that are vertically polarized. As can be seen from the diagram, the ground wave passes over and below the surface of the earth. There is a marked velocity difference for energy propagated through the ground and free space. The result is that the RF wavefront leans forward. As was stated before, the RF energy wavefront is composed of two components: the electric field (E) and the magnetic field (H). The E field is parallel to the surface of the earth. Because of the proximity of the E field to the surface of the earth, currents flow in the earth because of it. As a result, energy from the wavefront is lost as it is absorbed by the earth. Simultaneously, the H field component, which is vertical to the surface of the earth, induces eddy currents in the surface. That causes additional energy losses from

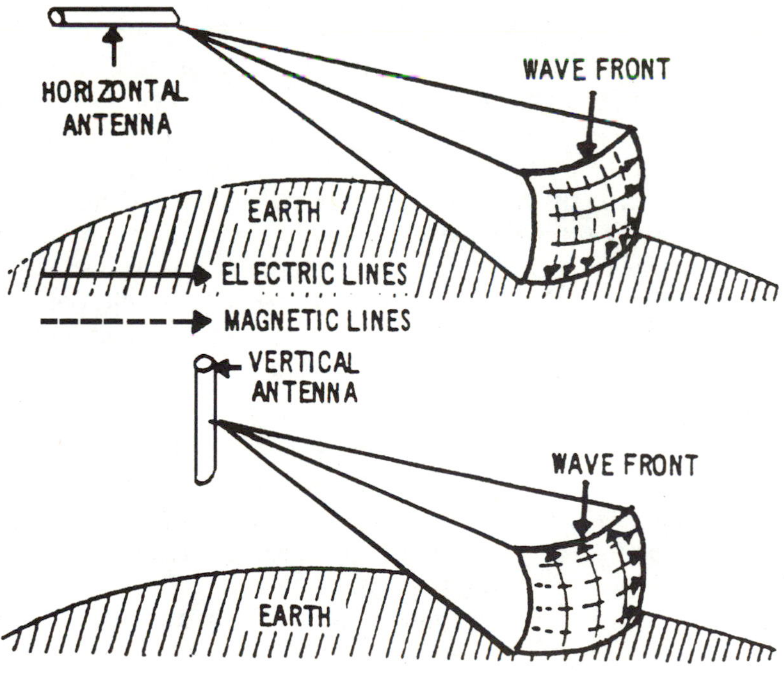

■ **5-4** *Vertically and horizontally polarized wavefronts.*

the wavefront. A horizontally polarized wavefront would have the E field vertical to the earth, causing almost total energy absorption a short distance from the antenna.

The amount of absorption loss with a vertically polarized wavefront due to ground loss, or *attenuation*, depends on the frequency utilized and the conductivity of the earth's surface. Signal attenuation is inversely proportional to the frequency of the transmitted wavefront. Currently, frequencies between 10 kHz and 2 MHz are suitable for ground-wave energy transmission. The U.S. Navy uses VLF communications in the 17-kHz to 23-kHz range to communicate with submerged submarines. Because the wavefront is in contact with the surface of the earth, it can be bent, or diffracted. The result is that the ground wave communications link can cover thousands of miles. It is possible that with a 1-MW transmitter worldwide coverage can be obtained. Disadvantages with this type of communications link would be the massive size of the transmitting antenna, the high powers involved, and the slow rate of data transfer. If a VLF antenna were constructed in the most efficient size, then it would be over 60 miles long. As shorter antennas are used, they are only 15% efficient.

The space wave is the component of the radiated energy that travels in a straight line from the transmitting antenna to the receiving one. Because of this fact, the transmitting and receiving antennas must be line-of-sight, or in other words not have any objects blocking the signal path. This is the most common form of propagation in use today. Typical frequencies encountered are 30 MHz and higher. In this frequency range the ground wave has little or no effect on the radiated energy. Also, as these frequencies are not normally refracted back to earth by the atmosphere, there is no sky wave component. Because of these factors, transmissions at more than 30 MHz are considered to be line-of-sight. Services found in that frequency range include TV, commercial FM, radar, VHF, UHF, and satellite communications.

A problem associated with the space wave propagation in communications is multiple-path reception. Figure 5-5 illustrates the problem. The radio or TV signal leaves the antenna in a 360-degree pattern. RF energy strikes every object in its path. That includes the receiving antennas, buildings, vegetation, large structures, and topographical features. The result is that the energy is reflected from its normally straight path. Some of the reflected energy, as well as the direct path energy, will be intercepted by the receiving antenna. The result could be multiple images or ghosts on TV, and echo with audio broadcasts. The audio echos are seldom if ever detected, as the time difference between the direct and reflected energy is so small.

Another problem (depicted in Figure 5-6) is a blind spot, or zone of silence. Notice that, in the diagram, the wavefront in the form of a ground wave or space wave is radiated from an antenna. A large hill or mountain blocks the energy from receiving antenna. This was one of the many problems that microwave transmission and cable TV were developed to overcome.

There is a phenomenon associated with line-of-sight transmissions known as *tropospheric ducting*. It is a well-known fact that line-of-sight propagation is a form of RF energy propagation that is limited by the height of both antennas and surrounding obstructions. As such, it is typically short-ranged. An exception would be air search radars and air-to-ground communication that encounter a phenomenon called a tropospheric duct. Often, surface communications links and shipboard radars will experience conditions of exceptional long-range communications and radar reception. This condition is caused by unusual circumstances in the troposphere, which is depicted in Figure 5-7. The *troposphere* is the region of the atmosphere that extends from the surface of the earth up to about 35,000 feet. Normally, the warmest air in the troposphere is found near the surface of the ocean. As altitude increases, temperature decreases as it should. However, in the tropics, pockets of warmer air become trapped between layers of cold air, and this condition is known as a *temperature inversion*. The layer of warm air then forms a duct, which is capable of trapping and guiding RF energy for

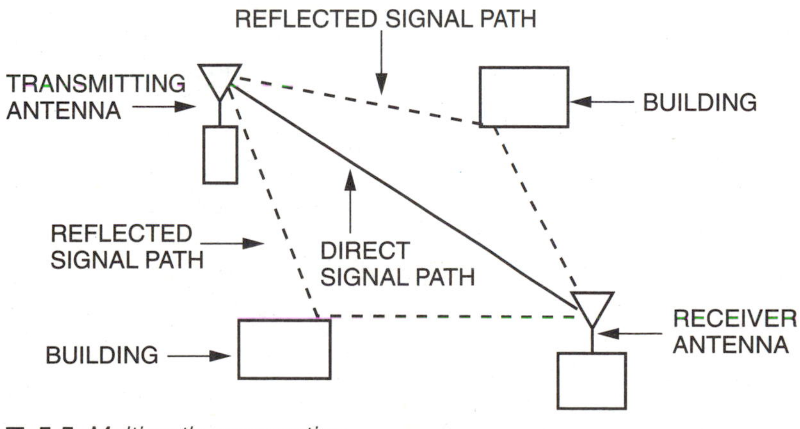

■ **5-5** *Multi-path propagation.*

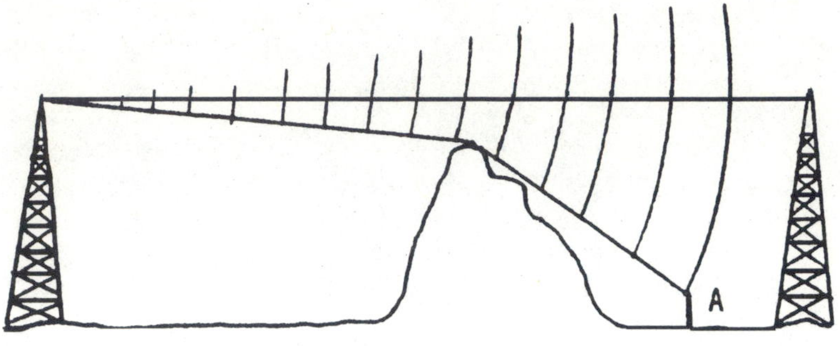

■ **5-6** *RF propagation blind spot.*

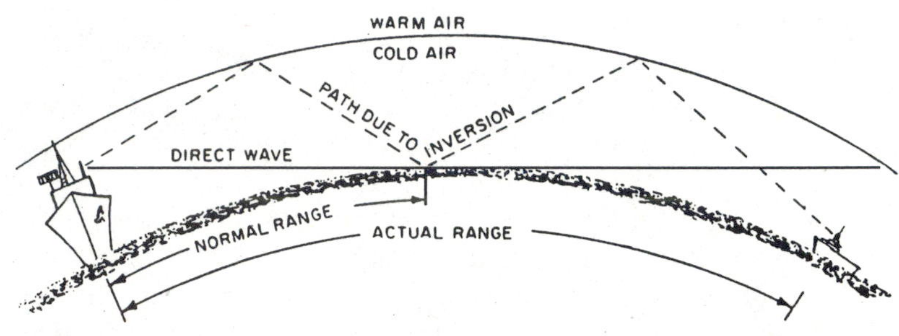

■ **5-7** *Tropospheric ducting.*

hundreds of miles. Typically, these ducts begin at heights of 500 to 1000 feet, and are 500 to 1000 feet thick. If a radiating antenna comes in contact with a duct, the transmitted RF will remain trapped within the duct until the temperature inversion ends, causing it to disappear. The result is that a communications range that should be measured in tens of miles is now measured in hundreds of miles.

Certain atmospheric conditions have been observed when ducts have formed. They include a slight wind blowing from land, a stratum of quiet air, high atmospheric pressure, clear skies, warm ocean surface with a slight cool breeze, smoke failing to rise, and received signals rapidly fading. Outside of the tropics, air temperature above the surface of the ocean initially increases, then forms an inversion. This results in RF energy being trapped near the surface of the earth in similar fashion.

Ducting is a well-recognized effect that cannot be used for reliable communications links, as its appearance is difficult to predict. Surface and high-level ducting are classified as anomalous propagation, as communications ranges far greater than normal are encountered.

The troposphere is very useful in providing long-range communications links. Communications through the troposphere function by taking advantage of the atmospheric conditions at an elevation of several thousand feet. Due to differences in temperatures and humidity in the troposphere, a scattering region is formed. Rather than a short-lived phenomenon such as the ducts, this is a permanent occurrence. This region is capable of refracting a fraction of the transmitted RF energy back to earth. Figure 5-8 illustrates the concept. Most of the transmitted energy continues in a straight line to outer space, with only a small portion being refracted back. Some of the refracted energy is scattered in directions other than that required for interception by the receiving antenna. As the discontinuity refracts only a small portion of the RF energy directly back to earth, high-powered transmitters and ultra-sensitive receivers are required. This type of propagation can be used from 300 MHz to 10 GHz. The wide frequency band is possible because signal attenuation increases slowly as the frequency increases. Fifty miles is the minimum range possible and is determined by the scattering volume. A maximum range is from 400 to 500 miles and is controlled by the average height of the scattering region.

Disadvantages associated with this type of propagation include the necessity of high transmitter power and sensitive receivers, multiple signal paths due to scattering, and severe fading due to atmospheric changes. Fading can be partially overcome through

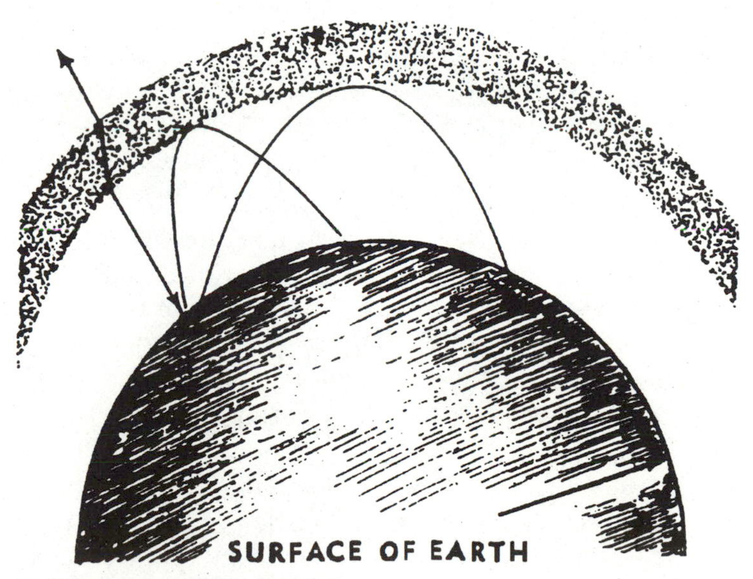

■ **5-8** *Tropospheric scatter.*

the use of *diversity reception*. Diversity reception is when several received signals are either combined to form one, or the strongest one is selected for processing. Frequency diversity is when the same intelligence is simultaneously transmitted on several frequencies. This takes advantage of the fact that fading will affect some frequencies more than others. Another form of compensation for fading is to utilize a technique known as *space diversity*. In this technique, a number of receiver antennas are spaced several wavelengths apart. Fading will have little or no effect on one of the antennas. The strongest signal is then selected and fed to the receiving system for processing. All signal selection is performed automatically to take advantage of rapidly changing atmospheric conditions.

The ionosphere is the region of the atmosphere that begins about 40 to 50 miles above the surface of the earth and extends up to about 350 miles. Bombardment by solar ultraviolet radiation and particles causes a very high number of positive and negative ions to exist in this region. The rotation of the earth and the development of sunspots affect the quantity of ions in the layer. This, in turn, affects the quality and distance of RF transmissions. Due to natural events, the ionosphere is under constant change. Various quantities of ions are always losing and gaining energy. The rate of variation between high and low levels of energy is determined by the volume of air present, the strength of the radiation received from the sun, and the positional relationship between the magnetic field of the earth and the propagated RF energy.

Above 350 miles, the atmosphere is far too thin to permit ionization on a useful scale for propagation. Below an altitude of about 40 miles, once again too few ions exist, this time due to ultraviolet radiation absorption. Therefore, the useable thickness of the ionosphere for radio communications is from 40 miles to about 350 miles.

The rate of ionization is not constant throughout the entire 310-mile thickness. Ionization densities vary with altitude. The result is that the ionosphere appears to be made up of layers. Any concentration difference between the layers is not an obvious dividing line; rather, it is gradual transition.

Figure 5-9 breaks the ionosphere down into its major layers. The first layer encountered is the D layer, at about an altitude of between 40 and 50 miles. Its ionization concentration is low, and it has little effect on the propagation of RF (with the exception of energy absorption). Its existence greatly reduces the strength of RF transmission. The D layer is evident only during daylight hours.

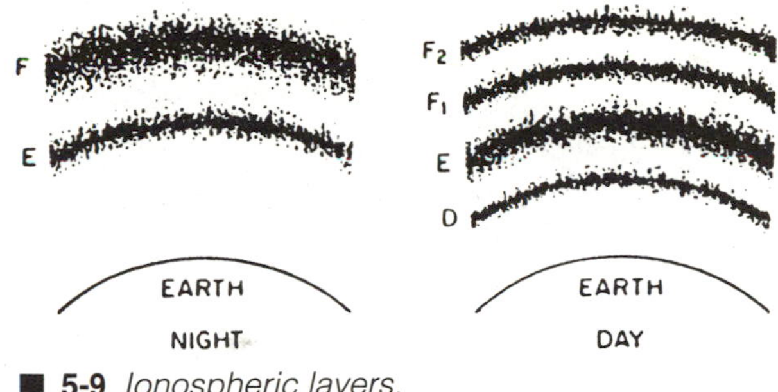

■ **5-9** *Ionospheric layers.*

From an altitude of 50 miles to 90 miles is the E layer. With its greatest density occurring at an altitude of 70 miles, it is very well-defined. This layer reaches its maximum density at noon, local time. Present constantly, it has its greatest strength during daylight hours and is very weak at night. The ionization of the E layer in the middle of the day is occasionally intense enough to refract frequencies up to about 20 MHz back to earth. Ranges of up to 1,500 miles are possible with daytime transmissions. The F layer is found from 90 miles up to 350 miles. This layer is unusual in that at night it is a single ionized band. During daylight hours, with the sun high in the sky, it breaks into two separate layers, F1 and F2. Usually, the F2 layer is greatest in the early afternoon. Shortly after sunset, the two layers recombine into one. In addition to the above four layers, erratic patches of ionized gas appear at the E level. Called *sporadic E ionizations,* they are often of a sufficient intensity to provide for reliable VHF transmissions over much greater than normal distances. They can appear at any altitude between 50 miles and 70 miles, covering an area of less than a mile to several hundred miles. Totally unpredictable in nature, they are useless for scheduled communications links. Other, unnamed sporadic ionizations appear at varying altitudes, and these are often strong enough to degrade communications.

The changes that occur in the ionosphere follow loose guidelines. The height and density of the layers changes between night and daylight hours. Also, there are seasonal changes that are linked to specific geographical areas. Another problem is *sudden ionospheric disturbances* (SIDs) or *Dellinger fadeouts*. This phenomenon is caused by intense ultraviolet radiation pulses that are produced by solar flares. A SID will cause a sudden massive increase in the ionic density of the D and E layers. As the D layer is highly absorptive and the E layer moderately so, communications are disrupted and maximum range decreased as transmitted energy is rapidly attenuated. Magnetic storms have a similar affect

ionospheric absorption, atmospheric noise, extraterrestrial noise, and receiver signal-to-noise ratio design considerations severely impact RF communications. LUF is different from MUF in that it can be controlled by improved receiver designs. Additionally, by increasing the *effective radiated power* (ERP) of the transmitter, LUF can be decreased. For each 10 dB increase in radiated transmitter power, there is a corresponding 2 MHz decrease in LUF.

In sky wave propagation, the electromagnetic RF signal is radiated toward the ionosphere at an oblique angle. Initially, it would appear as if the incident wave is then reflected back to earth. In reality, the wave is not reflected back, but rather it is bent by refraction, as one would observe with a light beam passing through a prism. Figure 5-12 will help to explain the concept. Refraction occurs because as the RF signal strikes the ionized layer at an angle, the top portion of the wavefront will experience a greater ionization intensity than the lower portion. As the RF signal's phase velocity increases with an increased level of ionization, the upper part of the wave is moving faster than the bottom, causing a bending, or refraction. For a given frequency, there is a maximum angle of incidence for which the RF signal will be returned to earth. The greater the angle at which the RF energy enters the ionosphere increases the distance it must travel through the ionosphere. An increased travel distance allows for the ionosphere to absorb more of the RF energy. That, in turn, limits the reception zone of the RF signal.

Figure 5-13 brings together all the factors of ground wave and sky wave propagation. The antenna is located on the left side of the di-

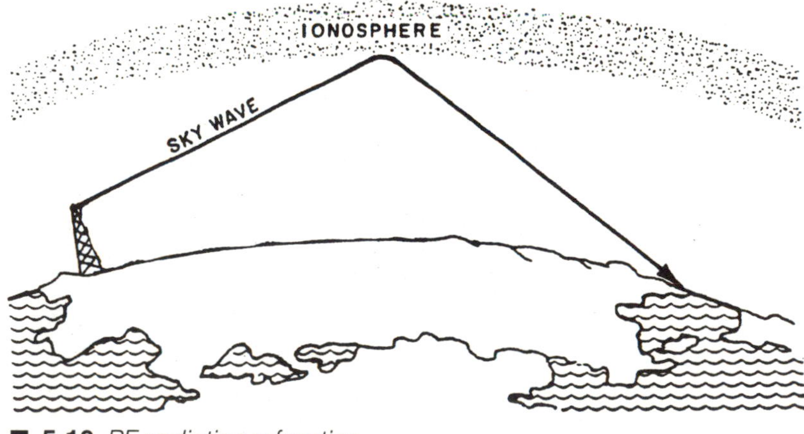

■ **5-12** *RF radiation refraction.*

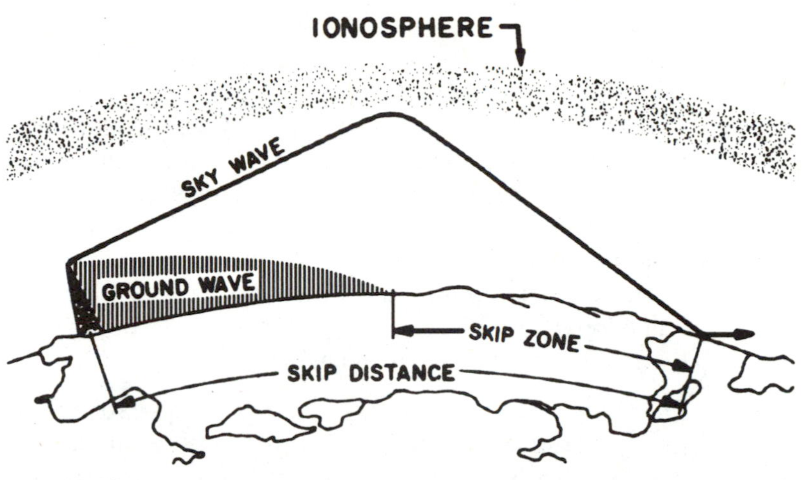

■ **5-13** *Antenna propagation.*

agram. Remember, the radiation pattern from the idealized antenna is in the shape of a doughnut.

The ground wave, as has been stated before, follows the surface of the earth. The sky wave, radiated more from the center of the antenna, leaves as a curved wavefront. As can be seen in the diagram, some of the RF energy is radiated at such an angle that it just passes through the ionosphere and is lost in space. Other portions of the radiated RF energy strike the ionized layer at such an angle of incidence that it is returned to earth. The skip distance is the distance that separates the transmitting antenna and the point on the surface of the earth where the first sky wave is returned and can be intercepted by a receiving antenna. The skip distance is dependent upon several variable factors. If the transmitted frequency is increased, the refraction in the ionosphere decreases, and the resulting skip distance increases. Other factors include the time of day, time of year, and physical location of the transmitter and receiver. As an example, the skip distance for lower HF frequencies is greater at night than during the daylight hours. If a station were located in New York City transmitting on a frequency of 5 MHz, it could be heard in Philadelphia during daylight hours, but not much farther. After sunset, however, it would be received in Chicago, but not Philadelphia. It is not uncommon for HF transmitters operating at frequencies up to 20 MHz to have a signal cover up to 2,500 miles in one hop. As a general rule, as frequency decreases, skip distance decreases.

Point A in Figure 5-13 is the location where the intensity of the ground wave decreases below receiver noise and can no longer be detected by a receiver. As the sky wave is not returned to earth at this point, radio reception is not possible, and this is the beginning of the zone of silence. The zone extends to point B, where the first sky wave is returned to earth. The physical size of the zone of silence can be reduced by decreasing the transmitted frequency. It is important that the frequency is not lowered to the point where the ground wave and sky wave overlap. If that were to happen, interference would result. As the ground wave follows a direct path to the receiving antenna and the sky wave follows a refracted, much longer path, the two signals will be out of phase. Fading often results, with a loss in signal quality.

There are several techniques that can be used to minimize the effects of fading. The easiest is to change frequencies. However, that is not always possible due to many factors, such as frequency assignments, equipment limitations, antenna designs, or because no such frequency is available. Another is to use one of the diversity techniques. The first, illustrated in Figure 5-14, is frequency diversity. In this technique, the communications link consists of two transmitters and two receivers. The transmitters are colocated and use the same antenna. The identical intelligence is transmitted with both transmitters. The signal is received by a single antenna that feeds two receivers. An electronic mixer is used to combine the two signals together, forming a usable output. As multiple frequencies are not subjected to the same level of fading at the same times, a usable signal is obtained.

Spacial diversity is depicted in Figure 5-15. In this arrangement, a single transmitted frequency is used. At the receiving station, two or more receivers are used, with three being the most common installation. The antennas are placed ½ wavelength apart. It has been observed that all of the antennas will not be subjected to the same degree of fading, due to their separation. In fact, as the signal strength decreases on one, it will increase on another. A common master oscillator is often used to keep all the receivers locked

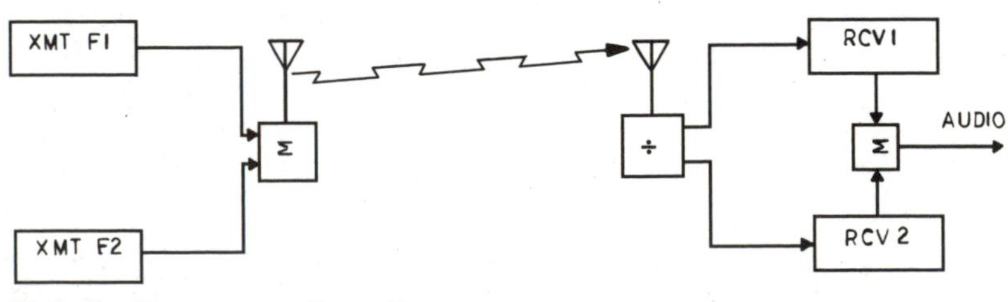

■ **5-14** *Frequency diversity.*

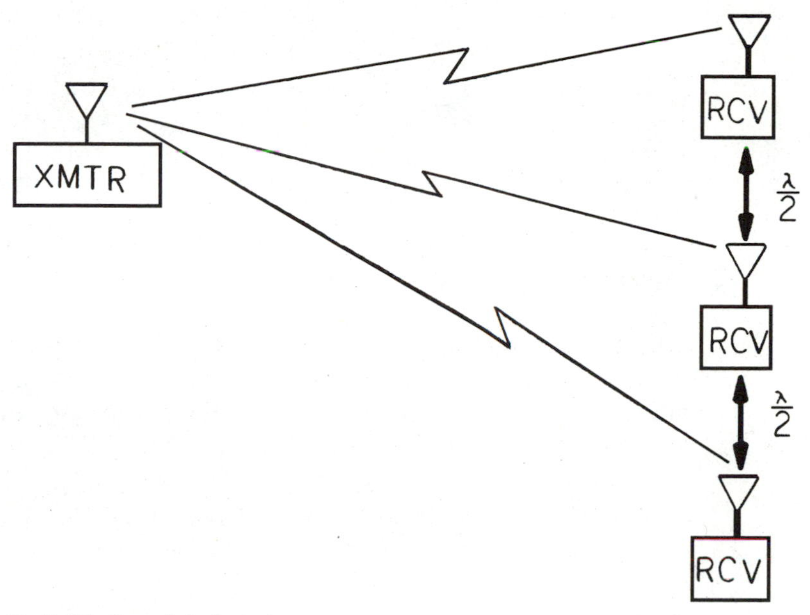

■ **5-15** *Spacial diversity.*

on to the same frequency. Once again, an audio mixer is used to provide a usable output signal.

The final diversity technique is polarity diversity, as shown in Figure 5-16. Notice that the receiving installation consists of two receivers and two antennas. One antenna is vertically polarized, the other horizontally polarized. This helps to reduce the effects of fading, as the action of fading often results in the polarity reversal of a radiated signal. The two signals are once more mixed to provide a composite audio output.

Multi-hop skip transmission is shown in Figure 5-17. The signal from the first sky wave strikes the surface of the earth, and then is reflected (not refracted) toward space. As it enters the ionosphere, it is refracted back to earth. Also notice that in the figure, the antenna is radiating two separate paths. A high angle of RF radiation causes a shorter skip. As the angle decreases, the skip becomes greater. A common technique with broadcasters is to use several different antennas with different radiation angles. That ensures that the correct skip distance is available to reach desired locations.

By using multiple skips, a greater distance can be covered, resulting in longer-ranged communications. However, a multiple

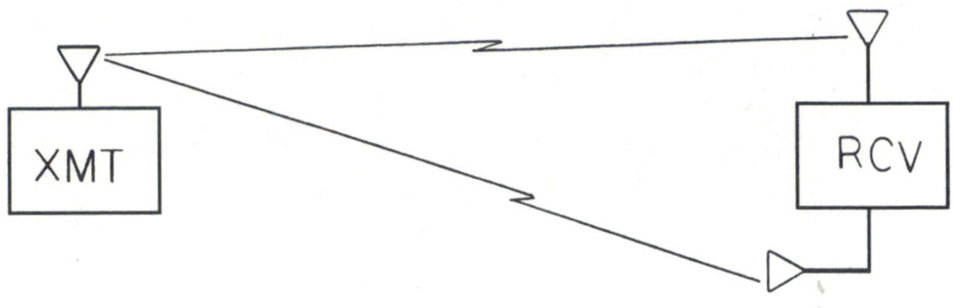

■ **5-16** *Polarity diversity.*

skip requires more power to have the same received signal strength. If the power cannot be increased, then that would result in a weaker signal at the receiver. The HF frequencies between 9 MHz and 30 MHz are used for long-distance transmissions. To minimize the number of skips a signal is subjected to, a lower vertical radiation angle from the antenna is used. However, due to atmospheric absorption, vertical radiation patterns below 3 degrees are impractical.

Another type of propagation is called meteor-burst or meteoric scatter. Although only a few meteors are bright enough to be observed from the ground, the earth is under constant bombardment by billions of objects daily. Of the small number of visible ones, even fewer are massive enough to reach the surface of the earth. All the others are completely burned up in the atmosphere.

The action of a meteor colliding with the atmosphere results in a cylindrical region of ions at the height of the E layer. The long-but-thin plume is sufficiently dense to refract radio waves. Depending upon the size of the meteor, the plume can last from less than a minute to several minutes. The frequency range of from 50 MHz to 80 MHz is best suited for meteor-burst transmission.

The system uses a two-way link. The message sent by the first station is returned by the second one to verify reception. The link begins with the transmitting station sending the first character. No other information is sent by the originator until the receiving station retransmits the character. At that point, the message is sent via *burst transmission*. A burst transmission is when the intelligence is electronically compressed and transmitted at a very high rate of speed. It has been theorized that meteoric atmospheric impacts act to strengthen the E layer. Although not in widespread use, meteor-burst transmission offers another potential propagation means.

Another form of propagation is *aurora propagation*. Due to the rarified atmosphere and strong magnetic fields found at the polar regions, the radiation from the sun causes ionization and air molecule ignition. The phenomenon is known as the aurora or Northern (Southern) Lights. It is very similar to the action contained within a neon light. It occurs at about the E level and can be seen up to 600 miles from the point of formation. The auroral zone sweeps across northern Europe and Canada, Greenland, Siberia, and Alaska.

{Typically the aurora is detrimental to communications links. It causes severe interference and absorption of HF communications. In the VHF range, the presence of an aurora can be beneficial for propagation. Frequencies from 100 MHz to 450 MHz are reflected by the aurora. Reflection properties can vary widely and change rapidly. They can cause distortion of VHF links through multi-path reflections. Voice modulation can be impacted. The best modulation technique for this form of propagation is CW. *As the aurora belt is shallow, east-west transmission paths are best.* Typical ranges that can be obtained are as much as 2000 miles. As the sun causes the auroras, they follow the sunspot cycle and usually peak in March and September. The propagation appears to peak twice a day—at about sundown and a couple of hours after midnight.

Transmission Lines

In small communications systems, such as a CB or handheld unit, the antenna can be connected directly to the equipment. However, when one is dealing with higher-powered communications systems, the antenna must be located some distance from the transmitter or receiver. The reason could be as simple as a high level of radiated power. Another could be the physical size of the antenna system. Many commercial and military communications

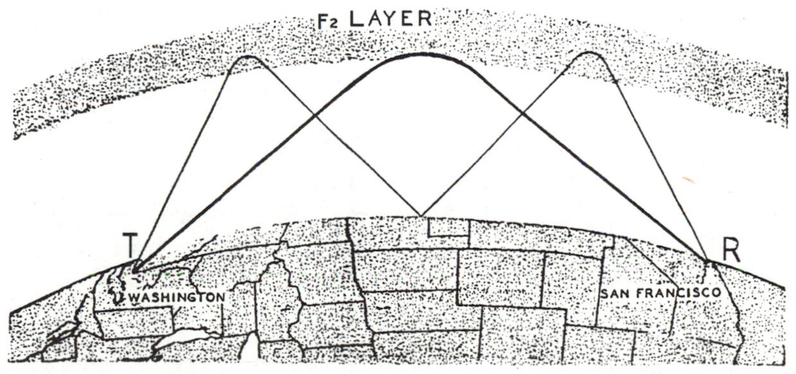

■ **5-17** *Multi-hop transmission.*

installations consist of dozens of receivers, transmitters, and antennas. To connect the equipment to the antenna, a means must be provided to accomplish the function safely and efficiently. The less power that is lost in the transfer process, the more that is available to radiate or process.

Any line or conductor that carries RF energy can encounter three types of energy loss—copper loss, dielectric loss, and radiation loss. Copper loss is determined by the physical resistive characteristics of the conductor or transmission line that carries the energy. Losses are least in a silver conductor, and increase if copper, aluminum, or gold is substituted. Gold has gained a great deal of popularity in connectors lately due to its corrosion-resistive properties (as opposed to a superiority in conductivity). Dielectric losses are due to heating of the insulating material between conductors in a transmission line or coaxial cable. This would be associated with coaxial cables. Radiation losses are caused by energy escaping to free space from conductor or line. Construction techniques such as shielding can limit or eliminate radiation losses.

Currently there are five types of transmission line in common use: parallel two-wire line, twisted pair, shielded pair, coaxial cable, and waveguide. The selection of a particular type of transmission line is based on operational frequency, power, and type of installation. An electrical characteristic that is often mentioned concerning transmission line is *characteristic impedance*. As a communications system must be efficient, any losses that *can* be eliminated *must* be eliminated. To ensure the maximum transfer of energy from a transmitter to an antenna or from an antenna to a receiver, the impedance of the antenna, transmission lines, and equipment must be as closely matched as possible. Any mismatch results in losses called *reflections*. As mundane and common as transmission lines might seem to be, their proper operation is vital to an effective communications system.

Parallel Two-Wire Line

One of the oldest types of transmission line is the parallel two-wire line illustrated in Figure 5-18. This transmission line has several advantages including simple construction, low cost, and efficiency. In addition to some communications applications it is used for electrical power lines, older telephone lines, and telegraph lines.

As can be seen from the diagram, it is constructed from two parallel conductors that are held at a fixed distance through the use

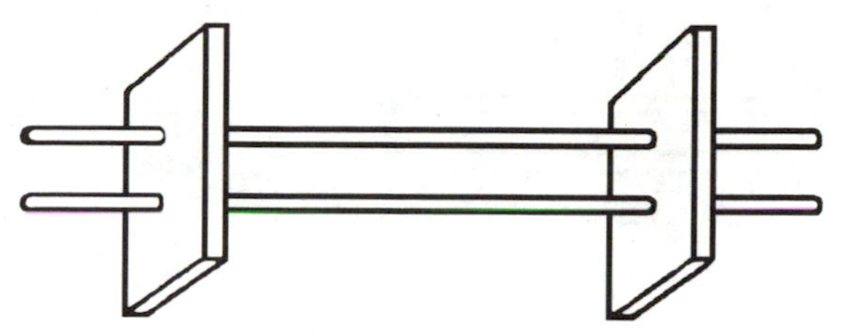

■ **5-18** *Basic two-wire transmission line.*

of insulating spacers. In this type of transmission line, the air separating the two conductors serves as the dielectric. The distance between the spacers is determined by the signal frequency that it will carry. For frequencies 14 MHz and lower, the spacers are from 2 to 6 inches apart. For frequencies above 18 MHz, the maximum spacing is only 4 inches. For best operational considerations, the wires should be separated by less than .01 wavelength of the RF signal.

This type of transmission line has the major disadvantage of a high radiation loss. As a result, it cannot be used in close proximity to metallic objects, because it is unshielded and dramatic energy losses will result. This problem can be overcome if a two-wire ribbon cable is used. Sheathed in a solid low-loss dielectric outer covering, it has the advantages of the two-wire line without the associated losses. This design is pictured in Figure 5-19. Familiar to many as either TV antenna lead wire or FM antenna wire, it is manufactured in two characteristic impedances—75 ohms and 300 ohms. The 300-ohm wire is wider, and is about ½ inch wide. Due to the size of the conductors and the fact that they are coated with only a thin layer of polyethylene, it features a dielectric that is partially air and partially insulator. Because of this, it is susceptible to weathering, moisture, or dirt changing the characteristic impedance. If that occurs, then a signal loss will result. The 75-ohm wire is narrower, which causes the electric field between the two wires to be confined to the dielectric. This wire is far less susceptible to the effects of weather, dirt, and aging. The two-wire ribbon cable was used for years in VLF, MF, and HF frequency applications. The higher characteristic impedance of the lines results in lower losses when compared with coaxial cable in high-powered applications.

The transmission line illustrated in Figure 5-20 is known as the *twisted pair.* Notice that it is nothing more than two conductors twisted into a flexible two-conductor line. This style does not use insulated spacers, so the insulation between the lines is provided

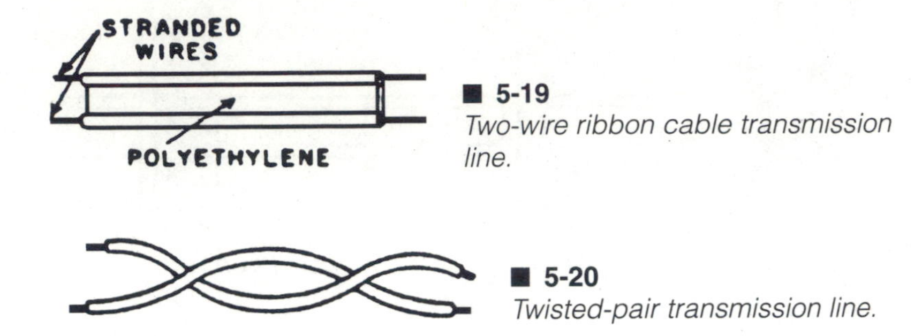

■ 5-19
Two-wire ribbon cable transmission line.

■ 5-20
Twisted-pair transmission line.

by the outer covering on each individual conductor. Twisted pair is suitable for only low-frequency applications. As with the two-wire line, it is affected by moisture, dirt, and aging. Typically, it has a characteristic impedance of about 100 ohms.

A type of transmission line that overcomes the deficiencies of twisted-pair and ribbon cable is the *shielded pair*, illustrated in Figure 5-21. A relatively complex design, it is constructed from two parallel conductors imbedded in a solid dielectric. The insulating dielectric is surrounded by a shield of copper braid. The entire cable structure is encased in a rubber or synthetic flexible sheathing. This type of line visually appears the same as a high-current power cord.

The prime advantage of this type of line is that it is balanced to ground. That means that the capacitance between each conductor and ground is consistent throughout the entire length of the cable. The balanced capacitance results from the construction technique that holds the grounded copper shield and two conductors a constant distance from one another. Another advantage is that the copper braid shielding prevents stray RF and electric fields from interfering with the desired signal carried by the cable.

The final type of transmission line commonly found in communications applications is the coaxial cable, drawn in Figure 5-22. It is the most common type of cable used for carrying RF signals. The line consists of a center inner conductor, held in place by a dielectric covering, such as polyethylene. This is in turn surrounded by a second conductor, or copper braid. The entire structure is covered by an insulating jacket. Various materials are used for the outer jacket, which allow for use in hazardous or harsh chemical environments. For some physically hazardous installations, a metal armor braid can be applied to the outside of the cable to provide protection from chafing and impacts.

Because the inner conductor is completely surrounded by the outer conductor, RF losses due to radiation are minimal. The main advantage of coaxial cable (in addition to the low losses) is the very slight electromagnetic field extending beyond the outer shielding. Because of this lack of EMF radiation, placing the cable in close proximity to metal objects or moisture does not increase losses. That means that the cable can be installed in metal structures or buried underground.

Coaxial cables are produced in several different power-handling capabilities and characteristic impedances. Because of this, coaxial cable is widely used in communications applications. As an example, RF-8/U cable, which has a characteristic impedance of 52 ohms, is suitable for frequencies up to 30 MHz and has a power-handling capability of up to 1 kW. The diameter of the cable is only ½ inch. The inner conductor is 13-gauge wire. At a frequency of 100 MHz, it has losses of 1.9 dB per 100 feet of cable. If the operating frequency is increased to 400 MHz, the losses increase to 4.1 dB. Cables such as RG-58 and RG-59 are very common in such communications applications as video, CB, radio amateur (ham), HF, VHF, UHF, mobile systems, and test equipment interconnections. RG-58 has a characteristic impedance of 50 ohms and losses of 4.5 dB at an operating frequency of 100 MHz. RG-59 (with a characteristic impedance of 75 ohms) is familiar to many as the cable TV line.

More common in radar applications than communications systems is a transmission line known as a *wave guide,* which is illustrated in Figure 5-23. Outwardly a wave guide appears to be nothing

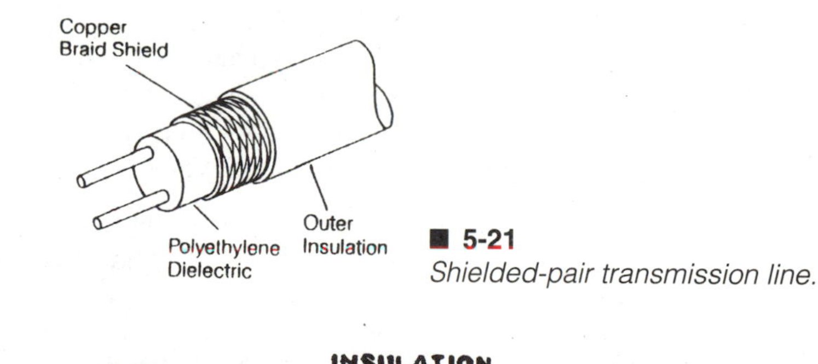

■ 5-21
Shielded-pair transmission line.

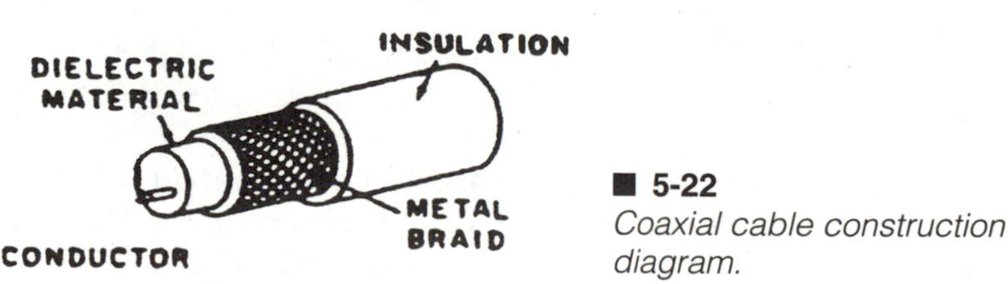

■ 5-22
Coaxial cable construction diagram.

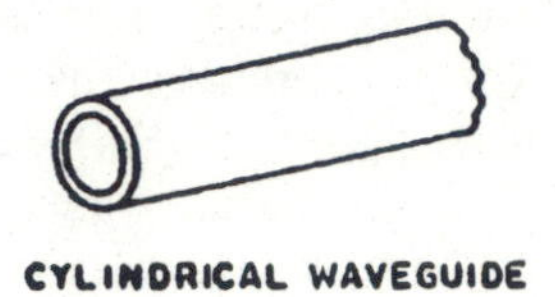

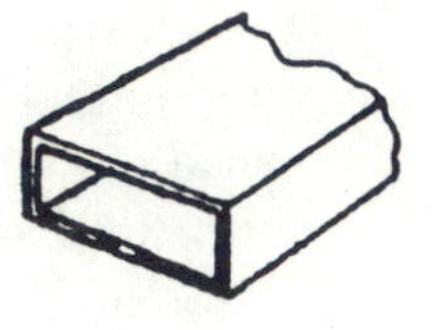

■ **5-23** *Example of wave guide.*

more than a rectangular or circular hollow metal tube with flanges on either end so that sections can be bolted together. However, this is a crucial electronic component that is manufactured to exact machining standards. The inner surfaces are carefully machined smooth with a uniform cross-section and are highly conductive. Common materials for guide construction include brass, aluminum, and copper. Rather than carrying liquid or gas, wave guides are designed to provide a sealed path for the movement of electromagnetic energy.

At microwave frequencies, RF energy escapes from conductors by radiating into free space, rendering transmission lines totally useless as conductors. Radiation parallel to transmission line conductors is confined by their close proximity, and not lost. In the perpendicular plane, the E fields produce a pattern very similar to that of an antenna. Due to the characteristics of current flow, any flow that occurs in a conductor is near the outer surface—hence the term *skin effect.* If most of the electron current flow is in the thin outer layer of a conductor, then the bulk of the conductive material contained within the conductor is wasted. Why not, then, use just the outer layer of metal as the means to transfer electromagnetic energy? The answer is the wave guide. Manufactured in rectangular and circular shapes, it is commonly used in communications and radar as an efficient means to safely transfer high-powered RF energy.

When compared to other conductors and transmission lines, wave guide is far superior for carrying RF energy. It has very low copper loss due to the large conducting surface area on the interior. Dielectric losses are also very low, as the guide is protected from the elements and can be pressurized with an inert gas or air—excellent insulators. Radiation loss is nonexistent, as the electromagnetic field is completely contained within the wave guide walls. For all its advantages, it does have disadvantages that limit its use: high cost, complexity of installation, a requirement for skilled installation personnel, and electrical characteristics that limit its practical use to only very high frequencies.

RF energy is propagated through a wave guide via an electromagnetic wavefront. For propagation to occur, the width of the guide must be one-half the wavelength of the energy that is to travel through it. Because of that, it is frequency-limited. If the applied RF has a frequency of 1 MHZ, the correct wave guide would have to have a width of slightly less than 500 feet. If the propagated frequency is increased to 100 MHZ, the guide width decreases to 4.9 feet—smaller, but still not practical. With an increased frequency to 1 GHZ, the wave guide width would only have to be a modest and practical 5.9 inches. The required wave guide width for a given frequency is easy to calculate, as for propagation to occur, guide width is inversely proportional to RF frequency. As a general rule, wave guide transmission lines are an impractical method of transferring RF energy below 100 MHZ due to the physical dimensions involved.

Regardless of the transmission used in a particular installation, an important consideration is impedance. At RF frequencies, the inherent capacitance and inductance in a conductor become vital design concerns. The inductive and capacitive components exhibited by a transmission line are controlled by several factors, including the dielectric material separating the conductors, the length of the line, the cross-sectional size of the conductors, and the spacing between the conductors. Also, as transmission lines are constructed from conductors, resistance is an electrical component that must be considered.

As these electrical characteristics are determined by the fabrication materials and construction designs, they are easily calculated. Figure 5-24 is the electrical equivalent circuit of a two-wire transmission line. Resistance, a factor of the material used in the conductors, is measured in ohms per unit length and is distributed equally along the line. In the diagram it is represented by R_1 and R_2. R_3, across the two lines, represents the leakage resistance between the two conductors. This results from the small leakage current that exists through the dielectric material separating the conductors.

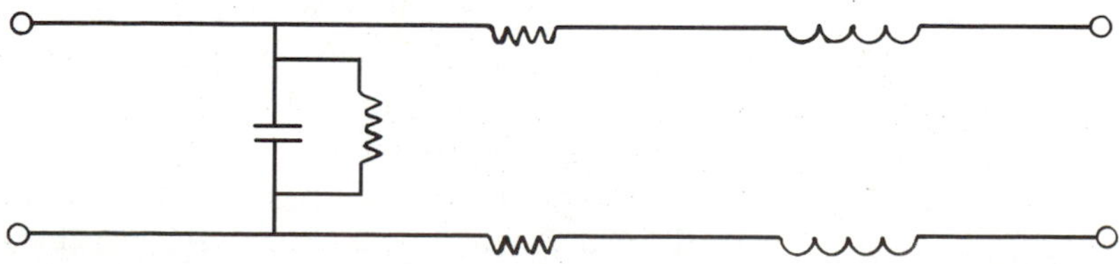

■ **5-24** *The electrical circuit equivalent of a two-wire transmission line.*

As all conductors have capacitive and inductive components, these values are known as *distributed constants*. The capacitive component, indicated by C_1, is the result of the electric field that exists between the conductors. It is shunted across the lines and is expressed as picofarads per unit length. The final electrical component is inductance. Current flow through a conductor results in a magnetic field surrounding the conductor. As you remember from Chapter 2, after the field is built, if the value of current flow changes, the magnetic field will resist the change. A decrease in current flow results in a collapsing magnetic field, which in turn feeds energy back into the conductor, preventing a rapid decrease in current flow. This induced inductance is measured in microhenries per unit length of conductor. In the diagram it is represented as L_1 and L_2.

A transmission line, in addition to the distributed constants, has a characteristic impedance. The characteristic impedance of a line is the impedance one would observe if the line had an infinite length. It is considered to be constant for a given transmission line. Examples would be RG-58 with an impedance of 50 ohms and RG-59 with an impedance of 75 ohms. Characteristic impedance is a very important measure of a transmission line, as it determines the efficiency (the ability of the line to transfer RF energy from the source to the load). Ideally, a transmission line with an infinite length would transfer all applied energy to a load with no losses.

Using the idealized transmission line in Figure 5-24, if a voltage is applied across the input terminals, current will flow through the line. Impedance is calculated in the same manner as resistance is in a conventional circuit.

$$Z = \frac{E}{I}$$

The impedance offered by the transmission line to the applied voltage is greater than one would measure if it were just resistance. The effects of the distributed inductance and capacitance are substantial. For simplicity, the equivalent circuit shown in Figure 5-25 can be used to illustrate the characteristic impedance of a transmission line. As the conductor resistance and the insulation leakage are both low, they are considered negligible. The capacitive component that exists between the two conductors is expressed as a single shunt capacitance. The distributed inductance is represented as two separate inductors forming the horizontal arms of a T. This should look familiar, as it is a simple T filter that was presented in Chapter 2. The characteristic impedance can be calculated by a simple formula.

$$Z = \frac{L}{C}$$

The characteristic impedance of a transmission line can be changed by the manufacturing technique and material selection. As an example, increasing the spacing between the two conductors will alter the impedance. An increased spacing has the effect of decreasing the distributed capacitance, because the two conductors function as the plates of a capacitor. The increase in distance leads to a weaker electric field existing between the two conductors. *As the distributed inductance is not controlled by the distance between the conductors, it remains unchanged.* By lowering the capacitance and not changing the inductance, the characteristic impedance of the line has been increased.

A change in the diameter of the conductors causes a change in the distributed inductance. If the conductor diameter were to decrease, then the characteristic impedance would increase. This action would affect the capacitance much more, as it would have the same effect as decreasing the size of the plates of a capacitor. *A change in the dielectric material can also cause a change in the line's characteristic impedance, by either increasing or decreasing the spacing between the two conductors.*

A very important fact to remember with transmission lines is that if the line is terminated in a purely resistive load, then it reacts exactly like an infinitely long conductor. All of the RF energy provided by the source is transferred down the line to the load. The load then completely dissipates the RF energy as heat.

In electronic communications, the antenna serves as a load that is connected by a transmission line to the transmitter. The function of the antenna is to radiate the RF energy as an electromagnetic

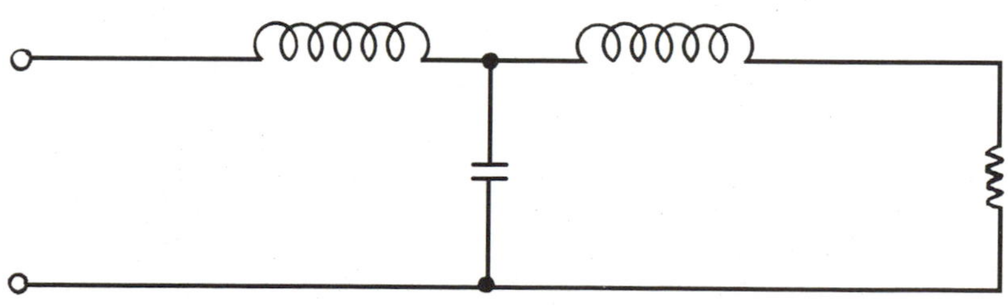

■ **5-25** *Transmission-line characteristic impedance.*

field, rather than dissipate it as wasted heat. On the receiver side, the antenna intercepts the radiated RF electromagnetic energy and conveys it with as little loss as possible to the receiver, which functions as the load. It takes the RF energy and converts it to a format suitable for extracting the intelligence.

From the previous paragraph it is obvious how important to a functioning communications system an efficient transmission line is. Ideally, a transmission line will act electrically as if it were infinitely long. One cycle is defined as one complete sequence of values of an alternating quantity. It includes starting at zero, reaching the maximum positive value, falling through zero to a maximum negative value, then returning to zero. A wavelength is the distance traveled by an electromagnetic wave during the time period of one cycle. The distance between two reference points, or one cycle can be calculated with the formula:

$$\text{wavelength} = \frac{V}{F}$$

V is an electrical constant for the velocity of light, which is 186,000 miles per second, or 300×10^6 meters per second. F is the frequency of the RF signal expressed in Hertz.

A traveling wave is an energy wave formed by the movement of energy along a conductor. For the movement to occur, time is required. Figure 5-26 illustrates the voltage and current waveforms that would be observed on an idealized transmission line of infinite length. Infinite length is an important concept, as it eliminates the possibility of an energy wave being reflected back from the end of the line. As can be seen, the current and voltage waveforms are in phase. Due to line losses, the amplitude of the current and voltage waveforms gradually decrease to zero. By the application of voltage on this line, an electric field (E) is formed between the two individual conductors. The application of voltage causes current to flow through the two conductors. That action in turn causes the building of a magnetic field (H) around each conductor. The E and H fields make up an electromagnetic wave that propagates energy along the transmission line. For an infinite length of transmission line to be valid, several characteristics are required. First, the voltage and current must be in phase along the entire length of the line. That results in minimal loss with maximum efficiency. Also, the voltage and current ratio, which is the characteristic impedance is constant over the entire length of the line. The input impedance of the line is equal to the characteristic impedance of the transmission line. Finally, as the line it terminated in a load that is equal to the characteristic impedance, it appears to be electrically infinitely long.

Transmission lines can be placed in one of two categories. The first is the *nonresonant* transmission line. A nonresonant transmission line is defined as one that is either infinitely long or is terminated in its characteristic impedance. As line losses are minimal, almost all of the electromagnetic energy is absorbed by the load, with only very small losses. Any losses are dissipated along the length of the line. The voltage and current waveforms are in phase along the entire length of the line. Transmission lines used to propagate RF frequencies have a characteristic impedance that is limited to the resistive value of the line. A nonresonant line would be found in applications where high efficiency is required in transferring RF energy to a load.

The other category, the *resonant* transmission line, is one that has standing waves, or reflections. To achieve this, the line is a definite length and is not terminated in its characteristic impedance. A resonant transmission line is very similar to that of a tuned circuit, as it is resonant at one definite frequency. The operational characteristics and impedance of a given line are determined by its length, the frequency of the applied RF energy, and if it is terminated as an

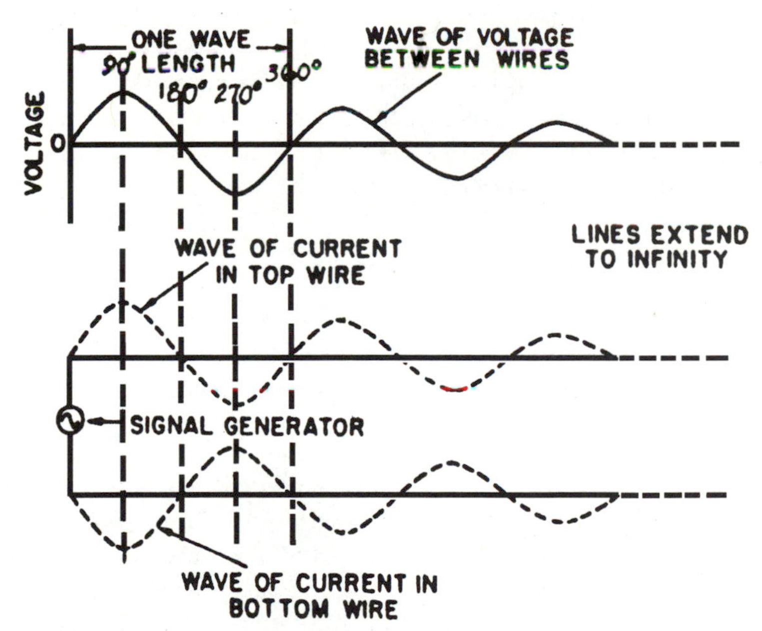

■ **5-26** *Current and voltage waveforms observed on an ideal transmission line of infinite length.*

open or a short. At different points along the line it can have resistive, capacitive, or inductive electrical characteristics. Resonant lines exhibit characteristics associated with a conventional resonant circuit constructed from reactive components. It can display a resonant voltage rise across reactive circuit components. One can also see a low impedance across a resonant circuit, such as that observed in series resonance. A resonant transmission line can also show a resonant current rise in reactive circuit components. It can act like a parallel resonant circuit by having a high impedance.

Based upon the length of the line, it can appear resistive, inductive, or capacitive. At the odd quarter-wavelength points, the current is at its maximum value and voltage is at its minimum. The impedance is low, with the transmission line acting as a series resonant circuit. At all even quarter-wavelength points, voltage is maximum and the current is minimum. The line exhibits maximum impedance and electrically is the equivalent of a parallel resonant circuit. At other than quarter wavelengths, the line appears to be either capacitive or inductive.

The resonance observed in a closed-end transmission line at even quarter-wavelengths from the shored end voltage is minimum and current is maximum. That is caused by the fact that the circuit is the electrical equivalent of a series resonant circuit. At the odd quarter-wavelength points from the shorted end of the line, voltage is maximum, current is at its minumum value, and measured impedance is maximum. That is because the line functions like a parallel resonant circuit. As with the open-end lines, the closed end can also appear to be purely capacitive or inductive, depending on the observed point on the line.

Resonant Line Applications

Standing waves are produced by two electromagnetic waves of the same frequency traveling in opposite directions along a conductor, such as an original wave and its subsequent reflection. This phenomenon is observed when a quarter-wave transmission line is shorted and has an RF electromagnetic wave of the correct frequency applied to it. A resonant quarter-wave line is the same length as a quarter wavelength of the RF electromagnetic energy that is applied to it. Figure 5-27a is a drawing of such a quarter-wave section being used as an insulator with a two-wire transmission line. At the resonant frequency, standing waves are present. At the shorted end, voltage is zero and current is maximum. That condition would indicate that impedance is low. At the input side

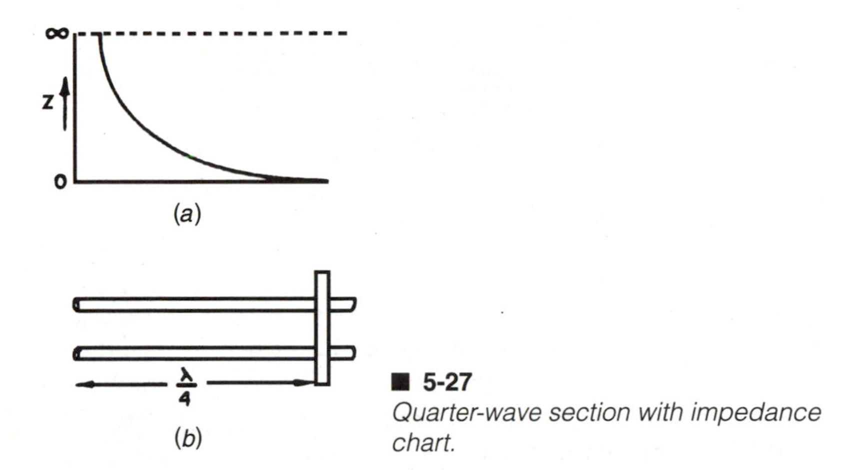

■ 5-27
Quarter-wave section with impedance chart.

of the quarter-wave section, voltage is maximum and current is zero. With those electrical conditions, the section would exhibit a high impedance, allowing the section to electrically function as an insulator for the transmission line at two open terminals. That concept is illustrated in Figure 5-27b. Notice that at the shorted end, the impedance graph is zero, indicating that it can function as a conductor. At the open end, impedance is infinite, which electrically is an insulator.

This electrical phenomenon based on resonance allows the quarter-wave section, or *stub,* to provide impedance-matching capabilities. In Figure 5-28 a quarter-wave section is used to provide impedance matching between a 300-ohm line and a 70-ohm line. To perform this task, the 300-ohm RF generator output is connected to the point on the quarter-wave section where it has an impedance of 300 ohms. This provides a perfect impedance match. The 70-ohm line to the antenna is connected to the point on the section where it has an impedance of 70 ohms. The result is two unmatched lines connected with minimum signal loss and reflections.

The versatile quarter-wave section can also be used to interconnect a resonant line to a nonresonant one, as depicted in Figure 5-29. This connection is slightly different from the one in Figure 5-28. If the line is to remain nonresonant, it must be terminated in its characteristic impedance, which should be purely resistive. The impedance of the shorted quarter-wave section is zero at the shorted end, and increases as you move toward the open end. To act as a matching device in this instance, the shorting bar is adjusted to produce a maximum voltage at the point marked cd. Point ab is manually adjusted to provide the least loss.

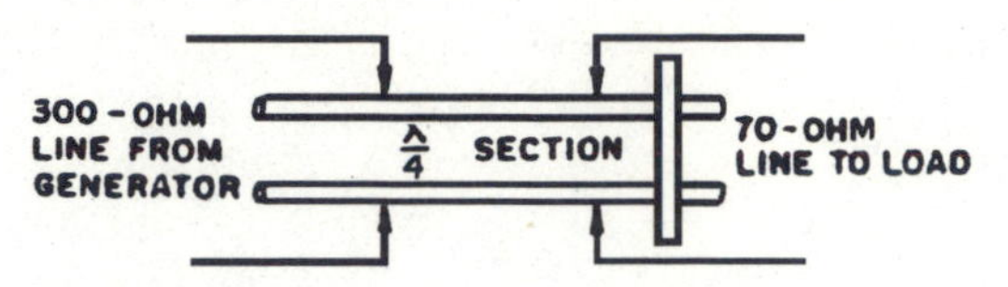

■ **5-28** *Quarter-wave section used to impedance-match from 300 ohms to 70 ohms.*

Another important use for the shorted quarter-wave section is as a filter. As the section offers a high impedance on the open end, it can be used to pass the desired frequencies. To other frequencies it will offer a low impedance, shunting undesirable frequencies to ground. As an example, a 2-MHz transmitter also produces an unwanted harmonic at 4 MHz. To the 2-MHz signal, the section operates as an open, not affecting the signal. At the 4-MHz harmonic, it is a short circuit, shunting it to ground. Quarter-wave sections are common in radar installations as filters, mixers, and signal-steering devices.

Standing Wave Creation

As has been previously stated, if a transmission line is infinite in length, or terminated in its characteristic impedance, then all the RF energy is propagated down the line. Any change in the impedance along the line will cause reflections of energy to occur. A drastic impedance change, such as a complete open or a short circuit, would cause a complete reflection of the electromagnetic energy. A less drastic change in impedance would cause smaller reflections to be produced. The magnitude of reflection is determined by the characteristic impedance of the line and the degree of impedance mismatch. The RF electromagnetic wave produced by the source is called the *incident wave.* A resulting reflected wave is known as the *reflected wave.*

In the open-end transmission line without a load attached, the impedance could be considered to be infinite. With the initial application of RF energy, the first wave propagated down the line would have the voltage and current in phase. This in-phase condition lasts until there is a change in the impedance of the line. When the wavefront reaches the open end of the transmission line, current flow is blocked. The result is that the current waveform at the open end of the transmission line attempts to instantaneously decrease to zero. As the current decreases, the magnetic field surrounding the conductor that was induced by the current flow begins to collapse. The action of the collapsing magnetic field cuts through the conductor, inducing an additional voltage across the line. The induced voltage acts as a generator, setting up new voltage and current waveforms that travel back through the conductors toward the input end of the line. A standing wave represents a power loss, as it is energy that is not delivered to the load. In this instance, a complete open would reflect all the RF energy back toward the source. A lesser mismatch, such as would be encountered in an antenna system, would reflect much less energy back.

When the antenna, transmission line, and transmitter impedances do not match, reflected RF energy develops standing waves on the line. Separate current and voltage waveforms are established. At the one-quarter wavelength point, the voltage is minimum and the current is maximum. At the one-half wavelength point, the voltage waveform is at its maximum value and the current waveform is at its minimum point. The ratio of the voltage at the maximum and minimum points is known as the *voltage standing wave ratio* (VSWR). It can also be calculated using the instantaneous minimum and maximum values of current.

When the load impedance matches the line impedance, no standing waves are formed. Under those conditions, the current is one constant value along the entire line. In that case, the SWR is said to be 1:1. In actual field conditions, a ratio of up to 3:1 is satisfactory for equipment operation. That means for every three watts of power sent down the line to the antenna, one watt is reflected back and lost.

Antenna Fundamentals

An antenna has the vital function of converting the RF oscillations of a transmitter into an electromagnetic wave of the same frequency. On the receiving end, the function of the antenna is to reverse the process by converting the electromagnetic wave into RF oscillations for processing in the receiver. Antenna designs are

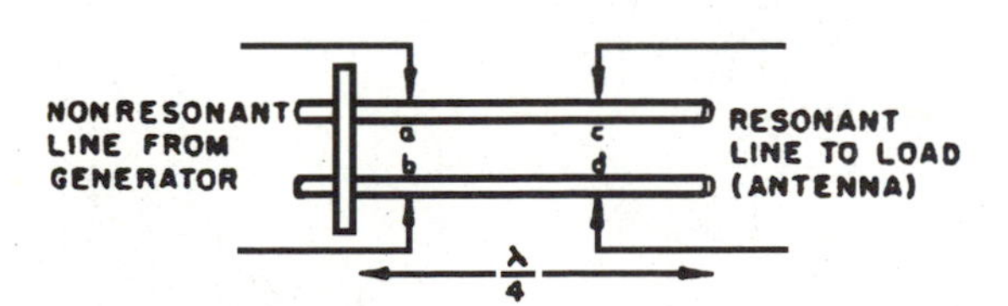

■ **5-29** *Quarter-wave section used to impedance-match a resonant circuit to a nonresonant circuit.*

based upon factors such as power to be radiated, operational frequency, and installation requirements. Because of this, they can range from the very simple and compact to the large and complex.

In a representative antenna system, the transmitter produces the RF electromagnetic energy modulated with the intelligence. The matching device is used to connect the output of the transmitter to the antenna. It can range from just an interconnecting transmission line to a "tuning" system. The RF energy is then coupled to the antenna for converting the RF energy into electromagnetic waves suitable for propagation through free space. On the receiving end, the antenna intercepts the electromagnetic waves and converts it back into RF signals to be processed by the receiver.

An antenna has a range of frequencies with which it is capable of interacting. This bandwidth is dependent upon the degree of impedance matching between all of the components of the communications system. A mismatch between the antenna and transmission line, or the transmitter and the transmission line, results in excessive standing waves and power loss.

Energy is radiated from an antenna through the formation of closed electric fields, which were first predicted by Maxwell in 1873. Figure 5-30 demonstrates how an electromagnetic wave is formed with an idealized antenna. The application of RF alternating energy on the antenna results in a linear displacement of electric charges along the antenna. Equal charges are repelled along the antenna by the expanding electric field, indicated by the solid line. The movement of the electrons causes the formation of a magnetic field. When the alternating RF wave begins to decrease, the charges are brought closer together along with the resulting electric field. The action of the charges moving close

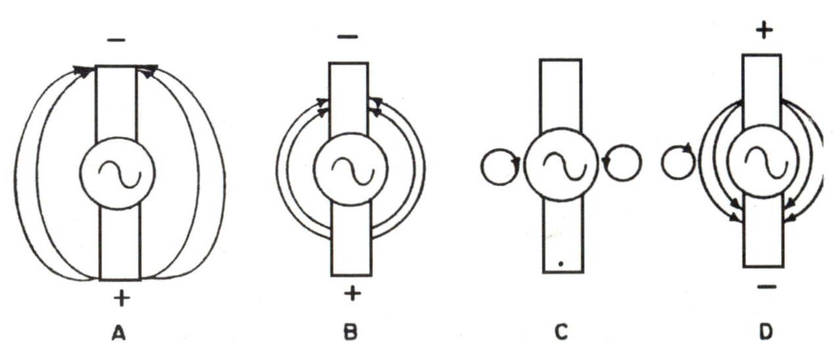

■ **5-30** *Electromagnetic wave formation on an idealized antenna system.*

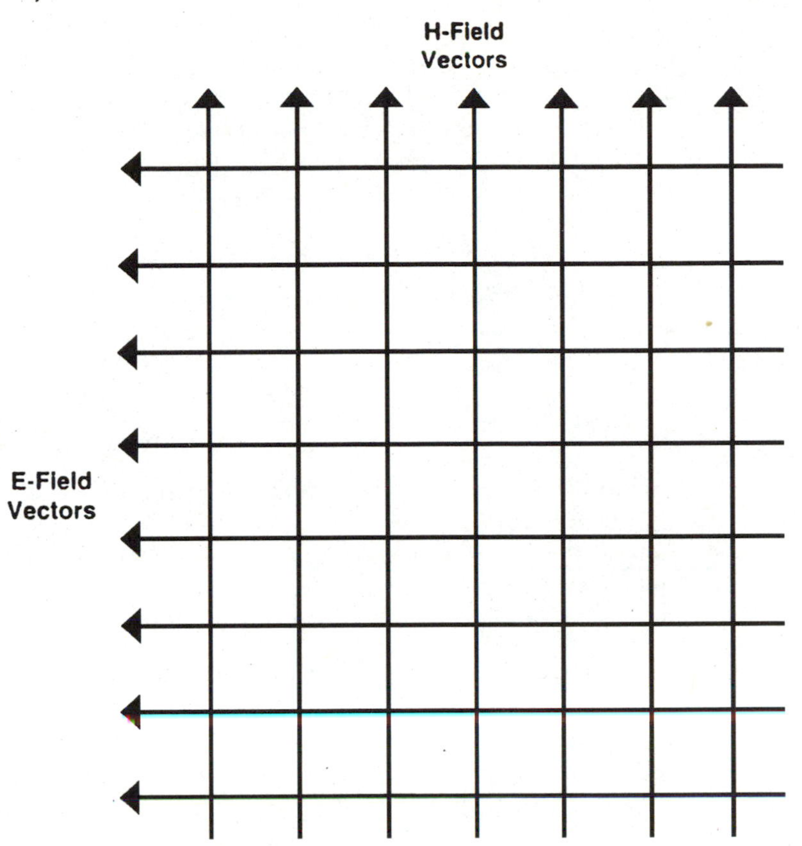

■ **5-31** *E- and H-field patterns in free space.*

together causes the lines of the electric field to close upon themselves, forming a closed loop. As the alternating RF is still applied to the antenna, a second charge displacement is formed. It repels the closed-loop electric field into free space. The movement of the closed-loop electric field causes the formation of a magnetic field. Figure 5-31 illustrates the resulting traveling electromagnetic wave through space.

Antenna resonance is a very important characteristic. The strength of the electromagnetic wave radiated by an antenna is determined by the amount of current applied to it and the size of the antenna. The highest value of current will flow through an antenna when the antenna is resonant at the frequency of the electromagnetic wave. Through experimentation, it was found that the shortest antenna that is self-resonant is one-half the length of the radiated electromagnetic wave. This concept is so important that the half-wavelength antenna is the base that all antenna theory is built upon.

The electrical length of a half-wave antenna is based upon the velocity of light and the frequency of the electromagnetic energy. This formula is a good starting point to calculate antenna length. Actual antenna length would be slightly different due to the physical size of the conductor used in its construction. The formula is as follows:

$$\text{half wavelength (meters)} = \frac{150{,}000{,}000}{\text{Frequency in Hz}}$$

$$\text{half wavelength (meters)} = \frac{150}{\text{Frequency in MHz}}$$

$$\text{half wavelength (feet)} = \frac{492}{\text{Frequency in MHz}}$$

Current is maximum at the center of the antenna. Voltage, on the other hand, is maximum positive on one end, and maximum negative on the other. The impedance is the lowest at the center and the highest at the ends. In this design, the center of the antenna is the feedpoint, or the entry point for the RF energy.

As antennas produce radiation patterns, they are directive in nature. Directivity is the ability of an antenna to concentrate electromagnetic radiation in a desired direction. Through the use of additional elements, the directivity can be enhanced. The elements can take the form of extra radiating elements, reflecting planes (panels), and curved surfaces.

The directive radiation pattern of an antenna can also be affected by nearby objects and topographical features. It has been found that structures within a few wavelengths of the antenna will have the greatest effect on antenna directivity. These structures can reradiate the energy in such a manner as to either reinforce or cancel the desired radiation pattern. Through additional elements on an antenna system, any effects can be either minimized or enhanced if beneficial. This problem is especially critical in military electronics. To minimize interference, high-powered land installations often separate the transmitting and receiving equipment by as much space as possible. In aircraft and shipboard installations this is a luxury that is rarely encountered. Interference minimization is an important area in ship and aircraft design. Antenna radiation patterns must be known so as to reduce this problem to bearable levels.

Antennas also have an effective power gain. The figure is actually derived from two power factors. The first is directive gain, which is a function of the radiation intensity per degree in the direction of radiation. The second is power gain. That is a ratio between the effective signal strength of an antenna and an idealized mathematical model. The two gains, expressed as a ratio, provide an overall efficiency measurement of an antenna. The antenna is the single part of a communications system that can provide the most improvement for the least cost and effort.

The elevation of an antenna is very important to its effective operation. The higher an antenna, the less impact surrounding structures and topographical features will have on the radiation pattern. Additionally, in communications frequencies above 100 MHz, the greater the antenna height, the greater the range. That is because in that frequency range the transmitting and receiving antennas must have an unobstructed view of one another. Another good point to remember with antennas is that good grounding can be critical to operation. A ground plane, which can consist of a grounding plate, an above-ground counterpoise, or an arrangement of buried radial wires is used in conjunction with antennas that depend upon earth as the return path for the radiated energy. A counterpoise is a system of conductors that is insulated from ground and which forms the lower portion of an antenna. It is used as a substitute for a direct ground connection.

Antennas can be broken down into several different broad classifications. The first is the *monopole antenna*. Examples of this would be the familiar whip antenna associated with car radios, CB communications, and military communications. This type of antenna is typically one-quarter wavelength or less. For operation it depends upon the ground plane. The ground plane is a ground point that is above earth ground and acts as the missing length of antenna. In this manner, the antenna is electrically the more desirable one-half wavelength. Figure 5-32 is a Marconi monopole antenna and resulting radiation pattern. The antenna can be mounted either vertically or horizontally. It has a lower input impedance than a corresponding one-half wavelength antenna. When mounted in the vertical position, the radiation pattern in the horizontal plane is a circle. In the vertical plane, it is an array of lobes, reflected from the ground up.

The *sleeve antenna* is a variation of the monopole antenna. It is basically a half-wave radiating element that has a lower half consisting of a metallic sleeve. The upper radiator is a one-quarter-wavelength antenna. The transmission line is connected to the base of the antenna through the sleeve.

Another type of monopole antenna is the *long wire*. This type of antenna is typically much longer than one wavelength and is found

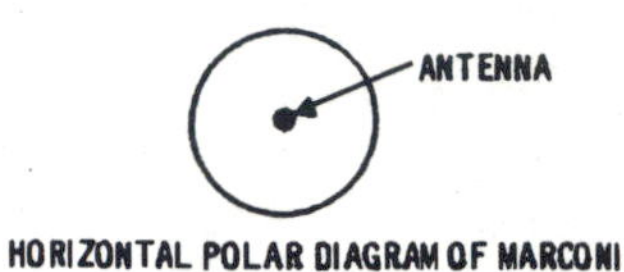

■ **5-32** *Marconi antenna with radiation pattern.*

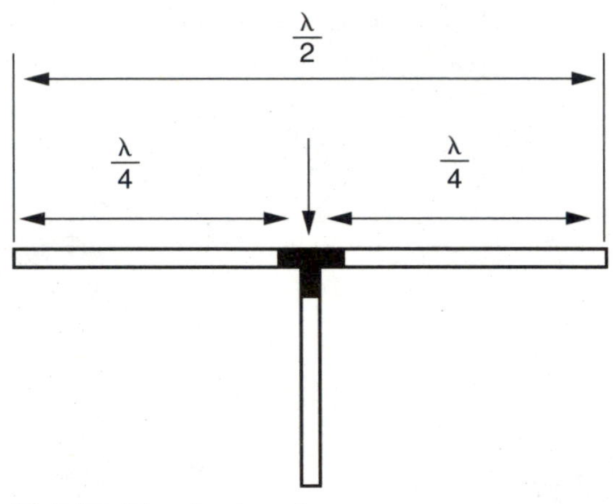

■ **5-33** *The dipole antenna.*

primarily in long-range HF communications. It does have gain, and is directional in nature. It offers simple design, low cost, installation ease, and good operational characteristics. A variation of this type of antenna is the *inverted L*. That is an antenna that consists of a long horizontal wire with a vertical transmission wire lead connected to one end.

The *dipole antenna*, illustrated in Figure 5-33, is a very common type of antenna. It typically consists of two metallic rods mounted horizontally, each of which is one-quarter wavelength long. The transmission line used to connect it to the transmitter or receiver is connected at the center of the antenna. Typical impedance at the transmission-line connection point is 75 ohms. This type of antenna has limited gain, but is low in cost, simple in design, and easy to mount.

A *folded dipole* is an improvement over the basic dipole antenna. It has a greater bandwidth and features a higher characteristic impedance: 300 ohms. Essentially it is two half-wave dipoles connected in parallel. The result is that the RF currents across both elements are in phase. This type of antenna is widely used in TV reception.

By adding to the basic dipole design, the more flexible logarithmically periodic antenna is obtained. Commonly known as the *Yagi,* it is presented in Figure 5-34. Although it was first developed by a Professor Uda, it was named after the Japanese physicist Yagi because he translated the theory into English. It is basically a single dipole with several reflectors and directors. The RF energy is applied to the dipole element, which is the driven element. Radiated RF electromagnetic energy radiated by the driven element toward the reflector is reflected back toward the front of the antenna. Perpendicular elements, called *directors,* are used to concentrate the energy in the desired direction. This type of antenna features high gain, high directivity, and a narrow bandwidth. Basically, the frequency bandwidth is inversely proportional to the antenna gain.

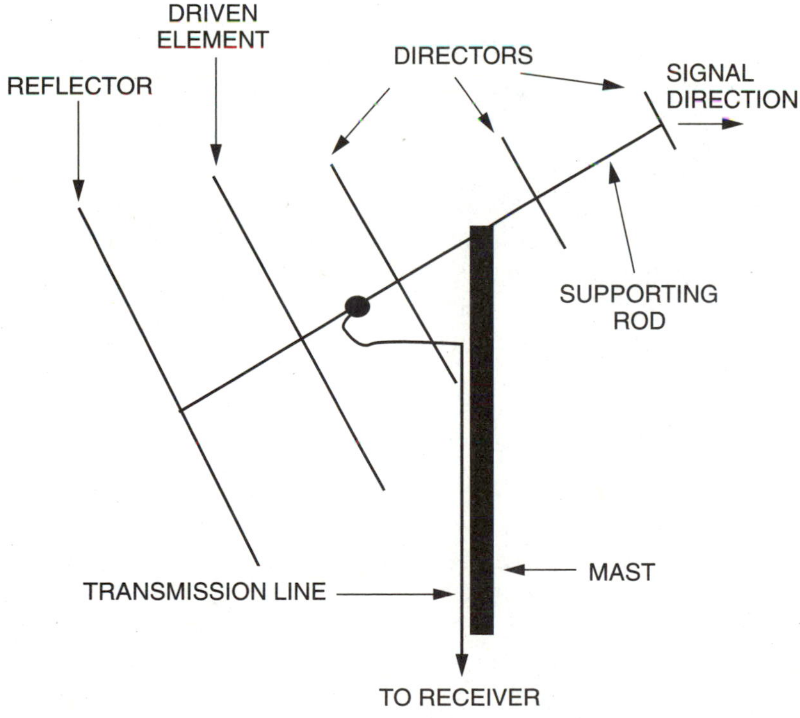

■ **5-34** *The Yagi antenna.*

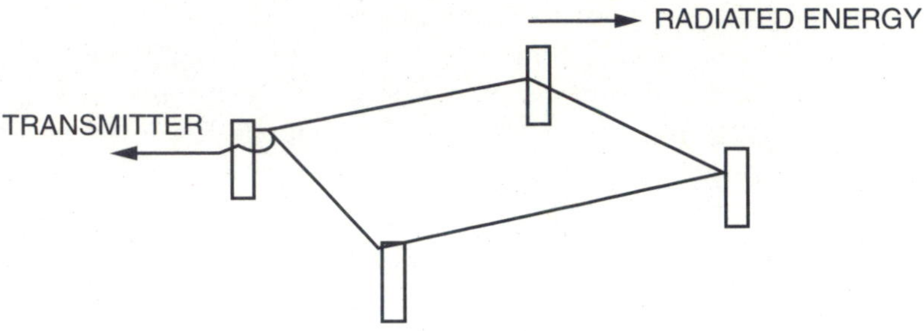

■ **5-35** *The rhombic antenna.*

The Yagi antenna can have an increased bandwidth by simply increasing the physical diameter of the antenna elements. The greater the diameter of the individual elements, the wider the antenna bandwidth. An additional improvement with the larger elements is a stronger antenna that is more resistant to the effects of wind and weather.

The *rhombic antenna,* shown in Figure 5-35, is essentially two longwire V antennas connected end-to-end. In this type of antenna, the transmission line is connected to the longer diagonal leg. As it is usually designed to be from three to ten wavelengths long, it is classified as a long-wire antenna. Terminal resistors are required opposite of the feed end to obtain proper impedance matching for the entire array. As you gain in experience you will encounter many other antenna designs.

VSWR should be as close to 1:1 as possible to provide peak system performance. By tuning, an antenna will be matched to the transmission line and antenna. There are many myths associated with tuning. In military and commercial applications, antenna couplers and tuners are used. Through the addition of reactive components, an antenna can be made usable over a wide band of frequencies as opposed to a small one. In small-scale installations, the antenna can be cut or designed for one specific frequency. In large installations, that would be impractical.

In military and large-scale civilian installations an antenna coupler is used. It can match the impedance of a 15-, 25-, 28-, or 35-foot whip antenna to a 50-ohm transmission line at any frequency from 2 MHz to 30 MHz. The tuning is accomplished by adding reactive components in series with the antenna to give it an electrical length suitable for the frequency it is to transmit or receive on. In doing so, the VSWR is reduced to the lowest level possible.

Another means to improve antenna performance is through the use of a *balun transformer* with center-fed antennas. A balun is a transformer that has a 1:1 impedance ratio. As such, it does not provide any matching. Rather, it balances the currents flowing through two radiators.

Now that the intelligence has been modulated, transferred to the antenna, and radiated in free space, it is time to receive it. The next chapter will introduce receiver theory associated with communications systems.

6

Communications Receivers

THE FUNCTION OF A COMMUNICATIONS RECEIVER IS TO take an input consisting of modulated electromagnetic radio waves and convert them into a format usable by an operator. Circuitry within the equipment will remove the intelligence from the RF carrier and restore the intelligence to its original frequency and format. High-quality receivers are designed to provide maximum intelligibility of signals under difficult conditions. Received signal strength varies from changing propagation conditions, and often has interference from other, unwanted signals. So, in effect, a receiver must reverse the process performed by the transmitter, and do so under many different conditions. The statement that "...a receiver is a receiver is a receiver" has always been valid. What it means is that the same concepts, and often the same circuitry, that are valid in communications receivers holds true for receivers used in radar, sonar, and other areas of electronics. Communications receivers are constructed for the reception of CW, AM, FM, SSB, narrow-band FM, *radio teletype* (rtty), and *facsimile* (fax) signals. Theories and concepts are identical for all forms of receivers.

Receiver Fundamentals and Characteristics

The variety and number of receivers are amazing. Ranging from small credit-card-sized FM radios to multimillion-dollar governmental and commercial installations, the design complexity and costs range widely. No matter what cost or level of technology, all receivers must be capable of several basic, but very important functions: *reception, selection, detection,* and *reproduction.*

Reception is when an RF electromagnetic wavefront passes through an antenna, inducing a voltage which is then routed to the input stage of a receiver. It marks the beginning of the receiver processes. Selection is the ability of a receiver to choose a single desired frequency from all the signals inducing a voltage on the antenna, without interference from the others. Regardless of physical position, frequency range, or format, every RF electromagnetic wave that cuts through an antenna is capable of inducing some voltage. The degree of selection is determined by the resonance of the frequency-determining circuits that comprise the receiver. The sharper the resonance of the reactive circuits used for tuning, the better the selectivity. It is a function of the engineering, design, and construction of the unit. Detection is critical, as it is the ability of a receiver to separate the low-frequency intelligence from the high-frequency RF carrier. The final function is intelligence reproduction. That is the action of converting the electromagnetic signals back into the original format. The type of intelligence that a receiver is designed to process controls both its complexity and cost. A CW-only receiver would be less expensive than a professional-level communications receiver capable of processing AM, FM, CW, and SSB signals.

All receivers must have the four basic operational characteristics to some degree. They are sensitivity, stability, noise, and fidelity. As a rule, the more expensive the receiver, the better these characteristics will be met.

Sensitivity is the capability of a receiver to reproduce weak signals. The minimum weakness of an intercepted signal that a receiver can have on the input circuits and still produce a usable output is determined by the receiver's sensitivity. Measured under standardized conditions with calibrated test equipment, it is usually expressed in microvolts or microwatts.

Stability is the ability of a receiver to remain tuned to a selected frequency under changing conditions. Constant changes, such as temperature, humidity, input AC power, and vibration can alter a receiver's tuning over a period of time. With the advent of low-cost digital components, this problem is becomming less prevalent. Pre-digital receivers are much more susceptible to noticeable drift. Drift is divided into warm-up drift and long-term drift. Warm-up drift is evident during the first few minutes of operation. Tubes and transistors take a definite amount of time to reach their normal operating temperature, stabilize, and maintain operating frequency. Long-term drift is apparent over a much longer period of time as components age. The stress from turning the equipment on and off, with the associated heating and cooling, causes electronic components to change in value, which in turn alters circuit operation. As the changes are gradual, it may be months or years before the changes become noticeable.

Noise is a very important characteristic, as it has a direct impact on the sensitivity of a receiver. All receivers generate internal thermal noise. Thermal noise results from the normal operation of resistive components, amplifiers, and PN junctions that form the receiver circuits. External noise is from outside the receiver and is in the form of atmosphere noise that can be from lightning, a magnetic storm, or solar activity.

The final receiver characteristic mentioned here is fidelity, which is the ability of a receiver to reproduce a detected output that is a faithful reproduction of the original modulating intelligence from the transmitter. These four characteristics must be balanced to produce a receiver that is satisfactory for its intended purpose. Each design is a compromise among the design criteria. As an example, excellent selectivity requires a receiver with a narrow frequency bandpass. Fidelity, on the other hand, requires that a receiver pass a broad band of frequencies so that the outer frequencies of the sidebands can be detected and amplified.

The sign of a quality receiver is its ability to reject spurious signals that fall outside of the desired frequency of the receiver. Additionally, it should not internally generate any spurious signals within the passband. Although the superheterodyne receiver is considered to be the best combination of circuits and compromises to ensure optimum signal reception, it is susceptible to various forms of spurious response. The most common culprit for spurious signals appears to be the frequency conversion process, when the high-frequency modulated carrier is reduced to the much lower intermediate frequency. Through the use of additional circuitry and careful alignments, the problem is controllable.

Another major advantage that the superheterodyne receiver enjoys over earlier receivers is the characteristic known as arithmetical selectivity. A common occurrence in HF communications is when a strong unwanted signal is close to the one of interest. As an example, the desired signal could have a frequency of 12,000 kHz. A strong unwanted signal with a frequency of 12,010 kHz could pose a great deal of interference, to the point where the desired signal would be totally unreadable. As the difference in frequency between the two is only 10 kHz, the high frequency tuning stages in many older receivers would pass both signals. In a superheterodyne receiver, the high frequencies are converted to a much lower IF. Using a standard IF of 455 kHz, the two signals are now

455 kHz and 465 kHz. After conversion to the IF, the two signals are now separated by a much larger frequency, and the unwanted one can now be easily rejected by the filters.

Phantom signals, or *birdies,* are caused by any possible combination of desired and undesired frequencies that are within the range of a receiver. These frequencies, when mixed with the LO frequency, produce an output of just noise. Often these spurious signals can be reduced or eliminated by shielding interference-prone circuits or providing additional filtering to prevent unwanted signal leak-through.

A very important consideration with a receiver is the *dynamic signal range.* The dynamic signal range is the range of input levels at which the receiver output is the exact replica of the input signal. Any signals that fall outside of this range are distorted and unreadable. The range is determined by several different factors. The low end is controlled by the internal receiver noise and hum. The high end of the range is controlled by *intermodulation distortion, gain compression,* and *cross-modulation.* Intermodulation is when the various frequencies that comprise a complex signal waveform are inadvertently modulated, causing spurious signals and interference. Gain compression is when for a given increase in input signal level, a receiver does not produce a further increase in output signal level. This phenomenon is the effect of saturation in a vacuum tube or transistor amplifier. The final limiting factor on the high end is cross-modulation. It can be characterized as a type of interference in which the carrier of a desired signal is modulated by the intelligence of an undesired signal with a totally different carrier frequency. Most often this is caused by a nonlinear amplifier acting as a detector for a strong unwanted signal. This problem is more common in locations where a number of strong signals are present. Testing for the dynamic signal range is performed under controlled laboratory conditions with specialized test equipment.

The dynamic measurements are made in terms of power. As was stated in Chapter 5, 0 dBm is 1 milliwatt. A good quality communications receiver should have a background noise level of –140 dBm. As input signal level is increased, it should show indications of blocking or cross-modulation at a signal level of –40 dBm. As the received signal would be lost in the background noise at –140 dBm and begins to show distortion at –40 dBm, it has a dynamic, or useable, range of 100 dB. State-of-the-art receivers today have a dynamic range that can be from a low of 70 dB to more than 120 dB.

Superheterodyne Receiver

Early receiver designs that were capable of tuning often used multiple tuning stages. Illustrated in Figure 6-1 is the block diagram of an early receiver known as the *tuned radio frequency* receiver (TRF). Notice that several of the amplifier blocks are mechanically linked, or ganged together. The receiver functions by having the input *radio frequency amplifier* (RFA) tuned to the desired frequency. This design has very poor selectivity. To compensate, several stages of RF amplifiers are used before the detector. Because of this, each individual amplifier is tuned by a variable capacitor on its input. To ensure that all the stages are tuned to the same frequency at the same time, they are mechanically linked, or ganged together. This type of receiver is suitable for only low frequencies, as it is difficult to construct a practical tuned-RF amplifier.

To overcome the selectivity problem, the *superheterodyne* receiver was developed. This type of receiver has a high degree of selectivity because it is easier to build highly selective tuned circuits that operate at low frequencies. Reducing the input frequency to a lower one for amplification, detection, and processing results in a more stable and selective receiver. Due to its flexibility, it is the most common type of communications receiver in use today. In a superheterodyne receiver, the very high input signal frequency intercepted by the antenna is mixed with a lower, internally generated variable frequency to produce a third fixed frequency, called the intermediate frequency (IF). The IF frequency is detected, processed, and amplified to convert back into the same format as the original modulating signal in the transmitter. A fixed IF frequency is used as it allows the engineers to design receiver circuitry for maximum sensitivity, selectivity, and gain.

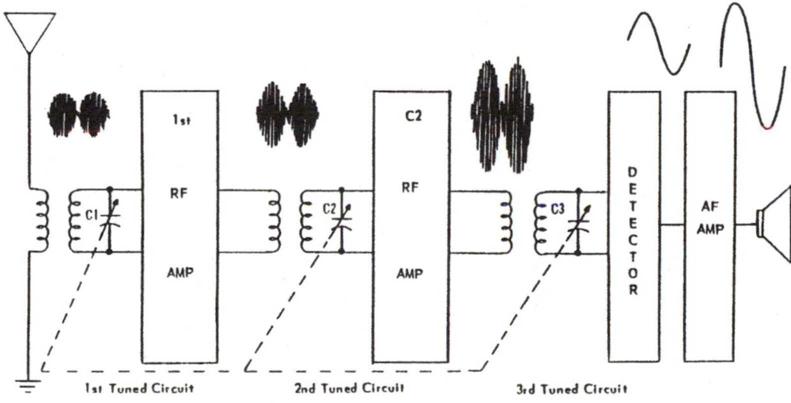

■ **6-1** *TRF receiver block diagram.*

Figure 6-2 is a block diagram of a basic superheterodyne receiver. Heterodyning is accomplished by mixing the input RF with the *local oscillator* (LO) frequency. To maintain system accuracy, the conversion process must be concluded without signal loss or distortion. The local oscillator is a critical receiver component, as any tuning or frequency adjustments required to lock the receiver to the transmitter frequency are actually controlled by it. *Automatic frequency control* (AFC) circuits are used to keep the frequency stable within system parameters, because needed adjustments occur too rapidly for a human operator to perform. The AFC signal is developed by comparing a sample of the transmitted RF with the LO frequency. If the transmitter frequency drifts, or propagation causes a slight shift, the AFC circuitry produces an error signal. Additional circuitry uses the error signal to develop a tuning signal. This is applied to the LO, which in turn changes the receiver frequency, which enables it to track the transmitter. By combining the input RF with the variable LO signal, the result is stable IF signal. A fixed frequency is desirable, as it simplifies receiver design and allows receiver circuitry to be optimized for maximum sensitivity and gain. After the input RF amplifier and the local oscillator stage, the frequency is fixed, eliminating the need for each receiver stage to be tuned to the input frequency.

The output signal of the mixer is now called the IF signal and is applied to the IF amplifier stages for amplification. The IF stages provide most of the receiver gain and determine overall receiver bandwidth.

From the IF strip the greatly amplified signal is applied to the second detector. The second detector actually rectifies the IF signal to derive the audio information. At this point, the high-frequency

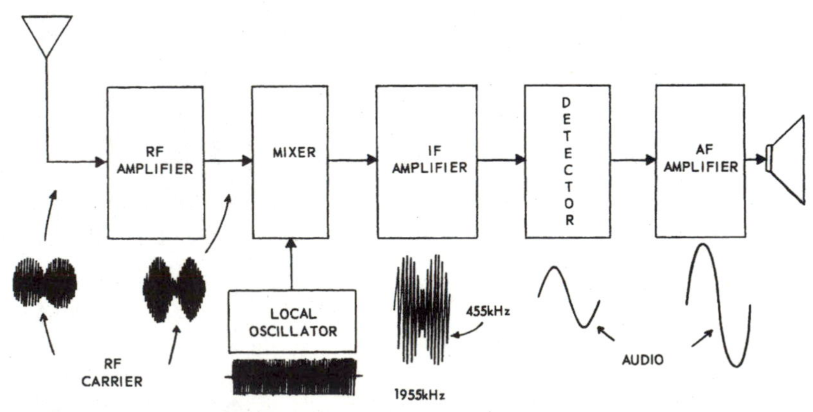

■ **6-2** *Superheterodyne receiver block diagram.*

carrier is removed, leaving only the modulation envelope. The resulting intelligence signal is then amplified by several stages of audio amplifiers. The audio signal is then routed to either external equipment for further processing or is available for the operator to listen to through either headphones or a speaker.

As has been previously stated, the mixing of the LO signal with the received RF results in four frequencies: the two original ones, the sum of the two, and the difference of the two. At any given selected frequency on the tuning control of the receiver, there are two frequencies that will result in the desired intermediate frequency. Below are three equations that explain the concept. In the first equation, the received frequency minus the LO frequency results in the IF. In the second equation the LO frequency minus the second frequency equals the desired LO.

$$F1 - LO = IF$$

$$LO - F2 = IF$$

$$F1 - F2 = 2 \times IF$$

In the third equation, the two signals are separated by twice the IF. One of the signals is the desired IF. The second one is an unwanted signal known as an *image frequency*. If the antenna intercepts the image frequency and it is mixed with the LO and converted to the IF, then there would be no way to separate it from the desired frequency. Filters can be used at the input of the mixer stage to block the image frequencies, but it is difficult to make them selective enough to provide for good image rejection. By increasing the IF to a high frequency, the interference from image frequencies is reduced. However, to provide for selective filters and amplifiers, the IF should be a low frequency.

To greatly reduce the problem associated with image frequencies, the *double-conversion receiver* was developed. Illustrated in Figure 6-3, the double-conversion receiver provides a high degree of image rejection along with excellent selectivity. Notice in the block diagram that the receiver features two local oscillators and two mixers. The first mixer provides a high LO frequency, such as 8 MHz, as the output. With the selection of this high IF, there would be a 16-MHz separation between the desired signal and its image frequency. This provides for a large separation between the desired signal and possible image frequency. The second IF frequency of 455 kHz is the typical IF. With the use of a much lower second IF, filters and amplifiers can be manufactured that provide for a high degree of selectivity. This design is the basis for almost all modern communications receivers.

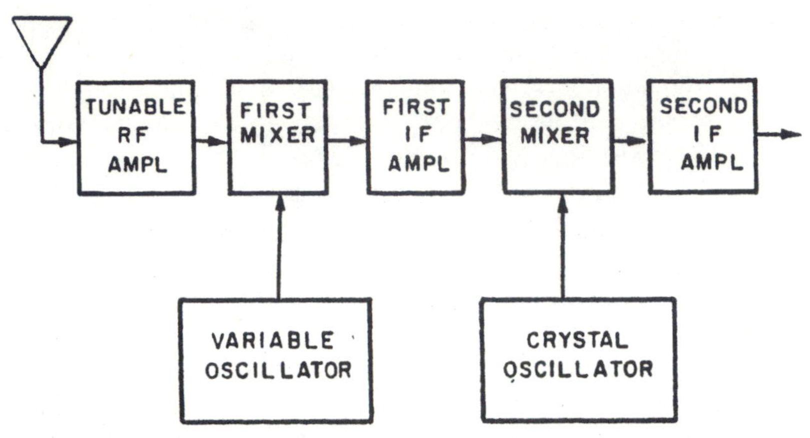

■ **6-3** *Double conversion receiver.*

Receiver Circuits

The primary difference between the superheterodyne receiver and the earlier TRF is that in the TRF receiver, the RF amplifiers are all tuneable, whereas in the superheterodyne, only one stage is tuneable. All input signal frequencies are converted to a much lower intermediate frequency through the use of signal mixing. If you refer to Figure 6-2, the simplified superheterodyne receiver block diagram, you will notice how the mixing process begins. An antenna intercepts the radiated RF energy and couples it into the RF amplifier. As the intercepted RF might be only a few microwatts, amplification is required to provide enough signal level for processing. From the RF amplifiers, the now-amplified RF signal is applied to the mixer stage. Here, with a stable and accurate frequency developed by the local oscillator, the RF is mixed (heterodyned) to develop the intermediate frequency. Often several stages of IF amplification are required to obtain the required signal strength. The output of the IF section is applied to the detector, where the high-frequency carrier is removed, leaving only the lower-frequency audio intelligence signal. The final function is the audio amplifiers, which are required to boost the audio signal to a level that will drive the output speakers or processing equipment.

RF Amplifiers

As will now be seen, RF amplifiers are used in both transmitters and receivers. The circuits can be divided into two broad categories: tuned and untuned. The untuned RF amplifier is responsive over a wide frequency range. The tuned amplifier features a very high amplification factor over a very narrow frequency range. In most installations the tuned amplifier is preferred, as it provides both amplification and selectivity. In a receiver, the RFA has a third and very important function, which is to determine the signal-to-noise ratio of the equipment. As the receiver circuits generate internal noise, a good quality RF amplifier must be capable of responding to weak signals and boosting them to the proper level without adding excessive noise.

The RFA and intermediate amplifiers are very similar in design and operation. The major difference is in the operational frequency of the circuit. You will find that many receivers will have only one stage of RF amplification and several stages of IF amplification. When multiple RFA stages are encountered, it is to increase the amplification factor and improve equipment operation.

Figure 6-4 is a representative RF amplifier constructed from a transistor. As with other amplifiers, vacuum tubes, FETs, and other semiconductor devices can be used as the active element. Notice that the circuit has a tuned input and output tank circuit. In this manner, both tanks will be locked on the desired RF frequency to

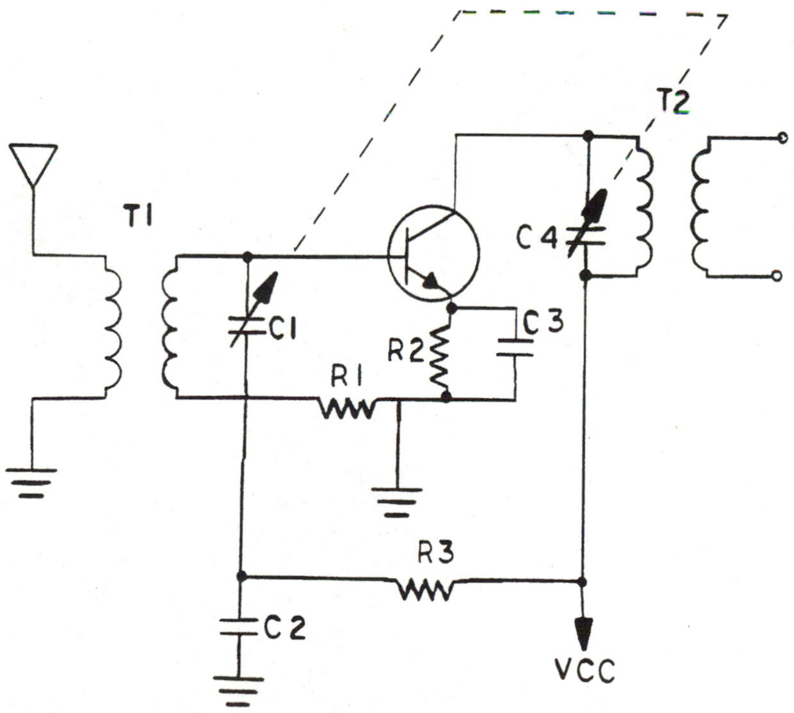

■ **6-4** *Representative RF amplifier.*

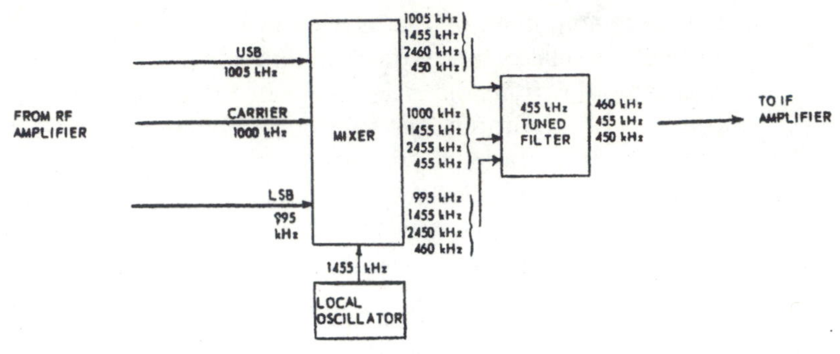

■ **6-5** *Resultant frequencies associated with the heterodyne process.*

be intercepted by the receiver. The function of C2 is to prevent AC variations from appearing on the power supply. C3 is to ensure that the emitter of Q1 is held at a DC level by shunting any AC variations to ground. R1 and R2 are for biasing. As the circuit is biased class A, any input RF signal is amplified and routed from this circuit to the mixer stage.

Receiver Mixer and Converter Circuits

The mixer stage operates on the principle of heterodyning. Figure 6-5 illustrates what happens within the mixer stage during the heterodyning process. In this example, a signal with a carrier frequency of 1000 kHz was received. It has a lower sideband of 995 kHz and an upper sideband of 1005 kHz. That means that the intelligence has a frequency of 5 kHz. In this receiver, the local oscillator produces a stable frequency of 1455 kHz. Within the mixer, the input RF is mixed with the LO frequency.

The local oscillator frequency is heterodyned with the input modulated signal. The LO frequency is combined with every input frequency that is found within the modulated envelope. The lower sideband is combined with the 1455 kHz LO, resulting in the sum, difference, and two original frequencies: 995 kHz, 1455 kHz, 2450 kHz, and 460 kHz. The carrier is mixed with the LO, resulting in the output frequencies of 1000 kHz, 1455 kHz, 2455 kHz, and 455 kHz. Also, the upper sideband is combined with the LO, resulting in the frequencies of 1005 kHz, 1455 kHz, 2460 kHz, and 450 kHz. In addition to all the sum, difference, and original frequencies, the mixer will also produce harmonics frequencies of all of them. The function of the 455 kHz tuned filter is to pass only the frequencies that lie between the two extremes of the sideband frequencies.

The characteristics of the output tank of the mixer stage are very critical. Figure 6-6 is the ideal bandwidth and response curve of a mixer tank circuit. Values are selected for ease of illustration, rather than typical receiver circuit characteristics. The filter is fabricated from components that will pass frequencies centered around 455 kHz. Gain should be uniform from the center frequency out to the half-power points. As the frequencies fall farther outside the half-power points, gain should fall off rapidly. Notice that the output of the mixer has an overall bandwidth of 10 kHz. If you refer to Figure 6-5, notice that the frequencies applied to the mixer have an overall bandwidth of 10 kHz. It is important that the mixer stage maintain the same relationship between the overall bandpass of the receiver RF and its heterodyned output.

The mixer stage has several important characteristics that determine receiver performance. One is the *conversion transconductance.* This is the ratio between the IF current in the output of the frequency mixer and the RF input signal voltage. Measured in micromhos, it is important in that it controls the gain of the stage. The higher the conversion transconductance, the higher the stage gain. Mixer-stage amplification is known as conversion gain. This value is the ratio of the IF output voltage to the RF signal input voltage. The circuit should also have as high a signal-to-noise ratio as possible. That is because all mixers do induce some value of noise into the receiver that tends to lower the unit's overall signal-to-noise ratio. Additionally, interaction between the mixer stage and the input RF amplifiers should be reduced to a minimum value. Any interactions can cause a shift in local oscillator frequency, which is known as *pulling.* Because of this event, the RF input amplifiers should be isolated from the LO to ensure the highest degree of frequency stability.

At this point a good question would be, "How does a mixer actually receive two separate frequency inputs and heterodyne them to

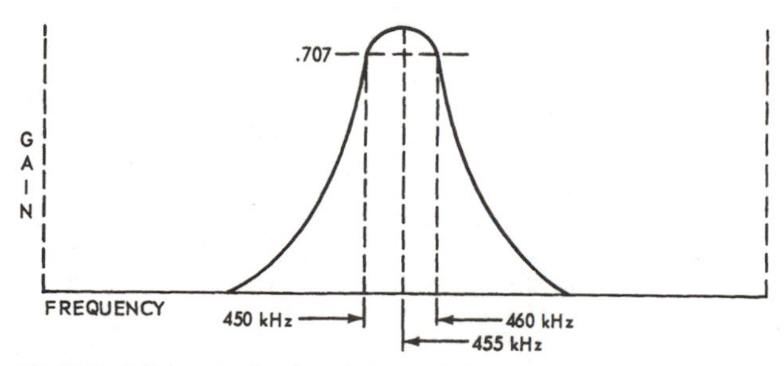

■ **6-6** *Mixer tank circuit bandwidth.*

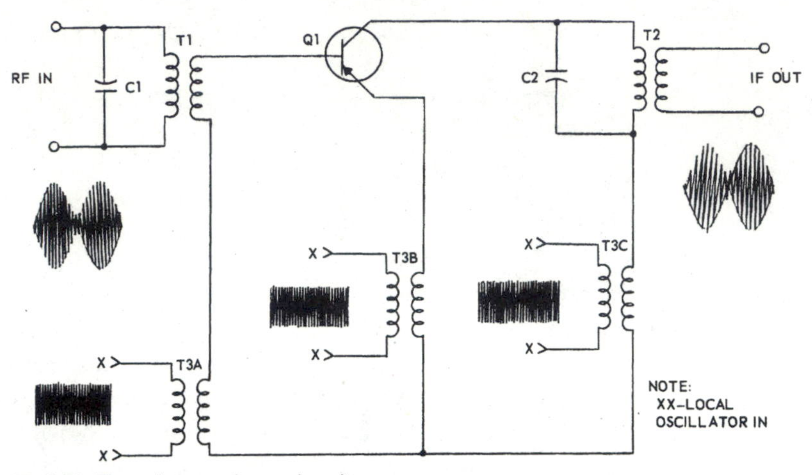

■ **6-7** *Transistor mixer circuit.*

produce a much lower IF frequency?" Figure 6-7 is a simplified schematic diagram for a typical transformer-coupled transistorized mixer. Notice that all signals are coupled into and out of the stage through the use of transformers. Also, this diagram illustrates the three methods of coupling the LO frequency into the mixer stage.

Circuit action begins with transformer T1, which couples the RF signal amplified by the RF amplifiers into the mixer stage. T2 primary and C2 form the tuned output circuit of the mixer stage. The component values are selected to resonate at the IF. The local oscillator frequency is coupled into the stage via transformer T3. In an actual circuit, only one LO input transformer would be used. Three are shown in this schematic to illustrate the different types of mixing available in the stage. T3A depicts a series connection with input transformer T1. In this connection, the LO signal is applied to the base of Q1 along with the RF modulated signal. Mixing is accomplished by controlling the base bias of the transistor. The T3B connection couples the LO signal into the emitter of Q1. In this configuration, mixing is accomplished by changing the emitter bias of the amplifier. The final connection, T3C, couples the LO signal into the collector circuit. Regardless of the LO input selected, the mixing process has the same effect: The high-frequency RF input is converted to the much lower IF frequency.

Figure 6-8 is the simplified schematic for a typical triode electron vacuum-tube mixer. Once again circuit action begins with transformer T1, which couples the RF signal amplified by the RF amplifiers into the mixer stage. T2 primary and C2 form the tuned-output circuit of the mixer stage. Notice that C2 is tuneable to provide for an accurate frequency setting of the tuned stage. To ensure proper operation, the component values are selected to be resonant at the IF. The local oscillator frequency is coupled into the stage via transformer T3 and, as with the previous schematic, three inputs are shown to illustrate the different types of mixing available in the stage. T3A depicts a series connection between the input transformer T1 and the LO input. In this connection, the LO signal is applied to the grid of V1 along with the RF modulated signal. Mixing is accomplished by controlling the bias of the triode. The T3B connection couples the LO signal into the base of V1. In this configuration, mixing is accomplished by changing the cathode bias of the amplifier. The final connection, T3C, couples the LO signal into the plate circuit. Regardless of the LO input selected, the mixing process has the same effect: The high-frequency RF input is converted to the much lower IF frequency.

Mixer design is very important to overall receiver operation. Signal injection of the LO frequency into the base or grid of a mixer is undesirable, as interaction between the two signals occurs. Another potential problem occurs if the antenna is connected directly to the mixer input. The LO signal can often be radiated out of the antenna into free space; this action can cause interference with other receivers. In military operations, it allows sensitive equipment to detect the operation of electronic receivers for many miles. The more preferable signal-injection method with LO signals is either the emitter/cathode or collector/plate technique. These connections minimize the possibility of the LO signal from interfering with the RF signal input. It also reduces the possibility of reradiation when the antenna is connected directly to the mixer stage. This is a form of interference that is annoying in civilian equipment and possibly fatal in military systems.

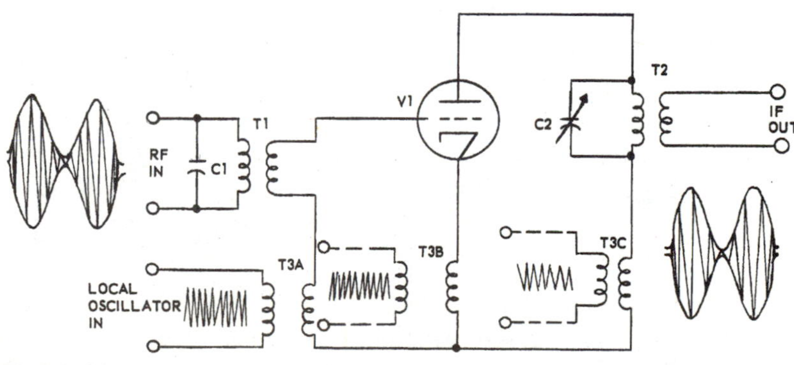

■ **6-8** *Vacuum-tube mixer circuit.*

Another type of signal coupling is the RC-coupled method, shown in Figure 6-9. In this configuration, the LO signal is injected in the emitter circuit of the amplifier. The RF input tank consists of T1/C1 and is tuned to the input RF frequency. The output of the tuned tank is applied to the base of Q1. Fixed bias to the transistor is provided by R1, R4, and the voltage input of V_{cc}. In this configuration, the transistor is biased to operate in the nonlinear portion of its operational curve. R2 provides base stabilization. The capacitors C3 and C4 are used to prevent any AC variations from causing a decrease in amplification. The output tank circuit is formed by T3 primary and C5, and it is tuned to the IF frequency.

Even with amplification, the RF input is often much smaller than the LO input. Because of this fact, the collector current of the mixer stage is controlled by the LO signal. Signals developed across the emitter-base junction of the mixer element are the instantaneous sum of the LO and RF signals. The result is signal heterodyning, and the collector current will be a very complex waveform consisting of the two original frequencies, the sum frequency, the difference frequency, and many harmonic frequencies. The output tank, the T3/C5 combination, is resonant at the IF frequency and therefore blocks any other frequencies. Because only the IF is passed, the overall gain of the stage is less than it would be if it were configured as a conventional amplifier. As the primary function of the stage is frequency conversion, amplification is performed by following stages more suited for that function.

Figure 6-10 shows the same type of RC-coupled mixer fabricated from a vacuum tube. Once more, input and output tanks are used. T1 secondary and C2 form the RF input tank. The cathode resistor

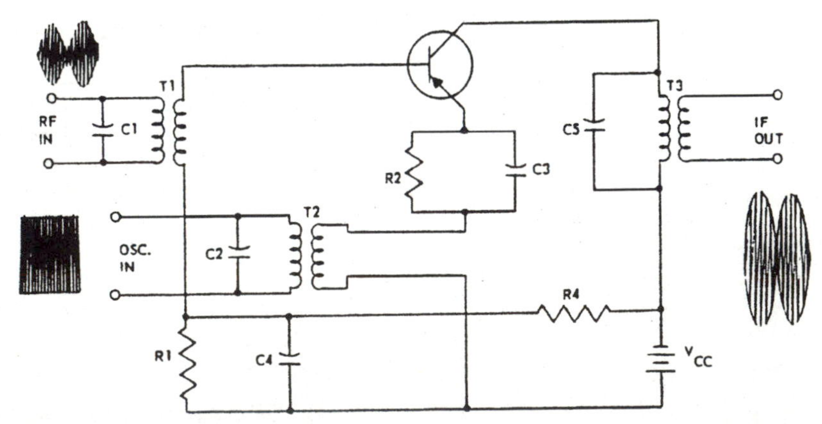

■ **6-9** *RC-coupled transistorized local oscillator schematic diagram.*

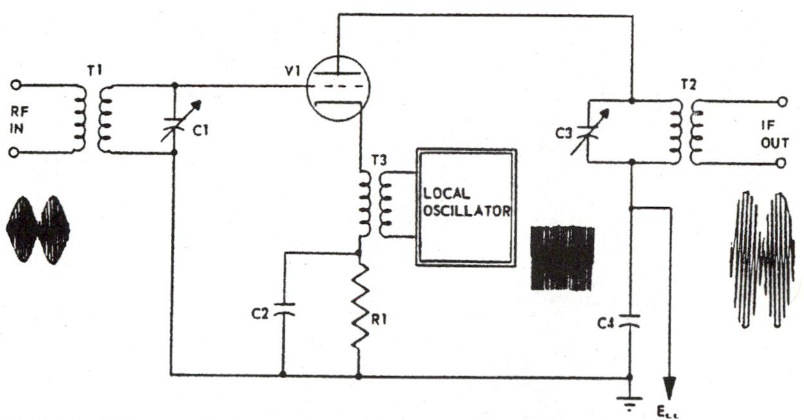

■ **6-10** *RC-coupled vacuum tube local oscillator schematic diagram.*

is R1, bypassed by C2 to prevent degeneration. The plate tank, consisting of T2 secondary and C3, pass only the IF frequency and block all others. C4 has the function of decoupling the power supply input to ensure that RF variations do not appear on it. If that were to happen, then receiver operation would be greatly impaired. As the RF input is on the grid and the LO input is on the cathode, both cause variations in the plate current. That action beats the two signals together, producing the heterodyne process. The result is the two original frequencies, the sum and the difference, appearing on the plate. The output tank blocks all frequencies with the exception of the difference frequency, forming the IF.

Occasionally, you will find the functions of both the local oscillator and mixer combined into one stage, called the *converter*. The single input to the stage is the RF input from the RF amplifier. Such a circuit is shown in Figure 6-11. It has the advantage of using fewer components, therefore it is less expensive to manufacture. It has a major disadvantage in that it is a compromise, which sacrifices some performance. An LO stage requires a high degree of frequency stability, which is produced by the circuit operating in the linear range of the circuit. A mixer requires operation in the nonlinear portion of the operational curve.

The functions of the various components in the converter stage are as follows: The receiver RF input is coupled via the input tank circuit, consisting of C1 and the primary of T1. The secondary of T1 applies the RF signal to the base of Q1. The local oscillator signal is generated by L2, C4, and C5. Notice that it is ganged to the input tank. That ensures that the proper separation is maintained between the input RF and the LO frequency. Voltage divider R1

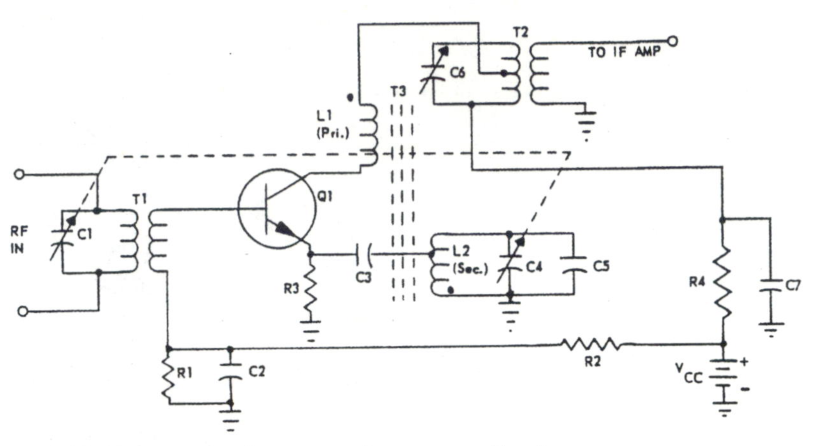

■ **6-11** *Representative converter schematic diagram.*

and R2 set the operating point of the transistor. R3 is installed for temperature compensation. T3, consisting of L1 and L2, is very interesting; this hookup is used so that it can also provide a path for regenerative feedback to the LO circuit. R1, which provides base bias, is bypassed by C2 to present degeneration.

Circuit operation is very straightforward. The input-modulated RF signal is transformer-coupled from the RF amplifiers by the action of T1. The quiescent operation of Q1 provides current to the LO tank so that it can oscillate at the correct frequency. The constant-amplitude LO frequency is coupled by C3 to the emitter of Q1. The voltage divider action of R1 and R2 bias Q1 in the nonlinear portion of its dynamic characteristic curve. This provides for proper heterodyning action. Transistor collector current is controlled by both the LO signal and the incoming RF. As the output of the transistor is a very complex waveform consisting of all the input frequencies, the resonant output tank consisting of C6/T2 primary passes just the desired IF. The final two components, R4 and C7, have the function of decoupling any of the undesired signals to ground.

A multi-element vacuum tube, such as the pentagrid, can also be used as a mixer and converter. Figure 6-12 is a pentagrid mixer stage. This configuration provides an excellent isolation of the LO from the RF input. Notice that the tuning element in the input RF tank and the LO frequency control are gang-tuned. This mechanical connection ensures that the frequency difference between the LO frequency and the input RF remains constant. As can be seen from the diagram, the tube consists of five grids, plate, cathode, and heater. Grid 1, the control grid, has a remote cutoff characteristic. Grid 3 is the injection grid, which provides for modulating the tube plate current. As the tube has a sharp cutoff, it produces a large effect on the developed plate current for a small value of LO voltage. Grids 2 and 4, internally connected accelerate the electron traveling from the cathode to the plate. Another function is to screen grid 3 from the other tube electrodes. Grid 5 is connected to the cathode to act as a suppressor grid, a conventional pentagrid connection.

Heterodyne action is provided by the plate current being varied by the combined effects of the RF and LO input signals. The RF signal is injected to the tube via grid 1. The electrons composing the tube current are accelerated by grid 2. The LO frequency is injected on grid 3. The LO frequency modulates the tube current, which is provided by the RF input. Because of the excellent isolation characteristics of a tube's grids, the RF signal has little or no effect on the LO frequency. In this type of mixer, pulling is rarely a problem. With the combined action of screen grid 4 and suppressor grid 5, the plate resistance is increased, providing the tube with an excellent amplification factor.

As with other types of mixers, the pentagrid can also be configured as a converter, as depicted in Figure 6-13. In this example the LO stage is a Hartley oscillator that is comprised of C5 and the oscillator coil. C4, a trimmer, is provided to allow alignment to ensure proper tracking of the LO. R1 and C3 provide grid-leak bias for the oscillator section of the vacuum tube. Grid 1 is the oscillator grid. Grids 2 and 4 function as the oscillator plate. As grids 2 and 4 are interconnected, they have a secondary function as the electronic shield for the RF signal input, grid 3.

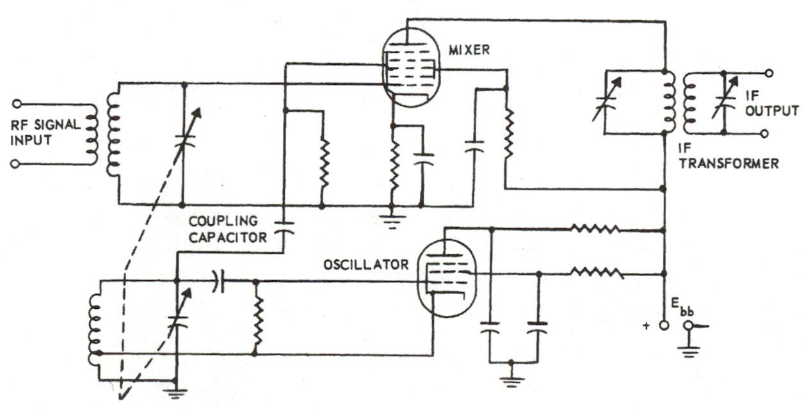

■ **6-12** *Pentagrid mixer circuit.*

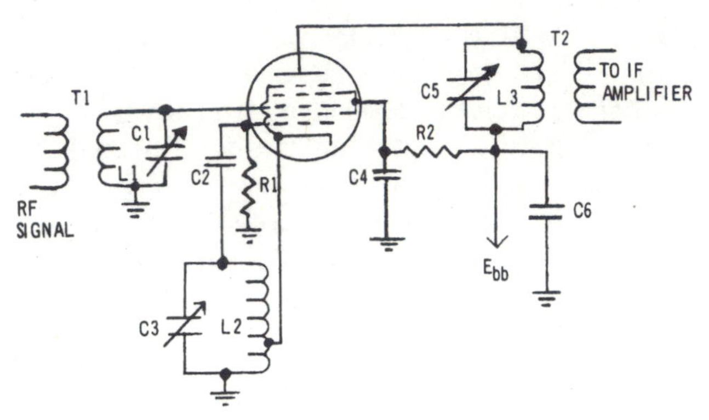

■ **6-13** *Pentagrid mixer circuit.*

Grid 3 in this configuration serves as both a mixer and amplifier for the stage. The tuned-RF input tank circuit consists of L1 and C1. For adjustment purposes, trimmer C2 is provided. The variable components of both input tank and the LO circuits are ganged together to provide for the proper separation between the LO frequency and input RF signal. The heterodyne action occurs within the pentagrid tube. Unwanted frequencies are shunted to ground via C6 and C7. The desired IF frequency is developed by the output tank circuit, C6 and L2. The output of the converter is routed to the IF amplifiers for further processing and amplification.

As any nonlinear device can function as a mixer, diodes can be used to construct simple mixers. Figure 6-14 illustrates a simple single-diode mixer and a double-diode mixer. An external LO is used to produce the LO frequency for the heterodyning process. The resistor in series with the input is to ensure that the RF input is low enough in amplitude so that it does not interfere with the LO. The double-diode mixer is superior in that it offers better isolation between the mixing signals and the output. This type of mixer is suitable for HF (30 MHz) and lower frequencies. In the VHF and UHF ranges, the interactions between the components would cause large-scale interference.

The double-balanced mixer illustrated in Figure 6-15 is found in HF, VHF, and UHF applications. This configuration offers a high degree of isolation, low distortion, and low intermodulation. The LO signal is injected via T2 and the RF input is T1. The output is taken from the center tap of T2.

Field effect transistors (FETs) and *metal oxide semiconductor field effect transistors* (MOSFETs) can also be used in receiver mixers. Figure 6-16 illustrates a common FET and MOSFET mixer circuit. These types of mixer circuits are superior to transistor mixers in that they can handle higher signal levels without intermodulation distortion. In the FET mixer, the RF signal and LO signal are both applied to the gate input. The heterodyne action occurs within the device. The two original frequencies, the sum frequency and the difference frequency, appear on the drain output. The function of the output tank is to pass only the IF signal to the IF amplifiers for further processing. The mixers presented here are just a sampling of what is available in the industry today.

Local Oscillator

The primary function of an oscillator is to generate a specific waveform at a constant amplitude and frequency. The resulting waveform must have amplitude and frequency stability. Amplitude

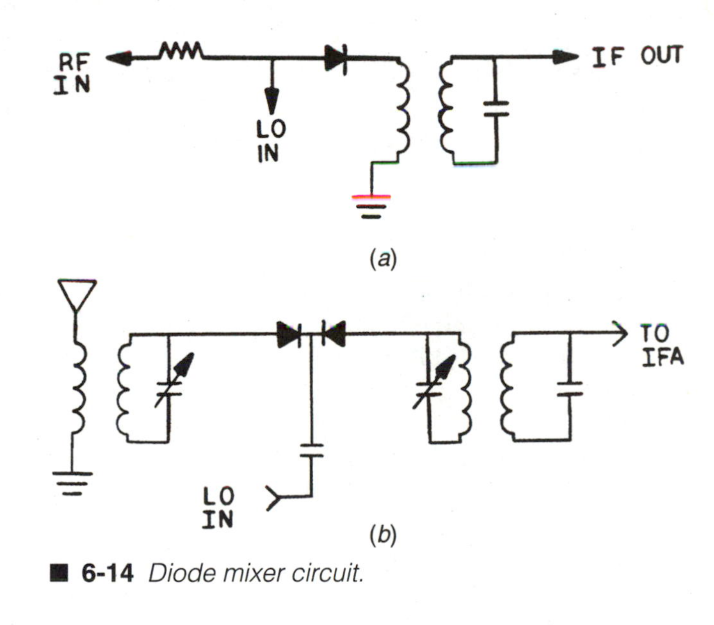

■ **6-14** *Diode mixer circuit.*

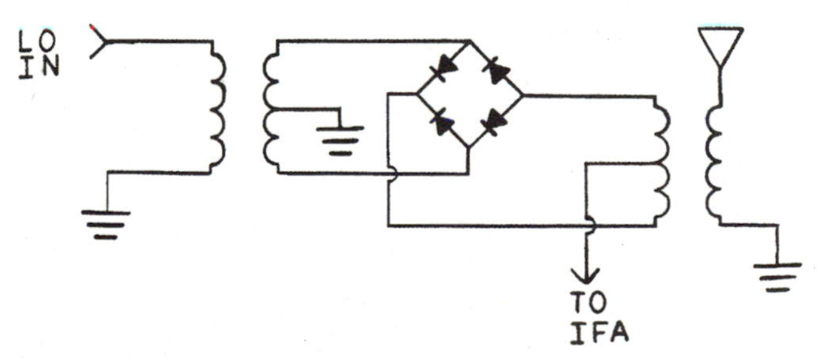

■ **6-15** *Double balanced mixer circuit.*

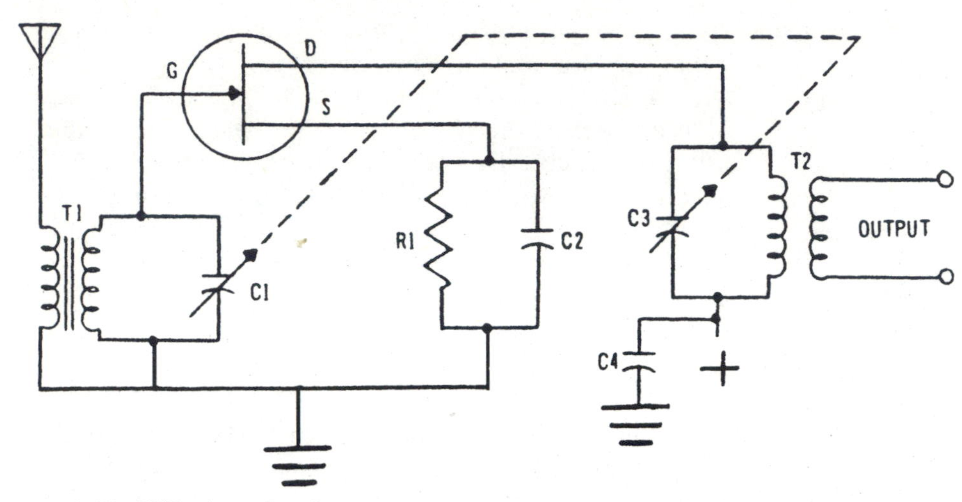

■ **6-16** *FET mixer circuit.*

stability is the ability of an oscillator circuit to maintain a constant-amplitude output waveform with changing voltages and system requirements. Frequency stability is the ability of an oscillator to maintain a desired operating frequency. In the case of a local oscillator found in a receiver, these two requirements are critical. Any deviation could noticeably impact receiver operation.

An oscillator can be divided into three main sections, as shown in the block diagram in Figure 6-17. The frequency control section of an oscillator is typically an LC tank circuit. Exact placement and design of a tank circuit can vary widely. Oscillations within the tank circuit determine the overall frequency. The transistor or tube serves both as an amplifier for the small oscillations from the tank circuit and as a feedback network. The function of a feedback network is to couple the energy from the amplifier back to the tank in the proper phase and amplitude to sustain oscillations. In actuality, an oscillator circuit is a closed-loop DC amplifier that is used to maintain constant-amplitude and constant-frequency AC oscillations. As oscillators have been covered in a previous chapter, only one example will be covered to serve as a review of oscillator operation. For the heterodyning action to be performed in the mixer, a stable, accurate local oscillator is a prime requirement.

Figure 6-18 is a Clapp oscillator that produces a stable sine wave in the RF range. It is typically found in test equipment and in high-frequency and very-high-frequency applications. This type of oscillator is based on a series-resonant LC tank to control the frequency of the circuit. Feedback from the amplifier to sustain oscillations is derived from a capacitive voltage divider network. In actuality, the Clapp oscillator is considered to be a variation of the Colpitts oscillator.

Notice capacitor C3 located in the tank circuit, as it is adjustable to allow for changing the frequency of the oscillator. Due to the small size of C1 and C2, the basic frequency-determining components are C3 and L1. C1 and C2 have the function of serving as the capacitive voltage divider network for feedback. R1 and R2 provide fixed bias to the amplifier. The *radio frequency coil* (RFC) serves as the collector load for Q1 and also to isolate the power supply from the oscillator.

Circuit operation is very straightforward. The first application of voltage allows all the capacitors to begin charging. C3's charge path is from $-V_{cc}$ through the RFC, L1, C3, C2, and then to ground. Charging current stops with C2 and C3 reaching full charge; however, the charging current also flows through the inductor L1. Due to inductor action, as the charging current begins to decrease, L1 collapses its magnetic field, returning energy back into the circuit.

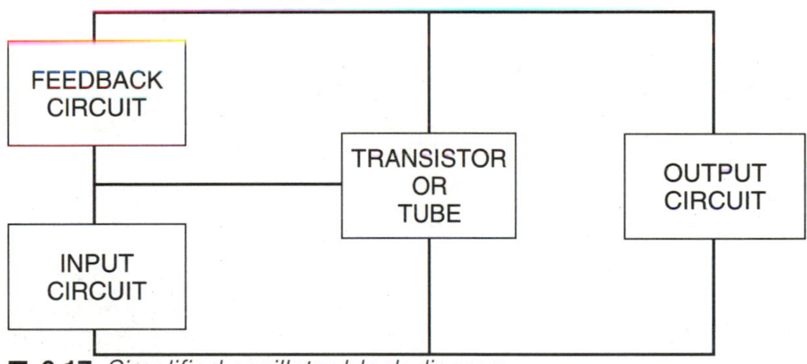

■ **6-17** *Simplified oscillator block diagram.*

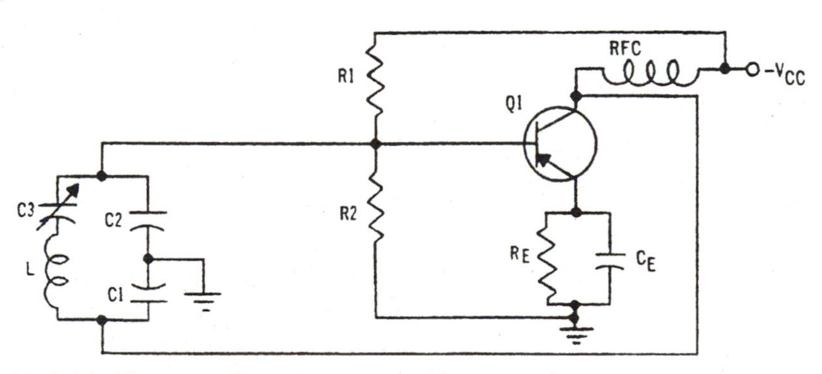

■ **6-18** *Clapp oscillator constructed from transistors.*

As current then decreases due to L1's collapsed field, the capacitors then discharge to maintain tank current. That allows L1 to recharge. Tank oscillations are thus maintained.

The tank also provides bias to Q1. The action of the tank circuit going in a positive direction reduces the forward bias felt by Q1. A decrease in bias causes a decrease in conduction through the transistor, which in turn drives the collector toward $-V_{cc}$. The increasing negative voltage is routed to the bottom of the tank circuit. That increases the voltage drop felt across the tank. That appears to drive the top of the tank more positive, further decreasing the bias on the transistor. That has the effect of acting as regenerative feedback, aiding in maintaining the oscillations in the tank circuit.

As the oscillation in the tank swings negative, the potential on the top of the tank goes negative. That has the effect of increasing the bias felt on the base of Q1. An increased bias causes the collector current to increase, causing the collector to swing positive. That in turn is fed back to the bottom of the tank circuit, acting as regenerative feedback and sustaining oscillations.

Figure 6-19 is the vacuum-tube version of the Clapp oscillator. Notice that with the exception of the capacitor in the plate circuit and the RFC in the cathode circuit, the oscillator is identical to the transistorized version. In this example, C3 and C4 provide a feedback path to the tank circuit. The capacitive voltage divider network is formed by C2 and C3. The RFC in the plate circuit of the tube functions has both a plate load component to develop output variations and to prevent oscillations from appearing in the power supply. Operation is the same as the solid-state circuit.

The *tunnel diode* is much smaller than either the transistor or vacuum tube; it is faster in operation, features low power con-

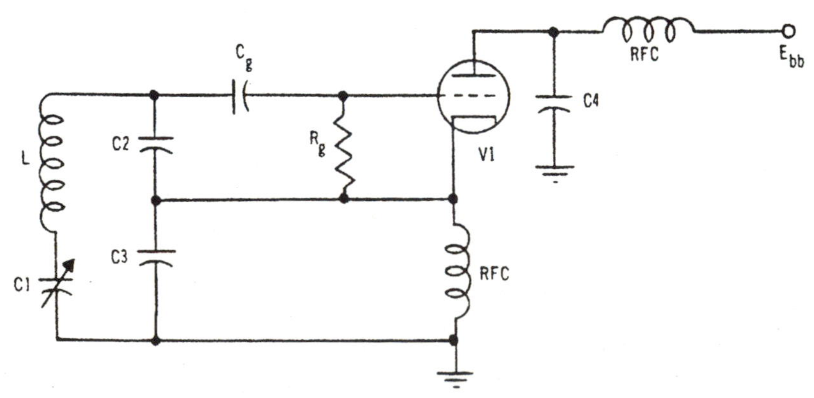

■ **6-19** *Clapp oscillator constructed from vacuum tubes.*

sumption, and is not affected by radiation or temperature variations. Simple in design, it is a very stable device. It is suitable for use as an amplifier, oscillator, and converter in microwave applications. The tunnel diode has only two connections, which is both an advantage and a disadvantage. Two terminals allow for small and simple circuits. However, as there are only two terminals, there is no isolation between the input and output points of the device. As can be seen from Figure 6-20, it is a single PN junction, as would be found in a conventional diode. Its tunnel characteristics come from the fact that it is more heavily doped than a conventional diode—from at least one hundred times to more than several thousands of times heavier. As a result of the very high impurity levels, it has a very thin barrier at the PN junction as compared to a conventional diode.

A tunnel diode is different in that it is capable of more conduction at near 0 volts of bias. A small increase in bias translates into a rapid increase in current flow through the device. Notice that the current flow through the device reaches a peak, and then begins to decrease. Current flow at this point is *valley point current*, which is controlled by a level of forward bias called the *valley point voltage*. After the valley region, a further increase in forward bias results in greater current flow. The current flow valley of a tunnel diode is considered to be a negative resistance region. Because of this curious negative resistance, a tunnel diode can be used to convert a DC voltage in AC—in other words, an oscillator. The governing factor on tunnel diode behavior is the physical size of the junction barrier and the doping of the semiconductor material.

No matter what type of oscillator is selected as the local oscillator, there must be a means of coupling it to the mixer stage. The two most common methods are *capacitive coupling* and *inductive coupling*. Capacitive coupling is simply connecting the oscillator with the circuit where it is to be used by a capacitor. A coupling capacitor should have a high value of capacitive reactance at the resonant frequency of the oscillator circuit. That allows the signal to be routed to another point without loading the resonant tank circuit characteristics. First, the amplitude of the output may be reduced. Then the bandwidth of the signal waveform would be increased, causing signal loss. Also, even if loading does not occur, adjustments of the oscillator might be required to compensate for frequency changes.

Inductive coupling is when the inductor of a tank circuit is also the primary of a transformer. The output of the oscillator is then taken from the secondary of the transformer. At times, the secondary is

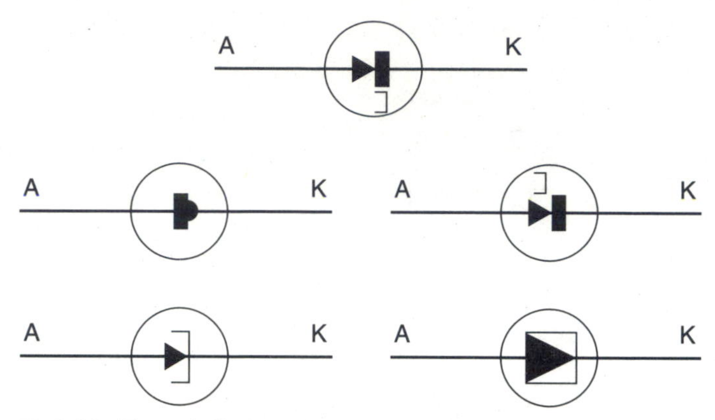

■ **6-20** *Tunnel diode schematic symbols.*

shunted with a capacitor to form a second tank circuit. This configuration is known as a tuned secondary. Inductive coupling has its own set of considerations.

Output amplitude from the coupling transformer is controlled by the degree of coupling that exists between the primary and secondary of the transformer. Tight coupling is when the primary and secondary are physically close together. This results in a high output amplitude with a large value of power being transferred from the tank. It has a disadvantage in that a high power loss is reflected back into the tank. It has the effect of loading the tank, lowering its Q, which in turn increases bandwidth. If a looser degree of coupling is used by moving the two windings farther apart, less power is transferred, decreasing amplitude. Loose coupling is used more often, as it has low reflected losses, which results in greater signal amplitude and increased frequency stability.

Two important tank oscillator characteristics, amplitude stability and frequency stability, are affected by many factors. Amplitude stability can be impacted by changes in bias voltages, amplifier gain, and reflected impedances. Each of these would affect the oscillator in a different fashion.

With most amplifiers, an increase in bias results in a decrease in feedback. As a tank circuit requires regenerative feedback to maintain oscillations, a reduced output results. A decreased bias would cause an increase in feedback with a corresponding increase in amplitude. Any changes in amplifier gain are the same as a change in the bias applied to the circuit. Changes in load impedance affect a resonant tank as well. A decrease in load impedance reflects an increased power drain back into the tank. If feedback amplitude remains constant, the reduction is not compensated for, causing a decrease in output amplitude.

Frequency stability, which is noticeable as a frequency drift, is most affected by changes in reflected losses, power supply voltages, tank Q, temperature, and vibration. Load impedance changes can cause a shift in tank Q, with resulting changes in frequency and bandwidth. By designing high-Q tanks with loose coupling, frequency instability can be minimized. Tank Q can be impacted by component value changes and equipment internal temperature. By ensuring adequate ventilation, clean filters, and equipment cleanliness, this problem can be virtually eliminated. Vibration can have the same effect on a circuit as temperature changes, except much more rapidly. To minimize this, equipment in locations that might encounter mechanical vibration are shock mounted. Finally, power supply variations can cause untold problems with oscillators. If internal equipment regulators cannot fully compensate for the problem, then external devices such as magnetic amplifier regulators or *uninterruptible power supplies* (UPSs) must be used.

A most vital concern in a superheterodyne receiver is resonant circuit tuning and tracking. Because a superheterodyne receiver uses a fixed intermediate frequency for signal processing, the input RF amplifier and the local oscillator must be tuneable. The LO frequency must be separated from the input RF frequency by the value of the intermediate frequency (IF). In reality, the local oscillator should be considered to be the heart of a superheterodyne receiver, as without its dependable operation the receiver would be useless.

The LO frequency can be tuned above or below the incoming RF frequency. The more common method is tuning above. A good example of how it functions would be as follows: A typical IF found in use today is 455 kHz. For an AM radio to tune to a station with a frequency of 1200 kHz, the LO frequency would be 1200 kHz + 455 kHz, or 1655 kHz.

A communications receiver with a range from 150 kHz to 30 MHz would use a double-conversion process. The incoming RF would be reduced to 45 MHz in the first mixer stage, and then to 455 kHz in the second stage. This design technique results in a greater operating stability in the higher frequency ranges.

As the LO frequency has to be a fixed frequency above or below the received RF, then it must follow, or track, the signal. The most common method in vacuum-tube and transistorized oscillators is to use tuning capacitors. Some oscillators are more readily made tuneable than others. As an example, a Hartley oscillator can maintain stability even with a variable capacitor in the tank circuit. A Colpitts oscillator is a little more difficult to modify. As the circuit depends upon an impedance ratio for operation, a variable tank coil or split capacitors may be used.

Voltage-controlled oscillators are gaining acceptance in communications tuning circuits. They offer low cost, accurate waveforms and ease of design. Often the entire circuit is contained within one small integrated circuit. The advantages offered by their adaptation is ease of construction, maintenance, stability, and accuracy. Even low-cost communications receivers now offer digital readouts, digital tuning, hundreds of memory storage channels, and continuous tuning throughout the range of the receiver.

Intermediate Frequency Amplifiers

The output of the mixer or converter is next routed to the intermediate amplifier. A point to remember is that a very small received RF signal is heterodyned with the LO to develop the IF signal. It is still very small and is not suitable for the detection process. To boost it to a level sufficient for processing, one or more stages of intermediate amplification are required. Figure 6-21 is a simplified receiver block diagram to accurately place the IF amplifiers. It receives its input from the mixer (converter) stage and has a single output to the detector stage.

IF amplifiers are very similar to other amplifier circuits that you have observed. The amplifiers operate at a fixed frequency. Tuning circuits are provided, but are used only to ensure that the amplifiers are aligned to the correct frequency. Specialized circuits are provided in the IF amplifiers to process the received signal and improve its quality.

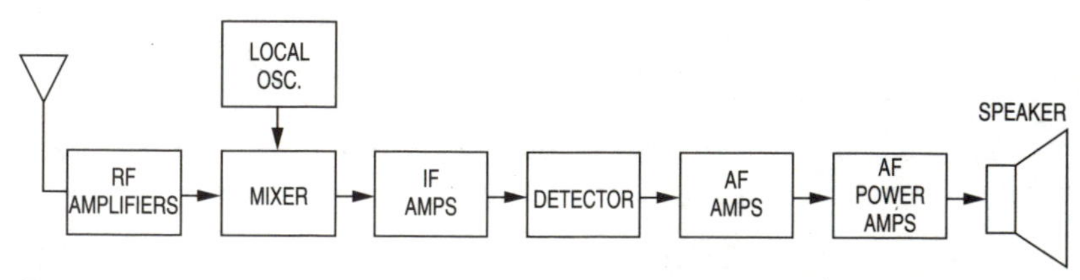

■ **6-21** *Superheterodyne receiver block diagram.*

The most important characteristic of an IF amplifier is its operating frequency. As with so many other electronic circuits, IF amplifier design is a compromise. The exact IF frequency will vary somewhat from designer to designer. Throughout the years, frequencies ranging from 130 kHz to 485 kHz have been used as intermediate frequencies for superheterodyne receivers operating in the broadcast band. Receivers using the lower end of the IF range are infrequent, even though it does offer some advantages. Selectivity, gain, and stability are all improved. However, the fact that a low-end IF is much more suceptible to image frequency problems has led to its lack of use. The majority of contemporary receivers use one of three frequencies: 455 kHz, 456 kHz, and 465 kHz. These frequencies appear to be the best compromise between amplification, stability, and image-frequency rejection.

Receivers operating in the shortwave and higher bands will use double conversion. The first IF will be much higher, ranging from 40 MHz to over 1 GHz. The second IF is then a more common frequency, such as 455 kHz. This enables high-frequency receivers to feature the same stability and sensitivity that used to be found only with crystal-controlled units.

Before covering the IF amplifiers, an important consideration should be discussed. The physical construction and characteristics of the transformers used in IF amplifiers are very important. In high frequency applications, such as an IF strip (several IF amplifiers in series), transformers are used as coupling devices. Often, as was introduced in local oscillators, the primary or secondary can be used as part of a resonant tuned tank.

The selectivity of a superheterodyne receiver is determined by the IF amplifiers. Because of this, the transformers used in this application should have a high Q. Therefore, usually transformers used in IF applications have a powdered iron core that is moveable. That means that the inductance of the transformer is tuneable. This type of component is said to be *permeability tuned.* When this type of transformer is used in a tuned tank, the capacitor is typically of a fixed value.

Transformers in IF circuits are also used for impedance matching between stages. To provide for impedance matching and to ensure adequate selectivity, the inductance of the primary is increased with a corresponding tapping. Transformers designed for IF applications are often very small and enclosed within a metal can or shield. In transistorized circuits, the metal shield often contains both the transformer and capacitor that comprise a tank circuit. That allows for both to be electrostatically shielded, preventing the magnetic lines of flux associated with the transformer from being induced in adjacent wires and component leads.

Figure 6-22 is a vacuum-tube IF amplifier. This circuit is slightly different in that it is double-tuned. Notice that on both the input and output circuits, a tuned tank is used to couple the signal into and out of the circuit. In vacuum-tube circuits pentodes are typically used as the amplifier. Another difference is in the input transformer. As a design consideration, the input transformer may have a lower coefficient of coupling than the output. That has an added benefit of reducing noise that is generated by the heterodyning process in the mixer stage.

A *single-tuned* IF amplifier is one that has only one tuned resonant circuit. A tuned circuit consisting of an inductor and a capacitor exhibits frequency discrimination. If you examine Figure 6-23, you will see the passband of a tuned tank. This compares the waveforms associated with a high-Q tank and a low-Q tank. The higher the Q, the sharper the waveform. This translates directly into selectivity. The sharper the slope of the response curve, the more frequencies outside of the half-power points are subjected to attenuation. That allows the equipment to select a signal frequency over another one that is very close to the first one. Single-stage tuning results in high amplifier gain and excellent selectivity. However, it can limit the amplification of frequencies that fall outside of the half-power points. That is a problem in receiving signals with a great deal of intelligence contained within the sidebands.

Wide passband is required in many instances, such as the need for constant amplification across the entire bandwidth of the amplifier. To maintain the selectivity and increase the passband, a tech-

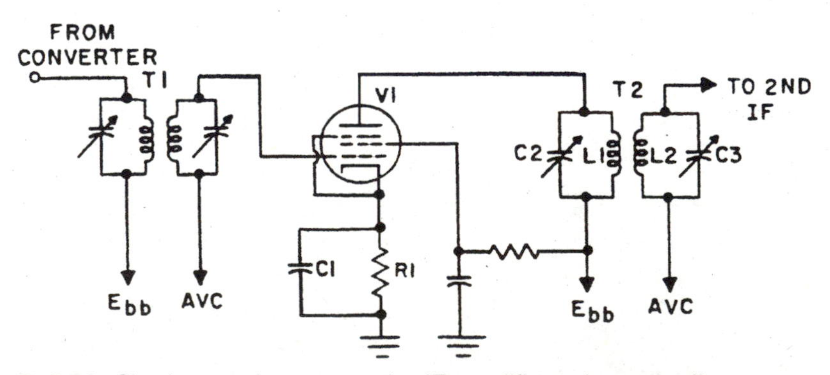

■ **6-22** *Single-tuned vacuum-tube IF amplifier schematic diagram.*

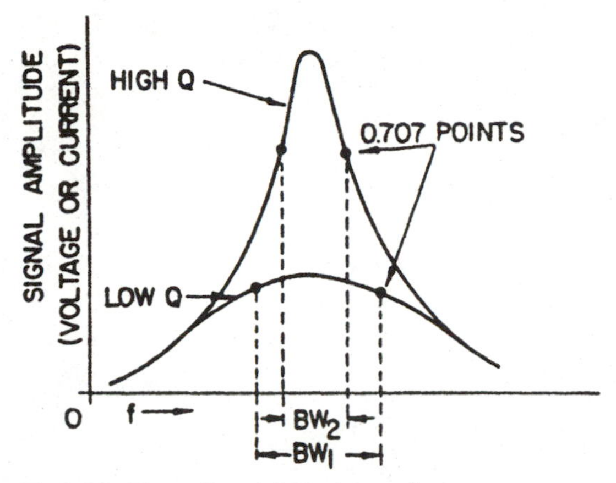

■ **6-23** *Tuned-tank bandpass frequency response curve.*

nique called *double-tuning* is used. Double tuning is when an amplifier has a transformer in which both the primary and secondary are part of resonant circuits. Figure 6-24 is the response curve for a double-tuned circuit. Three characteristics have the most effect on the passband of a double-tuned stage, and they are the coefficient of coupling between the primary and secondary windings of the transformer, the primary and secondary Q, and the mutual inductance between the transformer windings.

The response curves associated with a double-tuned IF amplifier are a compromise between the desired bandpass and the required and wave shape of the response curves. To be brief, this alignment of an IF strip is critical. The waveform in Figure 6-24 represents the degree of coupling between the primary and secondary. The first wave shape is loose coupling. Notice that the waveform is very sharp, with a narrow bandpass. This type of coupling would cause the loss of intelligence that might be found in the sidebands. An increase in the coefficient of coupling causes an increase in the value of mutual inductance. That in turn increases the value of voltage induced in the secondary winding, with the effect of increasing the bandpass. In this state, the impedance of the primary and secondary are equal, resulting in the maximum transfer of energy. The waveform labeled *critical coupling* has a wider bandpass and a rounded peak.

A further increase in the coefficient of coupling results in the optimum coupling waveform. Notice that the waveform has a slightly wider bandpass. The peak of the waveform begins to have a dip at the resonant frequency. As the coefficient of coupling is increased, then the *overcoupling* state is reached. The bandpass is too wide

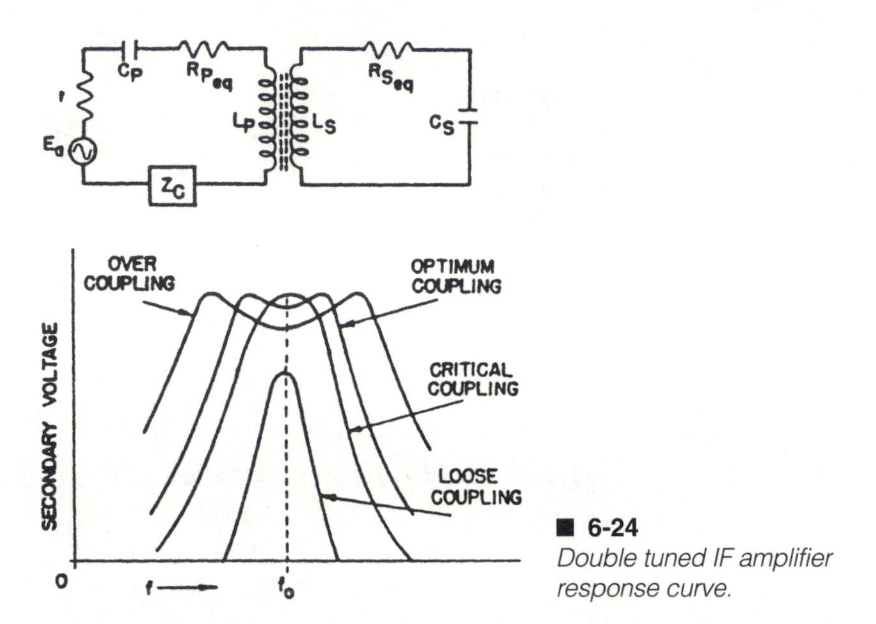

■ **6-24**
Double tuned IF amplifier response curve.

at this point, and the decrease at the resonant frequency is far more noticeable.

In double-tuned circuits, the bandwidth is controlled by the coefficient of coupling that exists between the primary and secondary windings of the transformer. The valley observed at the resonant frequency is determined by the interrelationship between the coefficient of coupling and the Q. If a circuit is operated with an optimum coefficient of coupling, an increase in Q results in a deeper valley in the waveform. Therefore, a large Q translates into a large valley and a small Q causes a small valley.

Stagger tuning is a technique that is used to develop a large value of gain and a wide bandpass. The concept is illustrated in Figure 6-25 with a highly simplified stagger-tuned IF strip. In this concept, the gain of a single-tuned IF stage is combined with a wide, uniform wave shape of a double-tuned IF stage. Notice that the two IF amplifiers appear identical, with the exception of the center frequency. In this example the receiver IF is 450 kHz. One stage is tuned to 443 kHz and the other to 457 kHz. Due to the lack of inductive coupling between the two stages, they will in effect have the same frequency response and operational characteristics.

Detection

Once the received RF signal has been amplified, mixed to convert it to a lower IF frequency, and amplified again, it is time to extract the intelligence from the modulated signal. *The action of detection is to remove the higher carrier frequency, leaving only the much lower intelligence frequency.*

Figure 6-26 is the schematic for a simplified series diode detector. This is considered to be a series detector, as the detection device is in series with the input signal. A diode is ideally suited for the function of a detector, as it conducts in only one direction. Notice that the circuit consists of a diode, input tank, load resistor, and filter capacitor. The function of the input tank is to pass only the desired IF frequency with adequate sidebands to contain the intelligence. The load resistor develops the intelligence signal developed by CR1. C1's value is chosen to allow it to operate as a simple low-pass filter.

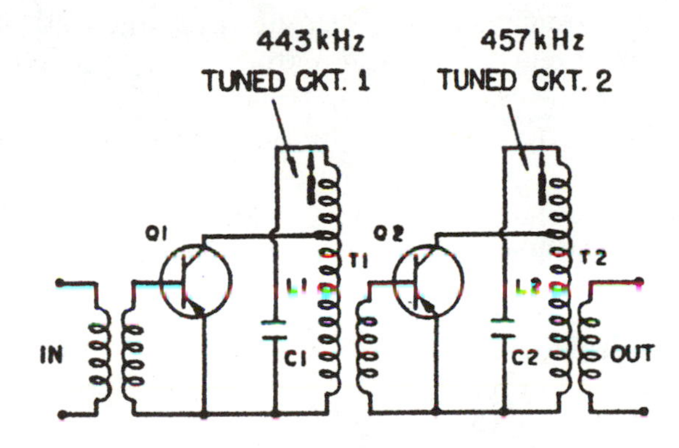

■ **6-25** *Staggered tuned response curve.*

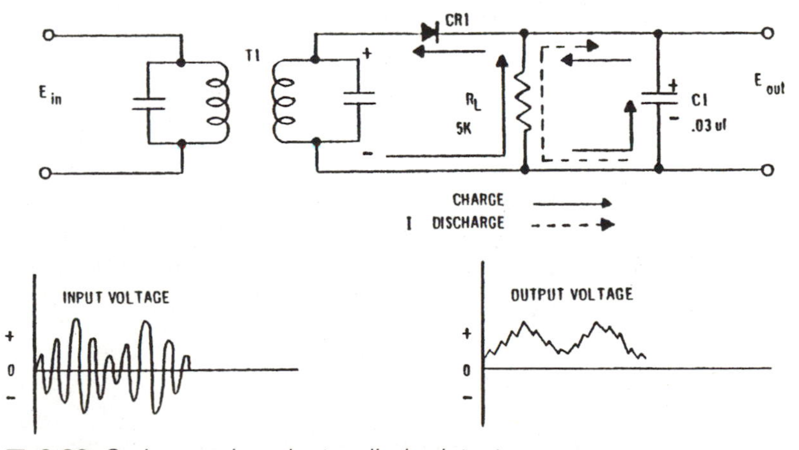

■ **6-26** *Series semiconductor diode detector.*

Detector operation is as follows: On the positive peak of the input signal at the top of the tank, CR1 is forward-biased and conducts. Current, indicated by the solid line, flows from the bottom of the tank to the first tie point. There it splits into two paths. The first one goes up through the load resistor. The second path is on to the second tie point, and into the bottom of C1. That allows C1 to charge. From C1 the current path is back toward CR1, where it adds with the current that flows through R1. As C1 has a low value of reactance, it rapidly charges to the peak value of voltage felt across R1.

As the input signal decreases past zero and continues on to the negative peak value, CR1 is cut off. The decrease in voltage felt across R1 and C1 causes C1 to begin discharging back into the circuit to maintain voltage. The discharge path, indicated by the dashed line, is from the bottom of C1, through the load resistor, and then back to the top of C1. As the discharge path for C1 is through the relatively high value of the load resistor, it only loses a small amount of charge before CR1 begins to conduct again, allowing it to charge again to the peak value felt across R1. The result is that the high-frequency oscillations of the carrier are removed, with only the much lower frequency of the intelligence left.

The selection of the load resistance and the filter capacitor are critical in proper detector operation, as they function as an RC circuit. The value of the resistor determines the time it takes to charge or discharge the capacitor. The values selected for the components is critical. If the capacitor is too large, it will take too long to discharge, resulting in distortion of the detected signal. Conversely, if the discharge time is too short, then unwanted high frequencies will be present in the output, once again causing signal distortion.

As has been stated in the explanation of detector operation, the capacitor must charge rapidly, and discharge relatively slowly to provide for an accurate extraction of the intelligence. When properly selected, the discharge time for the capacitor will appear long to the IF for efficient operation and filtering. At the same time, it will appear to be short to the modulation frequencies for good fidelity, or faithful reproduction of the intelligence signal.

The diode detector can also be connected in shunt configuration, as depicted in Figure 6-27. Notice that the diode is connected in parallel with the input signal to be detected. In this example an inductor in series with the signal is used as the filtering device. In conjunction with the load resistor it will form an LR circuit. As with the capacitor, the inductor must charge rapidly and discharge relatively slowly to provide for an accurate extraction of the intelligence. It must have a discharge time that will appear long to the IF for efficient operation and filtering. To the modulation frequencies, the discharge time must appear to be short for good fidelity.

Some older receivers still use vacuum-tube rectifiers. Figure 6-28 is the schematic diagram of a representative vacuum-tube circuit. As can be seen from the diagram, this is a shunt detector. It can also be connected in series configuration. Component functions are the same in vacuum-tube detectors as they were in the solid-state detectors. The input tank passes only the IF band of frequencies. The inductor, in conjunction with the load resistor, develops the intelligence signal.

Audio Amplifiers

Audio amplifiers perform the final major function that is common to virtually all receivers. Typically the circuits are designed to amplify audio frequencies, which are considered to be the band of frequencies from 20 Hz to 20 kHz. Often, to ensure adequate amplification of all frequencies of interest with sufficient power, a greater range is used, from 10 Hz to possibly as high as 100 kHz. Audio amplifiers can cover a wide range of complexity. The audio section of a high-quality stereo receiver would be more complex than that found in a communications receiver. Figure 6-29 is a block diagram of the representative superheterodyne receiver. Notice that the output of the detector stage is applied to the audio

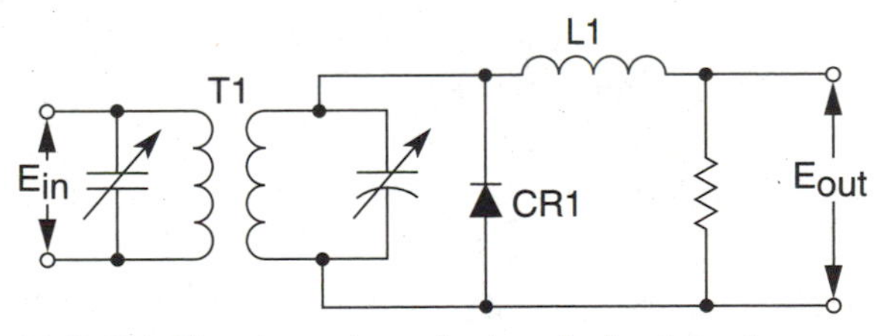

■ **6-27** *Shunt semiconductor diode detector.*

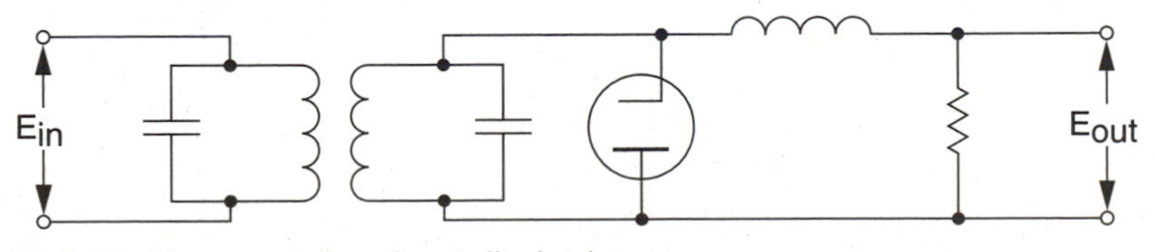

■ **6-28** *Vacuum tube shunt diode detector.*

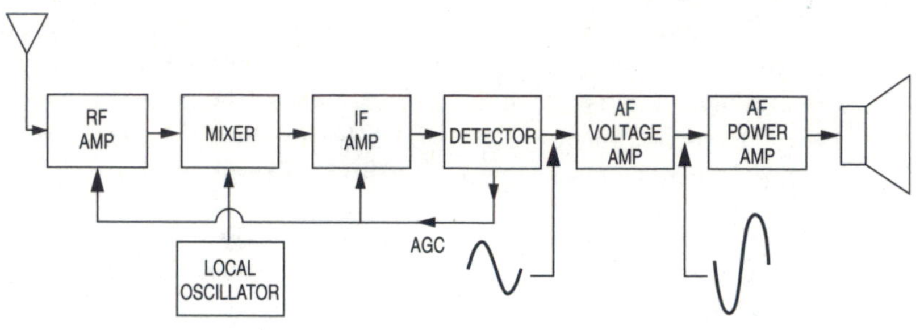

■ **6-29** *Representative superheterodyne receiver block diagram.*

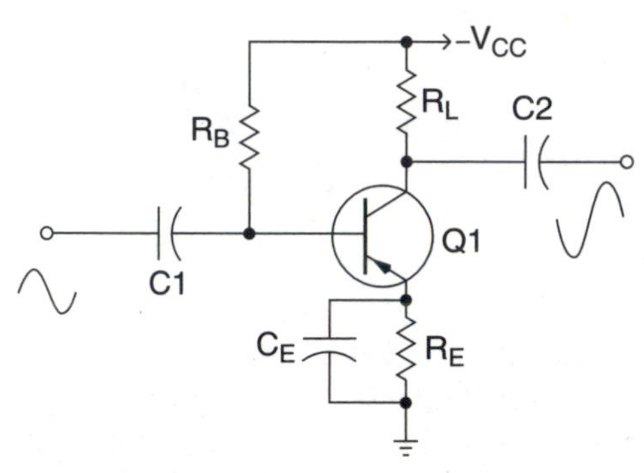

■ **6-30** *Transistorized audio amplifier schematic diagram.*

frequency amplifier stage. At this point, the audio frequencies are a very low level. Amplification is required to boost the signal to a level sufficiently high to drive either a set of headphones or speakers. As speakers often require a high level of power to be properly driven, then a power amplifier might be required.

Figure 6-30 is a simplified schematic of a typical transistorized audio amplifier. This would be operated class A. As was stated before, the input from the detector stage would be a very low-level signal, on the order of only a few millivolts. As can be seen from the diagram, this is a PNP transistor in a common-emitter configuration. The signal into and out of the stage is capacitively coupled.

Operation is as a straightforward amplifier. As it is a common emitter configuration, there is phase inversion. On a positive-going input signal, the transistor is driven toward cutoff. That results in the collector voltage swinging toward $-V_{cc}$ as conduction through the transistor decreases. As the signal then goes in the negative direction, forward bias increases on the transistor, causing the output to decrease toward zero.

Audio amplifiers can also be configured from vacuum tubes and other semiconductor devices. Figure 6-31 is an audio amplifier fabricated from a triode. Notice that the circuit appears to be very similar to the transistorized version. This is a high-gain amplifier that is operated class A. As signal is applied on the grid and taken from the plate, there is phase inversion. A positive-going input signal would increase the bias on the tube, increasing plate current. That in turn would cause a larger voltage drop over the plate load resistor, leaving less voltage on the plate. As the signal swings toward negative, plate current decreases, decreasing the voltage drop over the load resistor. That leaves a higher positive voltage on the plate—hence phase inversion. This type of amplifier has a gain factor of from 10 to 50.

An amplifier utilizing an FET is shown in Figure 6-32. This is a very common circuit, as it is a high-gain amplifier that is very similar to the characteristics one would associate with a pentode vacuum tube. The operation is very similar to that of the transistor and vacuum-tube circuits. As with the other two circuits, the source (emitter or cathode) resistor is bypassed by a capacitor to prevent any AC

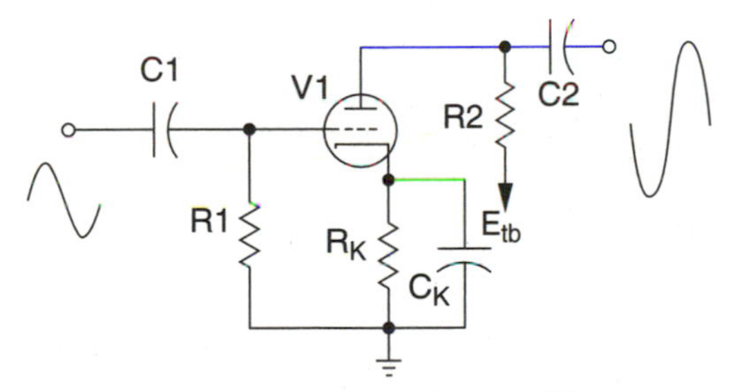

■ **6-31** *Vacuum tube audio amplifier schematic diagram.*

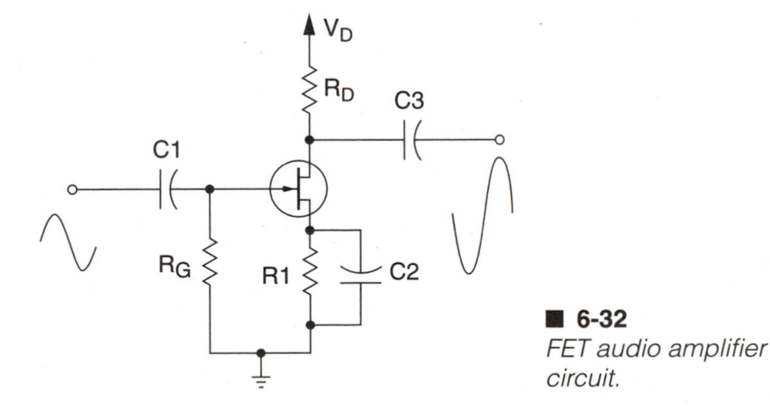

■ **6-32** *FET audio amplifier circuit.*

variations from appearing on the source, which would cause a decrease in the amplification factor. This device has a very high gain, on the order of 100 to 1000. Also, fewer components are required, it has a lower noise level, and allows minimal leakage currents. A high-impedance device, it is suitable for high-powered applications.

Audio amplifiers must be connected to other circuits for receiver operation to be possible. There are four different type of coupling networks that you might encounter in communications equipment. The four types are direct coupling, impedance coupling, RC coupling, and transformer coupling. All four types of coupling are illustrated in Figure 6-33.

Direct coupling is the simplest that you will encounter, as one stage is directly connected to another. In the case of a transistor, the collector of one is connected to the base of the next one. Although it offers simplicity and a flat frequency response, it re-

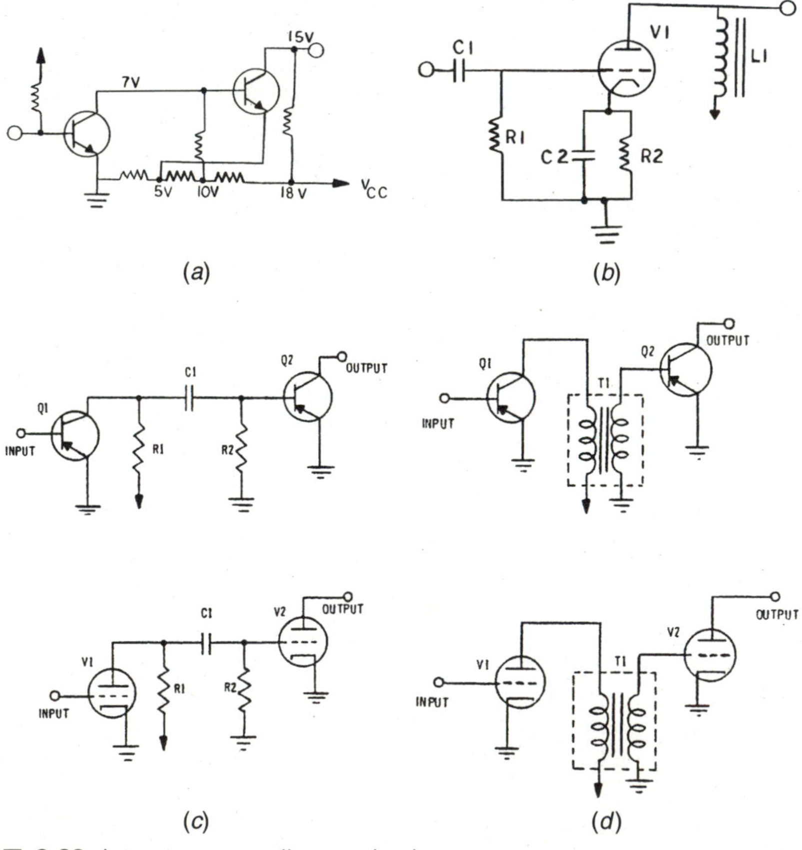

■ **6-33** *Interstage coupling methods.*

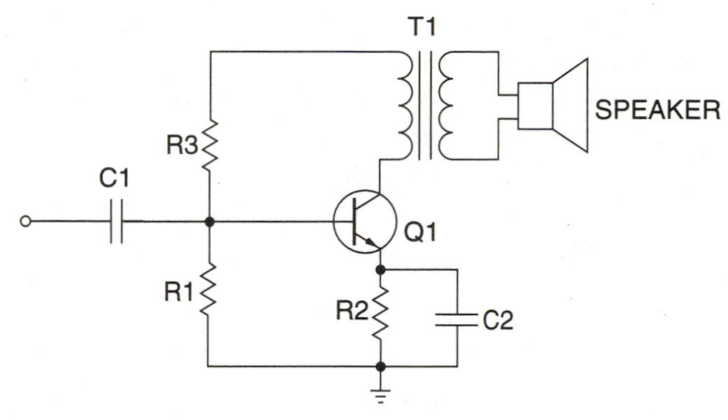

■ **6-34** *Transistorized power amplifier schematic diagram.*

quires each stage to have a larger power supply, and impedance matching can prove to be very difficult.

Impedance coupling is the least common of the four configurations. In this type, an inductor is used as the load for the output of an amplifier. It is very efficient; however, it has a less than optimum frequency response.

RC coupling, consisting of a Pi filter, is the most common form of coupling that you will encounter. It is very popular, as it features an acceptable frequency response, inexpensive components, and a small size. Its disadvantages include low efficiency and a limited frequency response. To compensate for a loss of low frequencies, additional components such as peaking capacitors are required.

The final type of coupling is the transformer. In this, the output of a stage is applied to the primary of a transformer. The secondary provides the input component to the next stage. It features a good level of efficiency. A problem is that it has a decreased frequency response when compared to the RC type.

The final amplifier is the audio power amplifier, as illustrated in Figure 6-34. This circuit, called a single-ended power amplifier, consists of one transistor. As can be seen, it is a conventional common-emitter configuration. C1 and R1 form the input RC coupling network. R1 and R3 form a voltage divider network to provide for fixed bias. R2, in the emitter circuit for thermal stability, ensures class-A operation. Notice that R2 is bypassed by C2 to prevent any AC variation from appearing on the emitter, thus preventing degeneration of the signal. The final component, T1, has the dual function of collector load and to couple the greatly amplified audio signal to the speaker. This type of audio amplifier can also be constructed from other semiconductor devices and vacuum tubes.

The single-ended power amplifier has the disadvantages of a lower power output, ability to operate only class A, and inefficiency. To overcome these problems, a dual-amplifier arrangement can be employed. Figure 6-35 expands the audio function block diagram of the representative superheterodyne receiver. To provide for a high-powered output, several additional circuits are required. The demodulated output of the detector is applied to the voltage amplifier. Its function is to boost the voltage level of the audio signal without increasing its power. This is required when the signal is used to provide a driving signal to a follow-on amplifier. Excessive power would lead to distortion and excessive power levels that could damage components. The increased signal is then used to drive a phase splitter. The purpose of the circuit is to have a single input and develop two output signals, each 180 degrees out of phase from the other. The two signals are then applied to a push-pull amplifier that uses two identical amplifiers operating in phase opposition to drive a high-powered output, such as a speaker system.

As a voltage amplifier operates in a manner very similar to a conventional amplifier, it need not be covered. The next circuit in the expanded block diagram is the phase splitter. Figure 6-36 is a phase splitter, which has the function of taking a single input and converting it to two output signals, 180 degrees out of phase. The key to the circuit is the use of a center-tapped transformer as the output load for the amplifier. Components have the same function as in other amplifiers that have been covered. R1 functions as the volume control. Fixed bias is provided by R2 and R3. R4 is the emitter resistor for thermal stabilization. To prevent any AC variations from appearing on the emitter, a bypass capacitor C2 is installed.

Figure 6-37 is the triode version of the same circuit. Components have the same function in this one as the transistorized version. The only differences are the components values, voltages, and active amplifier element. As you gain in experience you will note that there often is very little difference between the circuit designs using transistors and vacuum tubes.

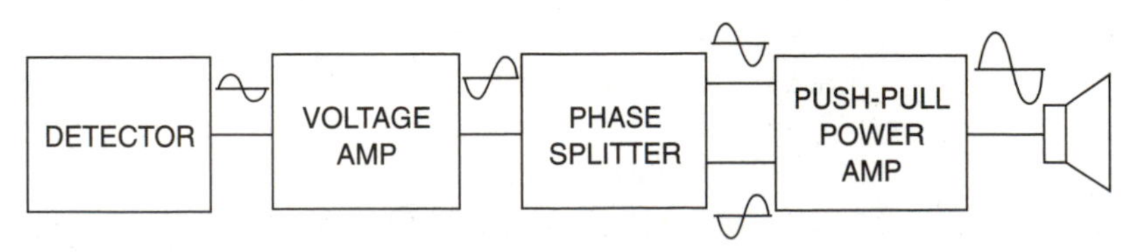

■ **6-35** *Expanded audio function block diagram.*

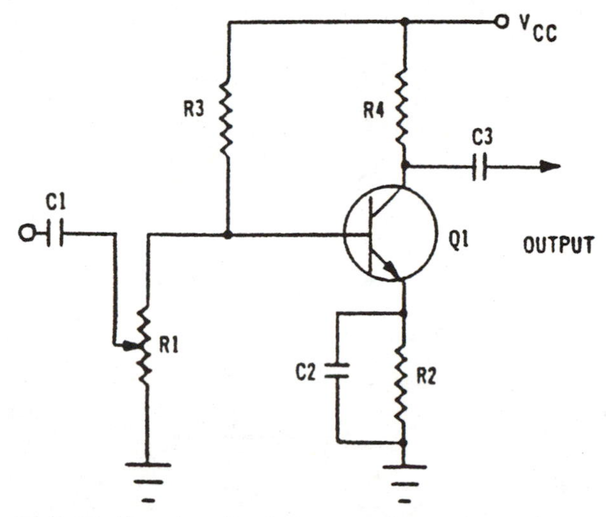

■ **6-36** *Transistorized phase splitter schematic diagram.*

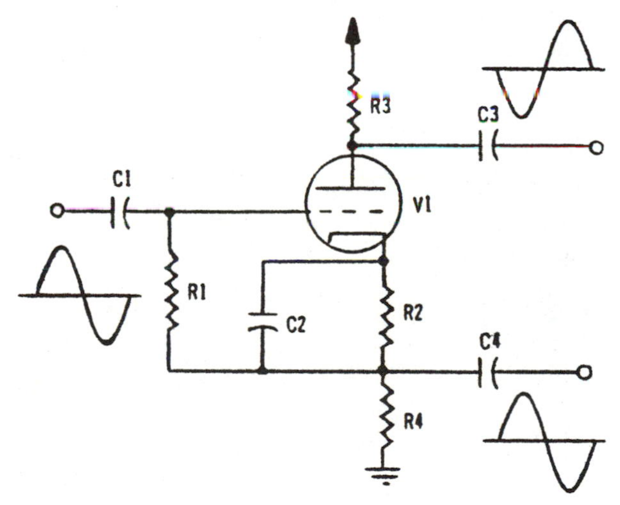

■ **6-37** *Vacuum-tube phase splitter schematic diagram.*

The same effect can be obtained without using a transformer. As the devices are bulky, expensive, and heavy, if a suitable replacement can be used, it would be a substantial savings to the manufacturer. Such a circuit, known as a *split-load phase splitter,* is illustrated in Figure 6-38. By taking outputs from both the emitter and collector circuits, two signals, 180 degrees out of phase, can be developed.

The circuit operates by having the output load resistors, R3 and R2, of equal value. It is possible for an impedance mismatch to

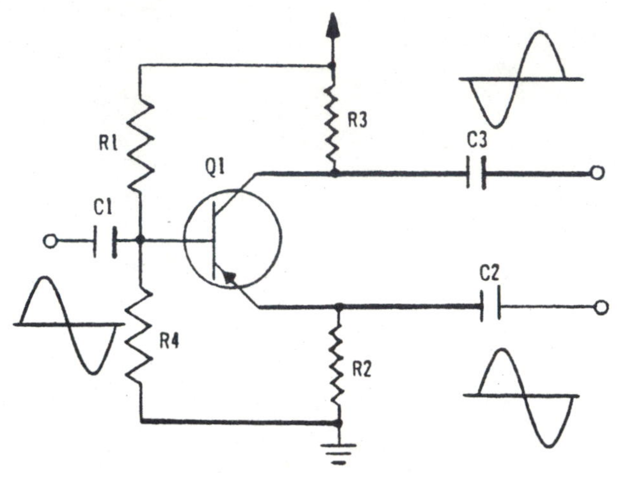

■ **6-38** *Transistorized split-load phase splitter schematic diagram.*

develop due to the output impedance of the transistor. If that is a problem, then a series resistor of a small value can be added between C2 and R2. This stage does not offer any gain, due to the degeneration involved with obtaining an output from the emitter circuit. In fact, the action of obtaining two outputs can result in a gain of less than one with this type of circuit. This type of circuit, just like all the others, can be fabricated from either transistors or vacuum tubes.

If phase splitting and voltage gain are desired out of a circuit, a *two-stage paraphase amplifier,* depicted in Figure 6-39, could be used. As with the phase splitter, two equal but opposite outputs are obtained. However, this circuit features a voltage gain on the order of 10 to 50 based upon the values of components selected.

The reason for developing two outputs is to be able to drive a push-pull amplifier. The purpose of it is to drive a speaker system with a distortion-free, high-powered audio signal. Figure 6-40 is the schematic diagram for a transistorized push-pull amplifier. This type of circuit can also be designed using vacuum tubes. As can be seen from the drawing, the circuit has two inputs from the phase splitter. A single output that is transformer-coupled is routed to a speaker. The transformer is center-tapped, which allows the upper half to function as the load for Q1 and the lower half as the collector load for Q2. Statically, current flows through the circuit without an input signal. Current flow is from ground to the transformer, where it splits. As the circuit is balanced, one half flows through T2 to Q1 and the other half through T2 to Q2. Current flows through both transistors back to ground.

Circuit operation is as follows: First, notice that the transistors are PNP—that has a bearing on the operation. On the positive alternation of the input signal, the top of T1 goes positive and the bottom swings negative. That causes the voltage on the base of Q1 to go positive, and on the base of Q2 to go negative. As the transistors are PNP, a positive on the base causes bias to decrease, decreasing conduction through the device, which happens in Q1. At the same time, the input to Q2 swings negative, which causes its bias to increase, increasing conduction through the transistor. That causes the top of the transformer to go in the negative direction while the bottom of the transformer goes toward the positive.

On the negative alternation of the input signal, the top of T1 goes negative and the bottom swings positive. That causes the voltage on the base of Q1 to go negative and on the base of Q2 to go positive.

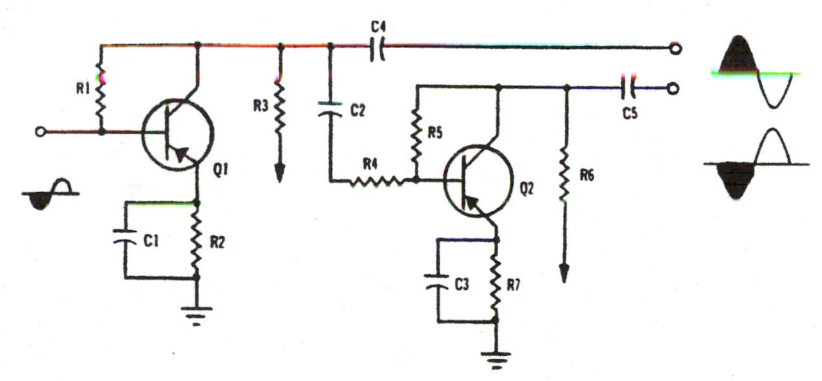

■ **6-39** *Two-stage transistor paraphase amplifier schematic diagram.*

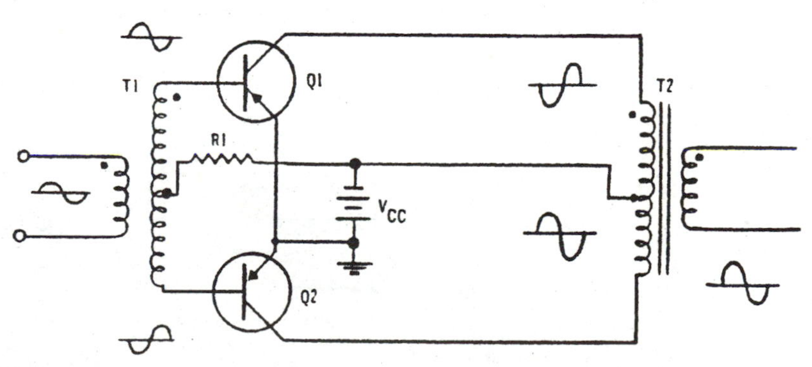

■ **6-40** *Transistorized push-pull amplifier schematic diagram.*

That causes conduction to increase through Q1. At the same time, the input to Q2 swings positive, which causes its bias to decrease, decreasing conduction through the transistor. That causes the top of the transformer to go in the positive direction while the bottom of the transformer goes toward the negative.

The effect is that the possible output power is double what can be obtained from a single-ended power amplifier. Also, and possibly most importantly, distortion is greatly reduced with a push-pull amplifier. A point to remember is that most distortion in audio amplifiers is caused by the second harmonic of the audio signal. Push-pull transformer action in a balanced amplifier cancels the second harmonics.

Receiver Control Circuits

Only the simplest of receivers lack a number of specialized control circuits. Everyone is familiar with the basic receiver functions, such as tuning, volume, and tone, that are controlled from front panel controls. Others, such as AGC and AFC, are internal and are automatic in nature. The more sophisticated the receiver, the more functions are user-controlled.

The first controls to be covered are gain and volume control. Figure 6-41 is a manual volume control. Although only a transistorized circuit is illustrated, this function could also be fabricated from a vacuum tube. The circuit is only one of the audio amplifiers before a phase splitter or push-pull amplifier. The only difference between this circuit and a conventional amplifier is the addition of

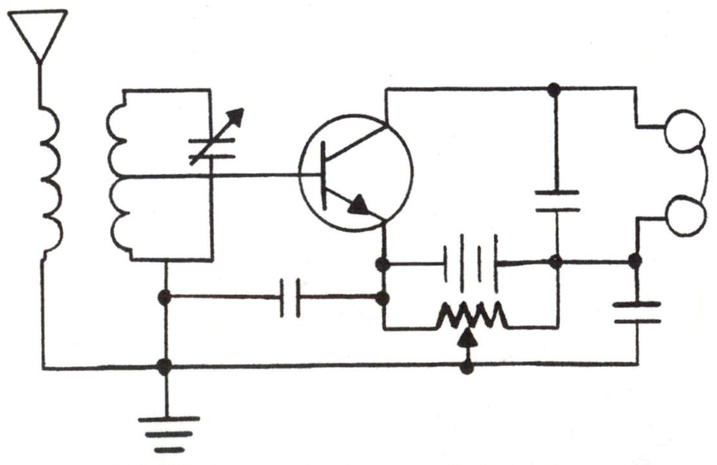

■ **6-41** *Simplified manual volume-control circuit.*

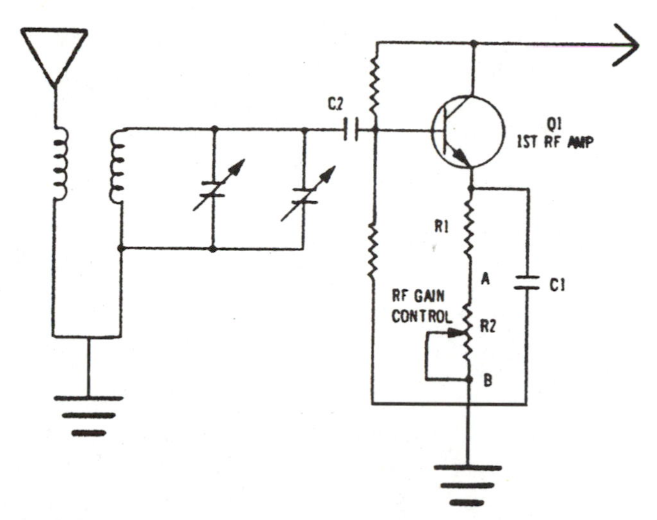

■ **6-42** *Representative RF gain control circuit.*

a potentiometer in the base circuit of the transistor. Volume is set by simply adjusting R1 to change the signal input to Q1.

Another common control often found in communications receivers is an RF gain control, illustrated in Figure 6-42. A frequency problem on shortwave bands is a high variation in signal levels caused by the constantly changing propagation conditions. Often the signal may be received so strongly that it distorts the audio signal. By reducing the RF input, it can be brought down to a level that results in good quality audio. A potentiometer is installed in the emitter circuit of the first RF amplifier. In effect, the variable resistor is only a manual bias adjustment for the stage.

All receivers that produce an audio output have its quality determined by several different factors. First would be the physical size and quality of the speaker(s), as it has a direct bearing on the frequency response. The frequency response of the audio amplifiers affects the audio output. An amplifier will have a different degree of amplification for different frequencies. Because of the two variables, some form of tone control might be part of the audio section.

Treble and bass are two terms that are familiar to you. You associate treble with the highs, and bass with lows. The accepted definition of treble is audio frequencies above 3000 hertz. Those frequencies below 300 hertz are considered to be bass.

The way that a tone control circuit functions is interesting. As an example, assume that an audio amplifier and speaker are producing two tones of equal amplification, one 5000 hertz and one 500 hertz. If one frequency is reduced, then the other one will seem to be louder or more pronounced. In this fashion, treble is enhanced by suppressing bass.

Figure 6-43 is the simplest type of tone control. In this example, only the speaker and output circuitry are shown. A fixed capacitor is wired in parallel with the primary of the output transformer. When S1 is closed, the higher frequencies are shunted to ground, leaving the bass tones. The value of C1 controls the lowest frequencies that are impacted.

Figure 6-44 illustrates a simplified selectable tone control. In this circuit, three capacitors are used to shunt the audio output transformer. The circuit is switch-selectable with three positions; one is for bass, one is for treble, and one is for normal. In this case, just select the switch position that gives the best audio output.

A control that many of you are familiar with if you have had experience with CB radio, SSB, ham radio, or scanners is the squelch control. In a conventional communications radio, the sensitivity is maximum when no signal is being received. This phenomenon is most noticeable to the operator when tuning between stations. As only atmospheric and circuit noise is present, it is greatly amplified. Two types of circuit can have an effect on the problem. The first would be a noise elimination circuit of some type. Often known as a noise limiter, noise silencer, or noise suppressor, they do have an effect on the problem. In actuality, these three circuits just remove the peaks of the noise spikes. It is still present and noticeable, just lessened. A circuit called the squelch eliminates noise. With it selected, you can tune from one station to another with total silence in the background.

Figure 6-45 is a representative squelch circuit. In this design, with no input signal present from the first IF amplifier, the input to the second IF amplifier is blocked. The key to circuit operation is diode CR1. Q1, an amplifier, is the control component for the cir-

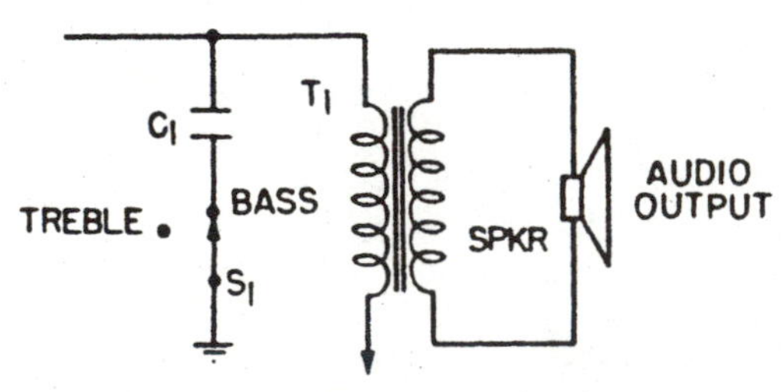

■ **6-43** *Basic tone control circuit.*

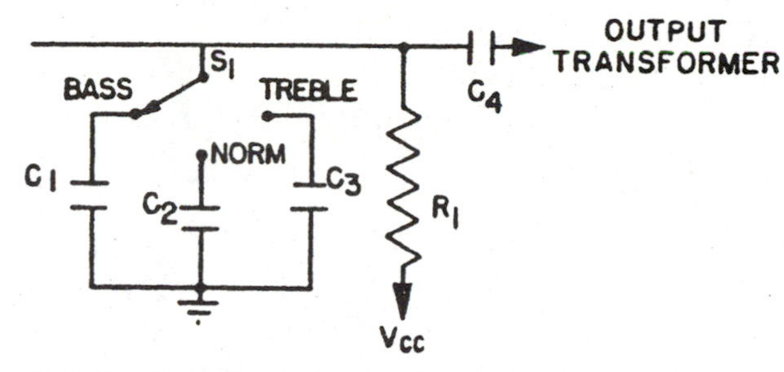

■ **6-44** *Selectable tone control circuit schematic.*

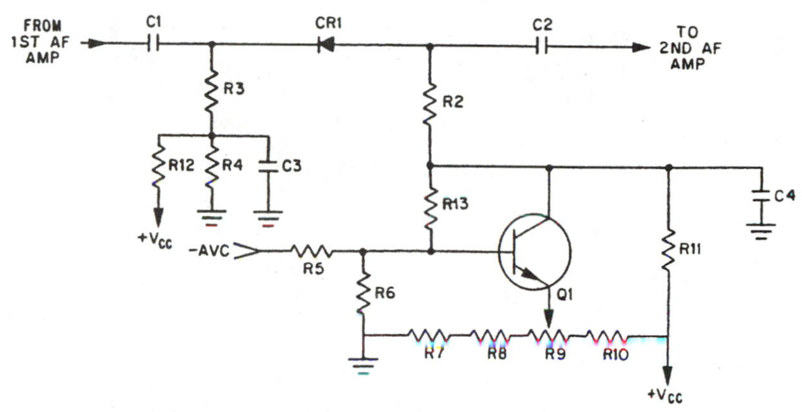

■ **6-45** *Squelch circuit schematic diagram.*

cuit. R9, in the emitter circuit of Q1, is the squelch control. With only background noise and no received signal present in the speaker R9, the squelch control is adjusted until there is silence. At that point, CR1 is reverse-biased and cut off.

The way that CR1 is forward-biased so that a received signal can be heard is through the automatic volume control (AVC). Any time that a signal above background noise is detected by the receiver, the AVC input swings negative. That drives Q1 into cutoff, which causes CR1 to be forward-biased and conducting. That allows the received signal to pass through the IF amplifiers.

Another common control, but one that is not operator-controllable, is the *automatic volume control* (AVC) or *automatic gain control* (AGC). The simplified schematic is illustrated in Figure 6-46. Communications receivers often suffer from output volume variations. These can be caused by rapid changes in the received signal strength, which leads to signal fade in and fade out. Many variables, such as receiver motion, rapid changes in atmospheric conditions, or transmitter fluctuations can cause distortion, signal

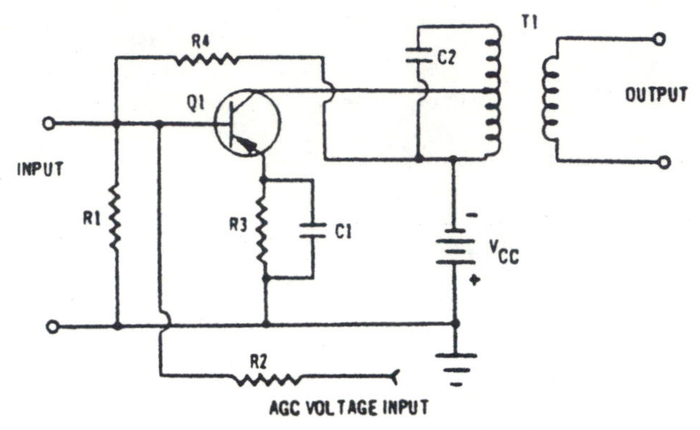

■ **6-46** *AGC signal development in the detector circuit.*

loss, or listening discomfort. Transmitters operating at different output power levels might be located at various distances and directions from a receiver. Without an automatic circuit, constant adjustments would be required to maintain a readable signal. Sometimes variations happen so rapidly that an operator would be hard-pressed to keep up with them.

AGC can be developed from the detector circuit, as illustrated in Figure 6-46. In this circuit, the detector is formed by T1, CR1, C1, and R1. AGC voltage is developed by R2 and C2. When the detector is forward-biased and conducting, both C1 and C2 charge. The function of C1 is to develop the DC voltage level that will be the demodulated audio signal. C2 develops the DC AGC voltage. When the input signal to the detector swings negative, CR1 is reverse-biased. That allows the capacitors to discharge. Due to component values, C2 has a very slow discharge time so that it will have a DC level across it. The discharge path for C2 is from the bottom of C2 to ground, up through R1, down through R2, and then back to C2. As C2 is charged to the average DC signal level, any instantaneous variations will have little or no effect on it. The developed AGC signal is a degenerative signal that is routed to the IF amplifiers.

Another specialized receiver circuit is a *beat frequency oscillator* (BFO). Of the four types of modulation, CW is the simplest in terms of use and hardware. A general-purpose communications receiver cannot pick up CW without additional circuitry, as it is not modulated by an audio signal. Once the CW-modulated carrier is detected by the receiver, there is no intelligence signal to process. To provide an audio signal, the CW signal must be heterodyned with an appropriate RF signal. Figure 6-47 is the block diagram for the location of a beat frequency oscillator (BFO) in a superheterodyne receiver. In effect, a BFO is an oscillator that is tuned to within 1 kHz or 2 kHz of the IF frequency. In that way, when it is mixed or beat with the IF, an audio signal results, allowing the operator to hear CW modulation.

The final receiver circuit is *automatic frequency control* (AFC). The function of AFC is to provide an automatic and accurate method of controlling the frequency of a receiver. To do this, the AFC circuit must sense the difference between the desired frequency and the actual frequency of the local oscillator. Then it must develop a signal or control voltage that will tune the LO back to the desired frequency. AFC circuits are found in radio receivers, transmitters, frequency synthesizers, radar, sonar, and large-scale x-ray systems.

Figure 6-48 is the block diagram of a receiver equipped with an AFC function. Notice that the diagram does appear to be very similar to the conventional superheterodyne that you have been introduced to before. A major difference is the addition of a second output off of the IF amplifier that is applied to the discriminator circuit, which has the function of converting frequency deviations into amplitude variations. This in turn is applied to the varicap, which will develop a reactance based on the amplitude variations. Varicap changes are then used to tune the LO back to the desired frequency.

As example, take the common LO frequency of 455 kHz. If the LO frequency drifts high for some reason, the resulting IF signal from mixing the received RF would decrease below 455 kHz. That causes the discriminator to decrease the capacitive reactance of the varicap, decreasing the LO frequency back to normal. Now if the LO frequency drifts low for some reason, the resulting IF signal from mixing the received RF would increase above 455 kHz. That causes the discrimator to increase the capactive reactance of the varicap, increasing the LO frequency back to normal. With this automatic control, rapid frequency variations caused by thermal

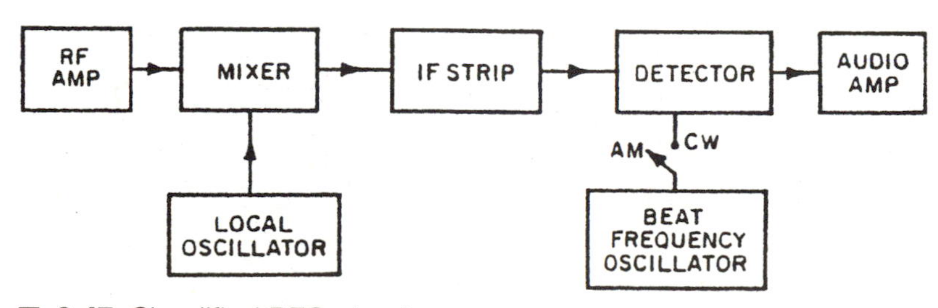

■ **6-47** *Simplified BFO circuit.*

changes, input power changes, and load variations can be corrected for so rapidly that only sophisticated laboratory instruments can detect them. The exact circuitry varies from equipment to equipment, but functions outwardly the same.

There are other specialized circuits, but they are beyond the scope of this book. Others include bandspread, which is a control that allows the frequency to spread out in a fine tuning. In that way, two very close signals can be separated. Another is frequency scanning. By using a VCO, the receiver can be set to automatically sweep between two limits, stopping if a signal is picked up. Many receivers use precision filters to eliminate noise, interference, and improve selectivity. As can be seen from this short section, there are many specialized controls to improve receiver performance.

Remember that SSB communications equipment must meet strict frequency stability requirements. Frequency deviations of ±50 hertz would be barely acceptable. A very small deviation in the carrier reinsertion oscillator could cause an IF signal that is too far off. As the received SSB signal has only one sideband and no carrier, a small IF error could cause loss of intelligence. That is because a small deviation in the carrier frequency moves the sideband containing the intelligence outside the bandwidth of the tuned circuits and filters, leading to its loss.

The basic frequency-modulated receiver block diagram is illustrated in Figure 6-49. In this type of receiver the RF front end appears to be similar to a conventional AM receiver. However, in the FM receiver the IF amplifier section is a wide-band amplifier. That is because the bandwidth of the circuits must be wide enough to pass all the sidebands of the modulated signal.

An FM signal is different in that the intelligence varies the frequency of the carrier when modulation occurs. This translates into a wider bandpass in the transmitter and receiver circuits. After the IF section, a limiter is used to reduce the amplified IF signal to the required level. The next block, the discriminator,

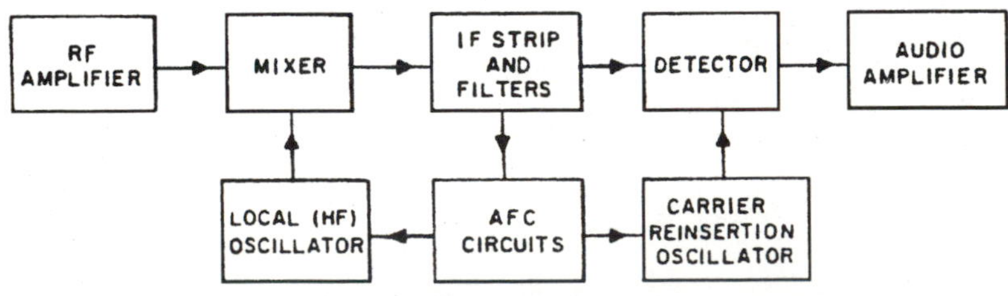

■ **6-48** *Superheterodyne receiver block diagram with AFC function.*

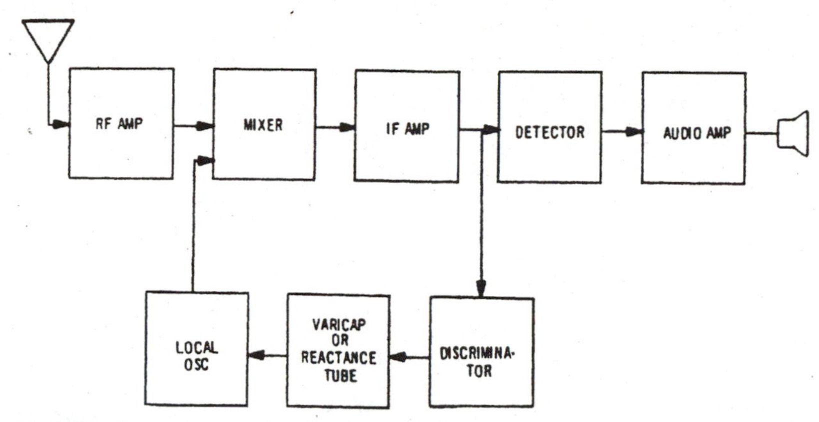

■ **6-49** *Superheterodyne FM receiver block diagram.*

converts frequency variations into voltage variations. This action detects the audio intelligence from the carrier. The receiver is completed with the required audio amplifiers. Not shown are blocks for the AFC function.

Now, the audio intelligence has completed its journey. It began in the transmitter where it was used to modulate the carrier signal. The antenna system then converted it into a wavefront that was propagated through the atmosphere. It was then intercepted by the receiving antenna, processed by the receiver circuits, and converted back into an audio format. The following chapters will cover important safety and maintenance information.

Safety

7

AN ELECTRONIC TECHNICIAN MUST BE PROFICIENT IN current electrical safety practices and follow them without fail, as electronic equipment often requires high voltage and current for proper operation. To complicate matters, many systems also produce significant levels of RF radiation, and in some instances, X-rays. To ensure personnel safety and prevent equipment damage, safety precautions must be known and practiced at all times. Electronic installations require more than just electrical safety. Other hazards encountered by electronic maintenance personnel include: electrical fires, chemical reactions, and working at heights, in trenches, and in confined spaces. As you gain experience, you will learn that most, if not all, safety precautions are nothing more than common-sense actions that can save your life.

Due to the hazardous nature of electricity, under no circumstances should any individual intentionally expose himself to an electrical shock. You would be amazed at how low a level of current flow can cause physical damage or death. A good point to remember is that every safety rule was paid for by someone's suffering.

This section is not a complete guide to safety, and does not replace safety organizations in any way. It is meant to introduce you to some hazards with which you might come into contact. You must be aware of precautions that are vital in your particular field. Many sources, such as equipment manuals, professional organizations, national electric code, EPA, OSHA, Underwriters Laboratories, ARRL, and others are excellent sources of safety recommendations. If you are with a large governmental or commercial organization, you will have a safety representative. Get to know the inspector, and follow any advice given. Your future depends upon it.

General Safety Considerations

Before contemplating performing maintenance, proper clothing is a must for any technician. Shirts and pants should not be overly lose or torn, in order to prevent them from being caught in moving parts, ladders, shut doors, cabinet edges, and antennas. Long-sleeved shirts and pants are preferable to shorts and short-sleeved shirts, as they can provide some protection from flash burn. There should not be any loose articles in pockets that could fall out and create either an electrical or mechanical hazard. It is important that safety shoes, not sandals or athletic shoes, be worn to prevent injuries to feet. Electronic components and subassemblies can be very large and heavy.

Safety glasses, goggles, or a face shield are another often-overlooked piece of safety equipment. Electronic maintenance personnel are frequently in areas where eye injury can occur. A good example is replacing defective electronic components. For proper installation, many small components such as resistors, capacitors, and semiconductors must have leads trimmed to fit them into the circuit. When leads are trimmed, small pieces of conductor can fly off and strike a person in the eye, causing injury. When soldering the component into place, hot solder can spatter, again with the possibility of causing eye damage. Equipment is often cleaned with compressed air and solvents, both of which require eye protection. As equipment cleanliness is many times ignored until overheating causes failure, dust and dirt can build to impressive levels. I have found that computer equipment is often installed as an afterthought, so placement is less than optimal. Many pieces of critical equipment are simply placed on the floor, where dust and dirt have easy access to the fans. In the cleaning process, face protection is a must due to the amount of material that has to be removed with compressed air.

A part of electronic maintenance is the proper cleaning and lubrication of equipment. This type of maintenance often has to be performed in a confined space with little or no ventilation.

You must exercise extreme caution when using any type of chemical agent. The problem is that in a confined space, chemical fumes displace oxygen, leading to oxygen deprivation, unconsciousness, and ultimately death. Always ensure that you have proper ventilation. If you have any doubts, have trained personnel verify air quality rather than trusting your nose. *Because carbon dioxide is odorless and colorless, instrumentation and a trained operator are required to verify its presence in excessive levels.*

Proper safety equipment such as gloves, goggles, and an apron might be required to prevent spilling chemicals on yourself. A good idea is to review the Material Safety Data Sheet for any new chemical that you encounter. Each chemical used in a work center must have an MSDS on file; this sheet lists all known hazards and safety precautions that are associated with it.

To ensure that equipment cleaning is a safe procedure, several steps should be followed. Ensure that the area where the cleaning is being accomplished has sufficient ventilation. If the cleaning agent is flammable, a handy fire extinguisher is a must. To prevent skin and eye contact, wear proper safety clothing such as rubber gloves, goggles, and a rubber apron. Importantly, when working with chemical cleaning agents, do not work alone.

The first rule of safety is to never perform maintenance without informing others. The list of possible accidents that could incapacitate you is almost endless. An excellent habit to develop is to never work by yourself. If you are working alone and without others knowing it, it could be hours before you were missed and someone looked for you. The other person does not have to have any technical knowledge of what you are doing, just be capable of de-energizing the equipment and summoning help. While having two people assigned to one maintenance task might seem extravagant, what is the cost if a person working alone is severely injured or killed? Even the safest job can become lethal if you slip on moisture or spilled lubricants, or fall off a ladder.

Electrical Safety

If you come into contact with an energized electrical circuit, injuries can range from a slight tingling sensation to burns, unconsciousness, cessation of breathing, ventricular fibrillation, and possibly death. Contrary to popular misconception, injuries associated with electrical shock are caused primarily by current flow, not voltage. Even the familiar 115-volt home lighting circuits can cause injury and death under the right circumstances. If you come into contact with an electrical current, the effects are determined by the intensity of the flow. Table 7-1 lists the effects of current flow on the human body. Notice that AC and DC have different levels of intensity. Also, a man, due to body mass, is capable of sustaining a higher level of current flow than a woman. With dry, unbroken skin, the smallest current flow that a man would feel is about 5 ma DC and all it would do is produce a slight sensation. An increase to only 9 ma is intense enough to paralyze your muscles.

If you had a conductor or hot component in your hand, you would be unable to release it. To break contact, outside assistance would be needed. Unless you were removed from the circuit, severe burns and injury would result. An increase in current flow to 90 ma is extremely painful, and fatal if the victim is not removed. Only half an amp for 3 seconds leads to death.

For AC the values are a little different when a male has dry unbroken skin. The level of perception is about 1 ma. Difficulty in releasing is only about 16 ma, and a current flow of only 100 ma leads to death. Death from electrocution is not easy to observe, nor is it a quick and easy way to die.

The value of current that flows through your body is determined by the applied voltage and your skin resistance. Dry, unbroken skin has a resistance of several hundred thousand ohms and offers protection from many shocks. If your skin is damp or cut, resistance drops to less than five hundred ohms. By cuts, I mean anything as insignificant as a barely visible paper cut. If a person with a skin resistance of only five hundred ohms comes into contact with an electrical circuit at a potential of 120 volts, a current flow of 240 ma will flow through his body: more than enough to be lethal. Deaths from electrical shock have been recorded when an individual has come into contact with a circuit at a potential of only 24 volts. If the victim has a pacemaker to control his heart-

Table 7-1 The effects of current on the human body*

	DC		60 Hz AC	
Effect	**Men**	**Women**	**Men**	**Women**
Slight sensation	1 ma	.6 ma	.4 ma	.3 ma
Perception	5 ma	3.6 ma	1.1 ma	.7 ma
Shock, not painful	9 ma	6 ma	1.8 ma	1.2 ma
Shock, painful	60 ma	40 ma	9 ma	6 ma
Shock, let go threshold	75 ma	50 ma	16 ma	20.5 ma
Shock, painful and severe, muscular contractions, breathing difficult	90 ma	60 ma	23 ma	15 ma
Shock, possible loss of life from ventricular fibrillation	500 ma	500 ma	100 ma	100 ma

* All values are estimates. Actual current levels will vary from individual to individual based upon skin resistance, humidity, skin moisture, and other factors.

beat, then an even smaller voltage could cause the device to stop, leading to death.

When working on electronic equipment, there are minimum safe distances. As an example, if the equipment has a working voltage of 600 volts, a minimum separation to prevent a shock hazard is six inches. A higher voltage, such as 10 kV, requires a minimum safe distance of two feet. For each additional 10 kV, another foot of separation is required.

To ensure the safety of electrical personnel, first aid and CPR training should be given to as many people as possible. The first problem in an electrical accident is removing the victim from the circuit. The safest method for both the victim and rescuer is to de-energize the equipment. If that is impossible, a nonconductor such as a wooden handle or rope can be used to remove the victim. Do not under any circumstances allow yourself to become part of the problem by coming into contact with the victim. If that happens, then there will be two casualties instead of one for the emergency response team to deal with.

To prevent yourself from becoming a statistic, there are several easy-to-follow safety precautions. If you are performing any maintenance that requires opening a cabinet or removing panels to the point where you could possibly be exposed to voltages, tag the equipment out and lock the power source. By placing a tag on an equipment's control panel or breaker, you inform others that a hazardous task is being performed and not to energize the equipment or change any settings. As a typical equipment installation can have the transmitter in one room, circuit breakers in second, and the operator control panel in a third, inadvertent operation is a possibility unless steps are taken. Governmental agencies, commercial installations, and commercial entities are required to have a *lock out/tag out program*. For civilian organizations, several industrial safety supply houses offer the complete program in one easy to purchase and setup package. Figure 7-1 is the currently acceptable tag and lock for a lock out/tag out program. Tags are serialized to maintain control. To facilitate reuse of tags, they are laminated, and information is recorded with a grease pencil. Locks, as with the example, can be color-coded for local use. If more than one technician must perform maintenance on a system, multiple lock fittings are available. That way the equipment cannot be placed back into service until the last person is clear.

By following the program, chances of injury while performing maintenance are greatly decreased. The steps must be performed

■ **7-1** *Recommended safety lock and tag.*

completely and in sequence for maximum effectiveness. You begin by ensuring that others are actively informed when maintenance is to be performed on an electronic or mechanical system that they operate. The next step is to place danger tags on all control panels and breakers to warn others not to operate the equipment. At this time, an entry must be made in the lock out/tag out log. To ensure compliance and to warn anyone that might have been inadvertently uninformed, the tags remain in place until the equipment is returned to service. If the system must be de-energized to troubleshoot, replace components, or align, the control panel, electrical breaker, gas supply, and water supply must be locked out to prevent accidental operation. Again, the placing of a lock on a piece of equipment must be logged to maintain control. You must ensure that all sources of compressed gases, liquids, and power required by the system are secured to prevent operation by using a lock. Only the person who installed the lock is authorized to remove it when repairs are completed. Upon completion of all maintenance, the tags and locks are removed and suitable log entries are made.

Many people would argue that a lock out/tag out program is more trouble than it is worth. It takes time to clear equipment, down time with all concerned parties, and time to physically install and remove the tags. From incidents that I have heard about, it is time well spent. One accident occurred in a clothing factory; a corduroy machine needed cleaning and time to schedule the outage was difficult to obtain. The maintenance person decided on his own to perform the task while the operator was on break. To save even more time, he neglected to tag the machine out, or flip the electrical breaker off and crawled inside the machine to clean it. While he was up inside the unit, the operator returned from break early and energized the machine to resume work. The screams from the maintenance person were the first sign that the operator had of a problem. As medical assistance was close at hand, injuries were limited. The misplaced desire to take several shortcuts resulted in months in the hospital due to severe leg injuries and unemployment for violating company policy. The moral of the story is "Follow safety procedures."

High-powered electronic installations such as radar and communications systems are provided with interlocks as a safety device. The idea is that if access panels are removed, the interlock opens, removing power and eliminating a shock hazard. To facilitate proper maintenance, the interlocks can be bypassed by pulling a shaft out. To accomplish this takes a conscious effort on the part of the maintenance technician. To make maintenance easier, some individuals have been known to bypass or wire around interlocks. The problem with that is that, when the affected panel is removed, power is still present. An unsuspecting person, believing that the interlock would provide protection, is now at risk of electrical shock. Never, never, for any reason bypass interlocks or other safety devices. Similar to the interlock is the Emergency Off button. High-powered systems are equipped with the red mushroom buttons in areas where maintenance personnel would be at risk if the system were active. When working on equipment, I use them as a backup to interlocks, lock out/tag out, and breakers. I want to make accidental operation as difficult as possible, which I feel is a good habit for all technicians to develop.

One of the easiest safety precautions to perform, yet one of the most frequently ignored, concerns equipment cabinets. Unless maintenance is in progress, all cabinet doors, access panels, and drawers should be closed and properly fastened. That prevents unauthorized personnel from gaining easy access to the inside of a piece of equipment. All nuts, bolts, screws, and Zeiss fasteners should be in place and tightened. It is advisable that all access panels require tools for removal. That makes it more difficult to obtain unauthorized entry. A bad habit exhibited by many technicians is to secure a door or panel with only few fasteners. Although it makes certain maintenance checks easier, safety is compromised. An additional hazard is if the equipment is mounted in an aircraft, ship, or vehicle. If the equipment were in motion, the stresses can cause an energized drawer to spring open, with catastrophic results.

When performing maintenance it is a good habit not to wear watches or jewelry. They could become caught in moving parts, pulling you into gear trains, shafts, or rotating machinery. As metal jewelry is a conductor, it is also an electrical shock hazard. Necklaces, rings, bracelets, and watches can provide an excellent conductor from an electrical source through you to ground. Any items that cannot be removed, such as rings, wrap in electrical tape to provide some protection. To prevent this hazard, I have not worn a watch or any type of jewelry for my entire electronics career.

A good habit to follow when taking any voltage checks is to use only one hand. That way, if you do come into contact with electricity, the path for current flow is through your hand, arm, and then ground. If you place both hands within the equipment, then the path to ground could be through your arm, heart, the other arm, and then ground. By becoming an electrical conductor across your heart, an electrical shock can cause cardiac ventricular fibrillation, or heart failure. As a further safety measure you should stand away from the equipment, so that no other part of your body can come in direct contact with equipment. This ensures that if you do come into contact with voltage, the chances of a fatal shock are greatly lessened.

Measuring high voltages presents additional hazards. If you place meter leads on energized contacts, you will draw an arc. To prevent injury and equipment damage, there is a safe and prescribed method to measure high voltages (300 volts or higher). Using the technical manual, determine the highest value of voltage that you should encounter in the circuit. Prior to placing your hands inside of equipment, turn the power off and be sure that all components capable of holding a charge have been grounded and discharged. Although high-voltage power supplies have safety devices and bleeder resistors to ensure total discharge, never trust them. It is impossible to tell if a capacitor is charged just by looking at it.

Therefore, de-energize the equipment and short out the area to discharge any residual electrical charge to ground. Any shorting must be accomplished only with an approved grounding probe, such as the one illustrated in Figure 7-2. The probe must be connected to a known-good ground for effectiveness. During the grounding process, touch the end of the probe to any points suspected of having a potential problem present. Never come into contact with the probe ground or ground strap; this makes sure that you do not become part of the ground path.

After grounding has been completed, insert the meter probes to the points to be tested, observing polarity. Remove the ground probe, pull your hands from the equipment, re-energize the equipment, and take the reading. If the meter scale is too high, DO NOT, repeat DO NOT change scales while the equipment is energized. Meters have been known to break down internally while measuring high voltages. If this were to happen and you were to attempt changing the scale, then a shock hazard would result. To change the scale you must de-energize the equipment, change scales, then re-energize. After the check is completed, to remove the probes, de-energize the equipment, ground components capable of retaining a charge, and then and only then remove the meter probes. Any time you are working on de-energized high-voltage equipment, it is a good practice to always leave a grounding probe in place to ensure safety. However, prior to applying power, please remove the probe.

The problem with electronic equipment is that, energized or de-energized, it often appears the same. The only way that you can be sure that it is de-energized and fully discharged is through the use of a grounding probe and multimeter. Reactive components such as capacitors, inductors, and transformers all can retain a residual charge for months. Although the steps seem time-consuming and overly cautious, to ensure your safety they must be followed in their entirety. Consider them the minimum level of safety that you will accept while performing maintenance on high-voltage equipment.

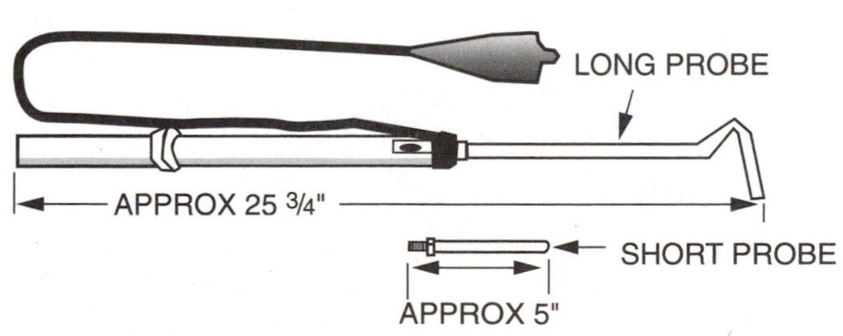

■ **7-2** *Typical grounding probe.*

These measures actually take only a few moments, but eliminate potentially life-threatening problems.

Static electricity is a problem that is often overlooked. It also has been linked to setting off explosive devices. All satellites with small solid-rocket kick motors have them x-rayed prior to launch. That is to ensure that solid fuel has adhered to the entire surface of the motor casing. Any cavities would result in uneven burning. To minimize the possibility of accidental explosion of explosive bolts, the motor must be grounded during all tests to prevent the buildup of static electricity. Any personnel coming in close proximity of the device must be physically grounded in some fashion.

Although you may not work around space hardware, you will come into contact with hazards associated with static electricity. Many semiconductor devices now have ultra-thin PN junctions and metallic contacts that are susceptible to damage from static electricity. MOSFETs are devices that are especially prone to this problem. A potential of 100 volts will cause an internal breakdown and component damage. Another consideration is that, if you are exposed to a static electricity shock, you could be startled as a natural reaction. It your hand is inside a piece of equipment, you're on a ladder or scaffolding, or you are near rotating machinery, a simple harmless involuntary reflex could become life-threatening.

To prevent this problem around sensitive electronic components, several precautions can be taken. The technician and equipment can be temporarily grounded to discharge any buildup. A soldering-iron tip should be touched to electrical ground before soldering to ensure there is not a buildup. Ensure that the electrical chassis is grounded.

If the component in question is a MOSFET, additional steps must be taken. Workbench tops, floor mats, and chairs must be adequately grounded. Technical personnel should be grounded through the use of wrist stats. Components and circuit boards constructed from them should be stored in electrically conductive Velofoam packaging. Clothing made from cotton, rather than synthetics, will minimize potential buildup.

Electrical Fires

Where you have electrical and electronic equipment installed, you have the danger of an electrical fire. Components short out, causing excessive heat; insulating oil can leak, and insulating material can catch fire. All electronics personnel should be familiar with fire safety. The first rule of fire safety is to practice good housekeeping. Properly store all combustible materials, keep the work site clean, keep all pathways clear, and always perform electrical repairs according to specifications. The repair of equipment with underrated parts can lead to an electrical fire, or at least increased maintenance. Also, the integral cleanliness of a piece of equipment cannot be overemphasized. Even in a seemingly clean environment, dust and dirt can build up to unacceptable levels within equipment.

If an electrical fire does happen, personnel safety is paramount. If in doubt, don't try to be a hero. What appears to be a small fire could rapidly explode into something much larger. With this type of fire, the easiest and safest way to fight it is to de-energize the equipment. With the removal of power, most electrical fires quickly go out on their own. If that does not work and the fire is small, use a carbon dioxide extinguisher. Do not try to fight a fire beyond that level—leave that to the professionals. Fires create a low-oxygen environment that requires breathing apparatus for survival. Burning electronic equipment gives off chemical fumes that can cause lung and air-passage injury. In the event of fire, follow the correct sequence. Personal safety permitting, turn off power to any equipment, close windows, and secure any ventilation. If the fire persists, try a carbon dioxide fire extinguisher. If the extinguisher fails, or the fire is too large, call emergency personnel and evacuate. I would only try to put the fire out myself if were very small. If in trying to fight an electrical fire, you were overcome with smoke, who would pass the alarm? In instances such as this, passing the alarm is vital. The paramount consideration for one is personnel safety. Never take risks of any kind because you might not get a second chance. Remember, equipment can always be replaced if destroyed, but can you be?

There are three classes of fires based upon the type of burning material. Class A fires are combustibles that leave ash, such as paper, wood, cloth, and trash. A Class B fire is confined to oils, fuels, paints, grease, and materials soaked with any of them. An electrical fire is Class C, and is limited to insulation and combustible materials found in electrical and electronic installations. A fire can start out Class C, but due to poor housekeeping can rapidly become an out-of-control Class A or B.

Due to electrical shock hazards and damage to delicate electronic components and contacts, carbon dioxide is the preferred method of combating an electrical fire. Carbon dioxide does not leave any chemical residue that damages delicate electronics and contacts. Also, it does not conduct electricity, so it is not a personnel safety hazard. Another type of fire extinguisher that can be used on electrical fires is the dry chemical. Potassium bicarbonate (purple K) is the most common chemical in use as a firefighting agent. It is desirable as it is not an electrical conductor and quickly smothers flames. However, it does leave a residue that is difficult to remove and can damage equipment. Other common firefighting agents are water and foam, both of which are not advisable for electrical fires as they can present a shock hazard to emergency personnel. Both foam and water conduct electricity. If either must be used on a fire, the equipment must be de-energized to provide for safety. Unfortunately, water or foam damages equipment, often to the point that it is unrepairable.

Portable Equipment Electrical Safety

A final electrical safety consideration is the proper grounding of all portable electronic equipment and power tools. Normally a three-wire conductor is used to provide power. Two conductors provide power, while the function of the third wire is to ensure that the equipment or tool case is at ground potential. If the equipment is ungrounded for any reason, an electrical potential is present on the case. In effect, the equipment has a floating ground. By coming into contact with the case while grounded, you will provide a convenient path for current flow. As was pointed out earlier, a potential of only 30 volts is sufficient to cause injury or death under the right circumstances.

To certify that all portable equipment is electrically safe, an electrical safety program should be instituted. Portable electrical equipment would be described as portable test equipment, power tools, and appliances. Equipment safety is verified by a few easy to perform steps. When checking the ground wire on a standard three-prong 120 VAC power plug, make sure that is has a resistance of zero ohms to equipment ground or metal case. The resistance between the ground prong and the other two prongs should be in the range of 100K ohms. The way I do it is to check the ground prong first. I then go from ground to each hot prong.

While performing the resistance test, wiggle the power cord to check for intermittent breaks in the conductor. Next, visually inspect the power plug and cord for damage and wear. If either show signs of wear or damage, replace it. Check the equipment case for loose and missing hardware, and tighten and replace as necessary. Whenever replacing a component that comes into contact with ground, ensure that all metallic surfaces are clean and have electrical continuity. Any signs of corrosion must be removed and cleaned. These few steps take very little time to accomplish, but can prevent personnel injury and equipment damage. The function of any electrical safety program is to verify that the checks are completed on a periodic basis and recorded.

Another useful safety device is the ground fault interrupter (GFI). OSHA and electrical codes call for the use of this device in an industrial or work environment where moisture may come in contact with the portable equipment or extension cords. A GFI is in effect a rapid-response power switch. Under normal conditions, it has no effect on the operation of the equipment. But, if a leakage of current to ground is detected, the device opens within a few milliseconds. Its response time is so rapid that you might not feel a thing. If you do, it might be painful, but not life-threatening.

While a GFI is a useful safety device, it is not a cure-all. It will not replace proper grounding techniques. It you inadvertently drill into a live conductor, it will not save you. Also, if you make contact from hot to neutral on a power line, it will not sense a short and trip. It protects only from hot line to ground. That in itself reduces the possibility for electrocution by almost 90%.

All GFIs fall into one of two designs: the differential transformer and the isolation transformer. The isolation type uses a transformer as a backup to its ground-fault-sensing circuit. This design trips when it senses a leakage current to ground as low as .2 ma.

The more common differential transformer has the power lines fed through a doughnut-shaped differential transformer. Under normal conditions the current through both lines is balanced. If a grounding situation occurs, a portion of the returned current from the load is shunted to ground. In 25 milliseconds or less, the GFI senses the current imbalance and shuts down the circuit. The typical trip level in a properly operating GFI is 5 ma. To ensure proper operation, any GFIs should be periodically checked in accordance with the manufacturer's standards.

Danger Signs

Danger and warning signs are an ever-present part of electronics. Major electronic installations are a hazardous location for the uninformed individual. High-voltage circuits, radiation hazards, ladders, and antenna platforms each have their own set of threats to personnel safety. To warn of potential dangerous locations, warning signs must be prominently posted and followed.

Any equipment cabinet or piece of equipment that contains high voltage must have a DANGER HIGH VOLTAGE sign displayed, as illustrated in Figure 7-3. This it to warn all individuals of the electrical hazard posed by opening the cabinet. Electronic equipment cabinets that have internal guards or screens to prevent access to high voltage points should have one posted there as well. Even the well-versed technician must be reminded of hazards.

Other signs should warn of radiation hazards from components and antennas. While an antenna is a familiar and respected piece of hardware to those that use it and work on it, it looks harmless to the uninitiated. There are several different signs (as depicted in Figure 7-4) that you might see. Sign 7-4*a* is normally found mounted directly on radar and communications antenna pedestals. Its purpose is to remind maintenance personnel of the dangers associated with RF-emitting antennas.

The sign in Figure 7-4*b* is a shipboard sign. This type of sign is used to warn nonmaintenance personnel of the fact that RF can be picked up by rigging and metal parts. If the ship has a poor ground, RF burns and shocks are a possibility unless safety procedures are followed. Another concern is that RF presents a hazard to flammable materials such as vaporized fuels. To warn of that hazard, the sign illustrated in Figure 7-6*c* is used. Even though mainte-

■ **7-3** *High voltage warning sign.*

(*a*) RADAR ANTENNA PEDESTALS (*b*) ON EXPOSED DECKS (*c*) FUEL HANDLING AREAS

(*d*) RADIO AND RADAR TRANSMITTER ROOMS (*e*) ANTENNA MASTS

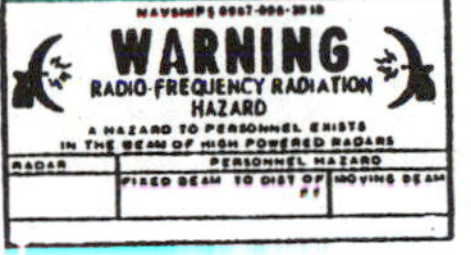

(*f*) ON RADAR CONTROL PANELS

■ **7-4** *RF radiation warning signs.*

nance and operations personnel are aware of RF radiation hazards, warning signs still must be displayed to serve as a reminder.

The sign depicted in Figure 7-4*d* is one such sign that is typically displayed in radar and communications transmitting spaces. Rarely are the antennas visible from the control spaces. Distance makes the threat seem less prevalent. For the safety of all individuals, transmitting antennas are often mounted on masts accessible only by a ladder. Signs such as the one pictured in Figure 7-4*e* are posted at the base of antenna masts and towers. The final RF warning sign is the one drawn in Figure 7-4*f*. This type is posted on transmitting control panels primarily for operations and maintenance personnel. It is usually prominently displayed close to the transmit controls as a reminder of RF hazards. When equipment is taken out of service for antenna maintenance, "MAN ALOFT" is often hung over this sign to serve as another protection against inadvertant operation. As a competent maintenance technician, part of your responsibility is to ensure that all signs are properly mounted, completely legible, and observed by all.

Heights

Many electronic technicians must work on equipment installed in high places. As radar and communications antennas must have a clear radiation pattern, they are mounted on towers and masts, often several hundred feet high. Antennas need periodic maintenance and visual examination to ensure trouble-free operation. When done correctly, working at heights is safe. However, there are many potential hazards associated with antenna installations. Often, the work space is very small, as illustrated in Figure 7-5. Installed guardrails are often very low. Unless the individuals performing maintenance are tethered with safety harnesses, a fall could easily happen.

The job begins before climbing a mast or antenna tower. Any maintenance to be performed at heights must be carefully planned, and in accord with ALL safety precautions. First, all equipment capable of transmitting energy must be secured to prevent personal injury. Surprisingly, because of the size and shape of the human body, it can act as a very efficient broadband antenna. Securing all equipment can be time-consuming in the case of shipboard electronic systems. Ships will tie up alongside one another to maximize pier space. That means if you are working on one antenna, another similar one may only be a few feet away. When working in a group of ships, all radiating devices must be secured to ensure safety.

■ **7-5** *Shipboard antenna platform.*

After obtaining permission and ensuring that all systems are secured, you can begin to go aloft. Never climb a mast or antenna without a safety observer. In addition to watching you, the observer can warn others away to prevent injury if any items are dropped. Once you begin to climb, keep both hands free. If the ladder is equipped with a safety rail, use the appropriate safety harness. Due to moisture, ladder rungs can be very slippery. Any tools that you require should be pulled up with a line. Even the unrestrained motion from a small tool bag can cause you to lose balance while climbing. Another point is that you must tether all tools, parts, and items that could fall. An article that is often overlooked is glasses. For safety they must be secured with a small piece of string. First, if you drop a wrench that means you must climb down, retrieve the tool, and then climb back up to complete the maintenance. Secondly, even a small wrench dropped a distance of 100 feet is lethal to any bystander. While working aloft, do not wear a hat, either soft or hard. Even a small breeze will blow it off of your head. If you try to grab it, you could lose your balance or grip and fall.

Most importantly, you must be tethered. Never work aloft unless you are in a secured area with handrails or are properly wearing an approved safety harness with manyard. The manyard is a relatively new safety device. Rather than a piece of rope or stitched material, it is literally a shock absorber. Once an individual has had his fall broken by one, it must be replaced. That is inconsequential, as the item costs less than $30 (a cheap price for safety).

If you must climb to a place where a harness cannot be used correctly, find another way to complete the task. A basket suspended from a crane, a cherry picker, or a bosun's chair might be more suitable. Risking your life to finish a job is not worth saving a few minutes. By following all applicable safety precautions, working at heights is safe.

Radiation

High-powered radars and communications systems are hazardous because of the presence of high voltages, currents, and RF energy. The RF energy radiated by an antenna can induce voltages into ungrounded metal objects such as antenna towers, ladders, handrails, and wire guys. Individuals can receive a severe shock or RF burn by coming into contact with an ungrounded metal structure.

RF is a hazard to us, as the human body acts as a broadband antenna due to length of trunk, arms, and legs. The effects from exposure to a high-energy RF field might not noticeable for hours, or even days. Because microwave energy is the cause, it has the same effect on the human body as a microwave has on food. When a human being is exposed to a high-powered RF source, the result is called *dielectric heating.* The body acts as an insulating material that has been placed in the presence of an RF field. The heating is caused by internal losses that are from the rapid reversal of polarization of the molecules composing the dielectric material in the human body. Any resulting damage is caused by heating from inside out. The main reason why the effects are not noted for some time is that the interior of the human body has relatively few nerve endings. The product is often very deep and penetrating third-degree burns that can often be as small as a pinhole.

When exposed to RF radiation, the eyes are the organs frequently affected first. Exposure can lead to cataracts and possible blindness. For safety's sake, never look into RF-producing devices, or stand in the path of RF-radiating devices. This safety rule dates from the early days of radar research. Because the hazards associated with RF energy were not known, several researchers viewed the interiors of waveguides and RF-producing devices to see if any visible phenomena were produced. Tragically, the result was that some of them went blind. The reason why we know of the risks is because someone paid the price for us. Never forget that radiation damage is cumulative. It might take years for you to absorb enough RF radiation to suffer noticeable damage. To ensure that you do not have future health problems, never expose yourself.

RF energy is lethal if absorbed in large enough quantities. The U.S. government, safety organizations, and professional associations have computed the lethal distances for virtually all high-powered RF-producing devices in use. To minimize risks, access to antennas and transmitter rooms is limited to prevent accidental exposures. Antenna systems are installed so as to limit the possibility of accidental exposure. Usually surrounded by a secure fence, the tower also has a locked ladder and appropriate warning signs. On shipboard antenna installations, red lines mark the closest point that one can approach an antenna without harm. Other, more dangerous systems have locked ladders and access points to prevent unauthorized entry. This is important, as radars and communications systems can often cause physical injury and death in distances measured in the hundreds of feet. The hazard associated with a particular radiating electronic system is determined by the frequency, power out, and beam pattern.

The amount of RF energy that you are exposed to is determined by your distance from the antenna and the length of time spent in the beam pattern. The longer you are in a beam, the greater the level of your exposure. The farther away from an RF producing source you are, the less RF energy is absorbed by your body. The radiation density decreases by a factor of the square of the distance. Remember, distance is your only protection from a radiating antenna. There is no safe way for you to be close to a radiating antenna. More importantly, there is no reason for you to be close to one. Table 7-2 is a relationship between peak power measured at an antenna and a safe distance for personnel exposure for an hour or more. This is a general table that does not take into consideration factors such as beam pattern, frequency, and height. As can be seen, even with a low output power of 50 watts, a safe distance is still at least 15 feet.

Another safety consideration with antennas is verifying radiation output and field intensity. There are a number of test devices on the market that can accurately measure field intensity without personal hazard. A test that I have observed old hands performing is to approach an antenna with a pencil. If the system is transmitting, a small arc will be drawn from the antenna to the pencil tip. DO NOT, I repeat, DO NOT ever do this foolish test. If you hear of anyone doing this, don't try it yourself, and discourage him. Radiation burns can easily result. If not, you have still received an RF radiation dose that you did not need for any reason.

Under the right circumstances, RF energy can also be a hazard to inanimate objects. It has been known to set off explosive devices such as explosive bolts and proximity fuses. Several explosions of ammunition loading sites were traced to electromagnetic radiation as being one of the possible causes. Large electromagnetic fields have also been known to ignite combustibles such as gasoline

■ Table 7-2 Required distances to protect personnel from the effects of radiation

Transmitter Power Output	Safe Distance from Antenna
Up to 50 watts	15 feet
51 to 200 watts	25 feet
201 to 1000 watts	50 feet
Greater than 1000 watts	75 feet

under certain circumstances. Normal equipment operation and refueling should not pose a significant hazard. For this to happen, three events must occur. A flammable fuel-air mixture must exist within the range of RF arcing. The RF arc must contain sufficient energy to cause ignition. And finally, the air gap across which the arc occurs must contain a sufficient volume of fuel-air mixture to ignite. Due to the potential hazard, there are areas on an aircraft carrier's flight deck where fuel tankers, gasoline-powered vehicles, and ammunition carts cannot be parked due to RF hazards.

When working in the vicinity of RF radiation hazards, there are several points you should always remember. Never fail to heed the warnings of radiation hazard signs. Do not ever visually inspect any component microwave device, such as a waveguide opening, feedhorn, reflector, or radiator while power is applied to the system. Under no circumstances should you climb a tower while the unit is in transmit mode. Finally, follow all local warnings and safety measures completely.

Electron Tubes

Electronic tubes present several hazards to the uninitiated. However, with information and a few precautions, there is nothing to worry about. First, if you are replacing an electron tube, never use your bare hands. If the equipment has been operated recently, a tube could still be very hot. Just like metal, it is impossible to tell if a tube is hot or cold. To safely remove it, always use either a tube puller or appropriate heat-resistant gloves. I recommend welder's gloves or asbestos gloves, as both have superior heat resistance.

Another problem associated with tubes is that some of them contain radioactive material. Typical radioactive tubes include TR, ATR, spark gap, gas-switching, and cold-cathode tubes. While not all of them contain appreciable amounts of radioactive material, some do contain significant quantities. Cartons for radioactive tubes are required to be appropriately marked to alert maintenance and supply personnel.

Defective tubes should be disposed of in accordance with EPA and OSHA regulations. As long as the tube envelope remains intact, no hazards exist. If for any reason the envelope is ruptured, radiation and radioactive material can escape. To ensure personnel and environmental safety, several precautions must be followed. Radioactive tubes should remain in approved cartons until installed in equipment. Defective tubes should be placed in secure cartons to prevent damage. Never place a radioactive tube in your pocket for safekeeping, as it could be broken there, complicating cleanup. If a tube is broken, then notify supervision and follow all local procedures for cleanup and disposal.

From this section, communications systems and other RF-producing devices sound dangerous. If approached correctly with knowledge and a good attitude, you have nothing to fear, as it can be a safe profession. If not, due to inattention, carelessness, or ignorance, then it is easily one of the more dangerous. First, you must know your equipment. By that I mean, does it have a lethal radiation zone you cannot enter? Where are the breakers, power disconnects, and switches? What is the layout of the control panel and what does each one do? Internally, know the values of voltage used in every circuit. Be aware of hazards such as large power-supply capacitors and pulse-forming networks. Use that knowledge in conjunction with all safety practices. You must know the rules and follow them. Never take chances, and never take shortcuts, because there is no second place or second chance. If there is an accident and someone is injured or killed, you can not just say you're sorry. The few moments saved are not worth the cost if you have an accident. Good safety practice begins and ends with a pilot's phrase: *situational awareness.* Always be aware of where you are and what you are doing. As safety is ultimately your responsibility, electronics operation and maintenance is as safe as YOU make it.

8

General Maintenance Considerations

Introduction to Maintenance

ALL ELECTRONIC EQUIPMENT MAINTENANCE CAN BE broken down into two broad categories: *preventative maintenance* and *corrective maintenance.* Preventative maintenance is any maintenance task that is performed on operational equipment with the expressed purpose of verifying operation and maintaining peak system performance. It can range from simple observation and measurement of operational parameters to alignments and adjustments. Corrective maintenance would be any maintenance task performed to correct an equipment deficiency by replacing a component or assembly and performing any adjustments or cleaning required to return it to peak operational performance. As an electronic maintenance technician, you can expect to perform more preventative maintenance than corrective maintenance on a major electronics installation. The effectiveness of any electronic system is based upon the professionalism exhibited by attention to detail and completing all necessary tasks. A misaligned receiver, a weak transmitter, or antenna system that is deteriorated has a direct and noticeable impact on the performance of a communications link. To ensure peak system performance, the entire communications system must be properly maintained.

General-Purpose Test Equipment

Before performing any maintenance, be it preventative or corrective, as an electronic technician you must be familiar with the use and operation of general-purpose test equipment. Also, there are many highly specialized pieces that are used with only certain types of equipment. As both corrective and preventative maintenance requires the accurate and effective use of several different types of test instruments, any time spent becoming competent in their use is time well spent, and should be considered an investment. The better a technician can operate test equipment, the easier and more professional the maintenance task will be.

One of the most common, versatile, and important pieces of test equipment is the *oscilloscope.* A simple representative oscilloscope is pictured in Figure 8-1. The primary function of any oscilloscope is to give a visual presentation of circuit action as it occurs. Through the use of this test instrument, a technician can observe and measure waveform frequency, duration, phase relationships with other signals, waveform shape, and amplitude. There are scores of different types of oscilloscopes in use today, and basically they all function the same. The horizontal axis of the CRT is the time base, and can be set in time increments such as seconds, milliseconds, and microseconds. The vertical axis is amplitude and is measured volts. Figure 8-2 is a photograph of a 60-hertz AC waveform. This allows for the accurate time and voltage level measurement of the waveform. To provide for more stable viewing, the instrument is usually triggered by the system under

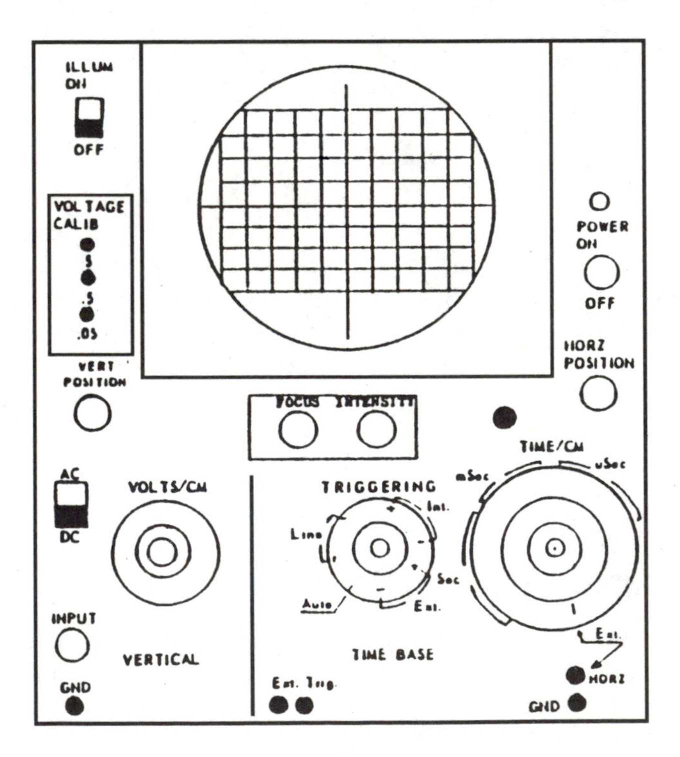

REPRESENTATIVE OSCILLOSCOPE

■ **8-1** *Basic oscilloscope front panel controls.*

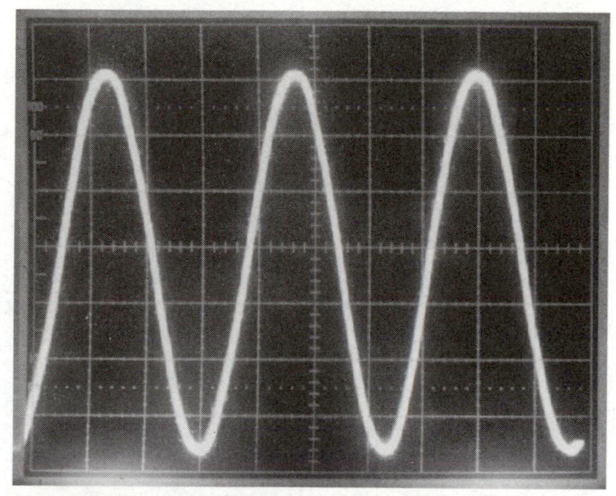

■ **8-2** *Representative AC waveform.*

test. For flexibility, most oscilloscopes are dual-channel, which allows for the viewing and comparison of two separate signals at the same time; this can be an invaluable capability for fault analysis and alignments.

The digital and microprocessor revolution of the past 20 years has had a tremendous impact on test equipment. The oscilloscope is not an exception from this today, as many oscilloscopes are microprocessor-controlled and have a multitude of special features in a small package. Figure 8-3 is an excellent example of the high-quality test instruments available on the market today. Manufactured by Tektronix, it is a high-performance test instrument that provides many advanced features in an easy-to-use package. It is actually a family of test instruments with a common front panel and controls. Equipped with a 3.5-inch floppy drive, it can store observed waveforms for many uses. A library of waveforms observed in a fully operational system can be saved to provide an advanced means of waveform comparison to speed failure analysis. When working on a problem, waveforms can be saved for future reference to record observations, thus speeding failure localization. Another good use, as the stored waveforms are in a computer-compatible format, is to forward them to other locations for sharing information. When performing preventative maintenance observations, waveforms can be recorded to provide long-term observations for detecting gradual system degradation.

To speed operation, test instrument settings can be saved and recalled to prevent errors. By using internally generated cursors,

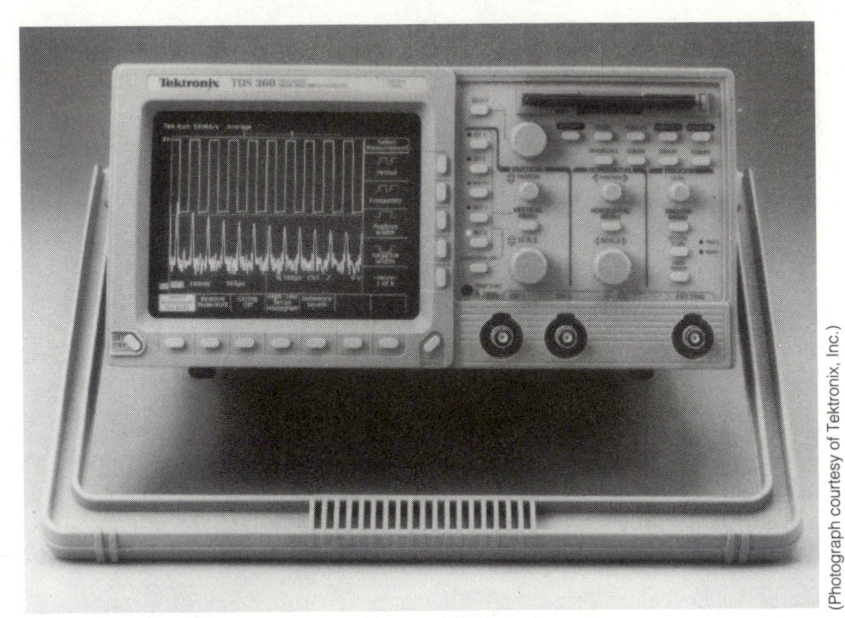

(Photograph courtesy of Tektronix, Inc.)

■ **8-3** *Tektronix TDS 300 digital real-time oscilloscope.*

different sections of a waveform can be expanded and measured. A digital readout right on the screen gives an accurate readout of waveform frequency and amplitude.

Multimeters are also important test instruments with which every technician should be familiar. There are two broad categories of multimeters today: *digital multimeters* and *analog meters.* At a minimum, a good multimeter should be capable of measuring AC/DC voltage, DC current, and resistance. Additional and highly desirable features are the ability to perform continuity tests and measure AC current. When a meter is configured to measure continuity, an audible beep is heard if the conductor under test is a short circuit.

An innovation of the '80s was the digital multimeter, which has virtually taken over the multimeter field. The modern DMMs are low in cost but high in quality and are extremely versatile test instruments capable of performing a wide variety of measurements. Some of the high-end DMMs are capable of providing readings of temperature, frequency, capacitance, and inductance. Other useful features found in many DMMs include continuity test, diode check, bar-graph readout, capacitance, transistor gain measurements, and logic probe function for digital troubleshooting.

Heavy industrial applications would call for the meter illustrated in Figure 8-4. Designed for production-line uses, it provides automatic testing of variable equipment parameters for quality assurance purposes. It can perform pass/fail tests, compare functions, or null measurements.

As flexible as these instruments are, there are still some tests that are performed better by the old-style analog meters. I have found that an analog meter is superior when testing the changing resistance of a potentiometer, or the charging action of a capacitor. Also, when performing antenna alignments on a precision approach radar, the analog meter provides a much better indication of antenna motion. However, with the rapid pace of change in electronic test equipment, the days of the analog meter are numbered.

The advent of low-cost digital and computer-based circuitry has also influenced the multimeter and oscilloscope in an unusual and beneficial fashion. The best features of both are combined in the *handheld oscilloscope.* Lightweight, accurate, and flexible, this new configuration will certainly affect the test instrument field in a positive way.

The signal generator has many applications in both communications and radar system maintenance. Due to the different characteristics associated with frequencies ranging from audio to RF, signal generators are classified by frequency range. To properly maintain a large-scale electronics system, you would use two or more separate instruments, as you might have to test circuits through *signal injection* from audio to RF ranges. Signal injection

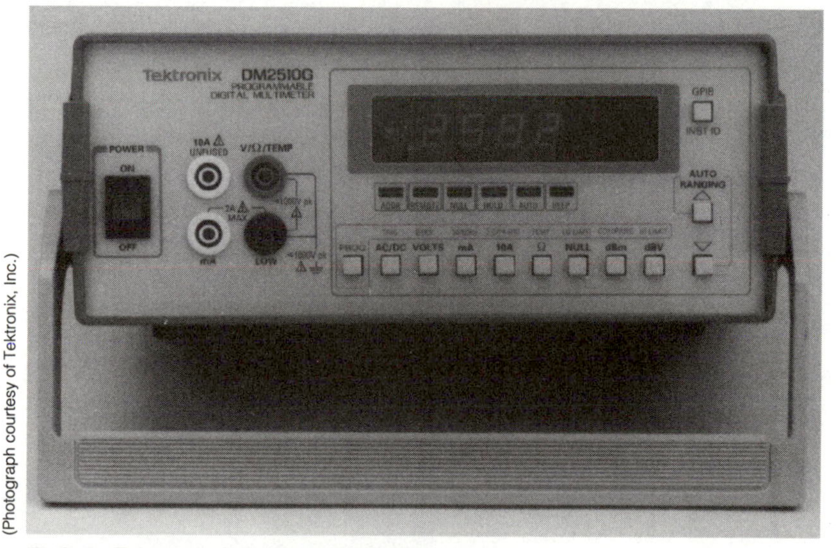

(Photograph courtesy of Tektronix, Inc.)

■ **8-4** *Tektronix DM2510 production test set.*

is used to provide a known-good test signal to verify the operation and alignment of AF, IF, and RF stages.

Another test instrument designed for maintaining the advanced equipment now in use is a signal generator, such as the Rohde & Schwarz SME signal generator shown in Figure 8-5. The purpose of this instrument is to provide complex signals associated with digital mobile radio networks. It can also produce analog signals to test AM, FM, and pulse modulation systems. As with other advanced instruments, as it is microprocessor-controlled, test sequences can be programmed to speed maintenance tasks.

A *spectrum analyzer* is a highly specialized piece of test equipment that is capable of observing the frequency spectrum produced by an RF generator, such as the final stage of a transmitter. The unit features a CRT display to give the technician a visual indication of the transmitter's output. By periodically testing the frequency spectrum produced by a communications system, you will have an indication of transmitter performance. With this test instrument, component problems and misadjustments that are normally very difficult to analyze are quickly isolated. Due to the high voltages and currents present in a transmitter, locating some problems can be challenging. However, with a spectrum analyzer, failures related to the RF generator and modulator section are unmistakable.

Figure 8-6 is of a state-of-the-art test instrument from Tektronix, the R2465 Digital Radio Transmitter Analyzer. This one very flexi-

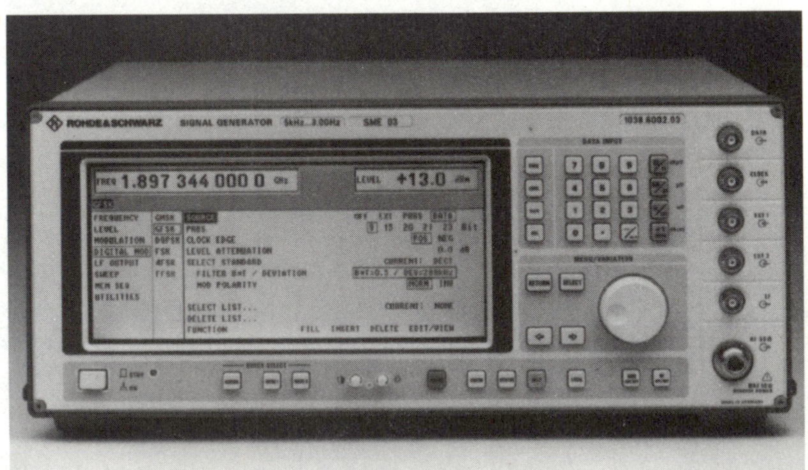

(Photograph courtesy of Tektronix, Inc.)

■ **8-5** *Rohde & Schwartz signal generator.*

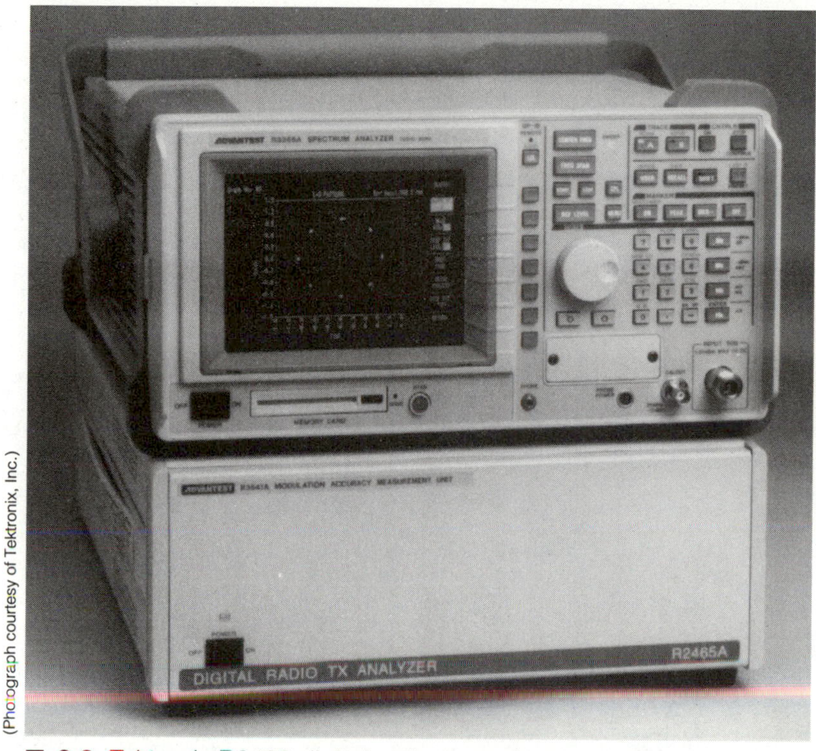

(Photograph courtesy of Tektronix, Inc.)

■ **8-6** *Tektronix R2465 digital radio transmitter analyzer.*

ble test instrument is capable of observing frequency accuracy, RF power output, harmonic signals, and spurious signal levels. With this type of instrument you can obtain precision modulation observations and measurements of a transmitter's output. By observing the transmitter output on a regular basis, you can document signal degradation and equipment deterioration before it becomes a problem.

A *time delay reflectometer,* as pictured in Figure 8-7, is a highly specialized piece of test equipment that is vital when there is a need to troubleshoot failed cables, wiring harnesses, and conductors. This instrument can be used to verify and locate conductors that have shorts, opens, crimps, or defective connectors. If an antenna has a coaxial feed, then it can be used to check for defective cables. As a technician, you have to check many cables and wires with continuity tests. While that type of investigation is satisfactory at low frequencies and short distances, there are times when it is inadequate, if not impossible. Rather than just testing for continuity, which is only a resistance check, a TDR observes the effect that a cable has on actual signals.

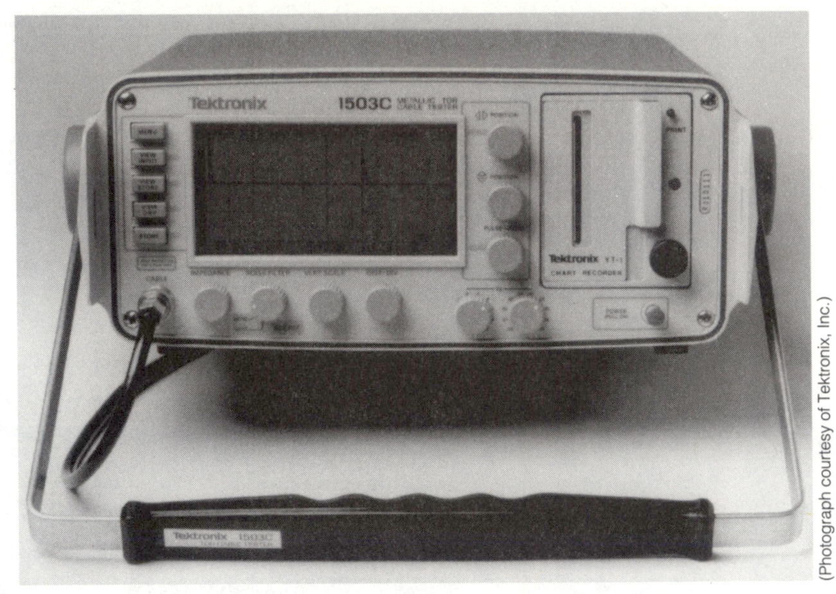

(Photograph courtesy of Tektronix, Inc.)

■ **8-7** *Tektronix 1503 time-delay reflectometer.*

A TDR functions by transmitting a short burst of energy into a cable under test. The instrument is then used to observe for reflections at the point of signal insertion. If no energy is reflected back to the point of insertion, then the cable has a uniform impedance and it is functional. Any defect in the cable is reflected back as a discontinuity, which can be positive, negative, or a fast-changing indication. The position of the discontinuity on the CRT display indicates where in the cable run the fault is located. When testing cables measured in thousands of feet, it is vital to know exactly where the fault is located. By using a TDR, you can determine the exact point on the cable run where there is a short or open. By using system drawings, the problem is isolated to a specific access point, facilitating rapid repairs. The unit can be just as useful on shipboard installations, where cable runs can be several hundred feet through several different decks, levels, and bulkheads.

This high-quality test instrument is capable of locating defects in metallic cables up to 50,000 feet in length. Capable of testing cables with a wide range of impedances, it can detect a fault as close as ten inches. As with the other advanced types of test instruments, it can store observed waveforms on a 3.5-inch floppy disk for future reference.

As rapidly as the communications field is evolving, so is the test equipment to support it. Figure 8-8, a Rohde & Schwarz CMS 54 Radio Communication Service Monitor, is a one-piece radio communications test instrument designed for use in maintenance and equipment testing. Specifically built to test transceivers, it features flexibility, small size, and light weight. Through the use of front panel controls, it can be configured for troubleshooting or automatic manufacturing quality tests. In one small package, it can perform spectrum observations, transmitter testing, receiver testing, and RF frequency counting. As with other advanced test instruments, it features a disk drive for saving observations for future reference.

The final test instrument is the *logic probe.* Advanced electronic systems are constructed from digital circuits. As the signals within the equipment are digital highs and lows, this device is invaluable for tracing failures. The advantage of this test instrument is its compact size and ease of use. LEDs provide a rapid readout of what a particular gate or digital circuit is doing. This would be ideal in very tight locations, where an oscilloscope would be too bulky and difficult to observe.

As can be seen from this short section, there is a wide range of test instruments available on the market today. All of them are lower in cost and far more flexible than what was in use just 20 years ago. Just as the equipment we maintain has become more complex, the manufacturers have provided us with the tools needed to keep it in peak operating condition.

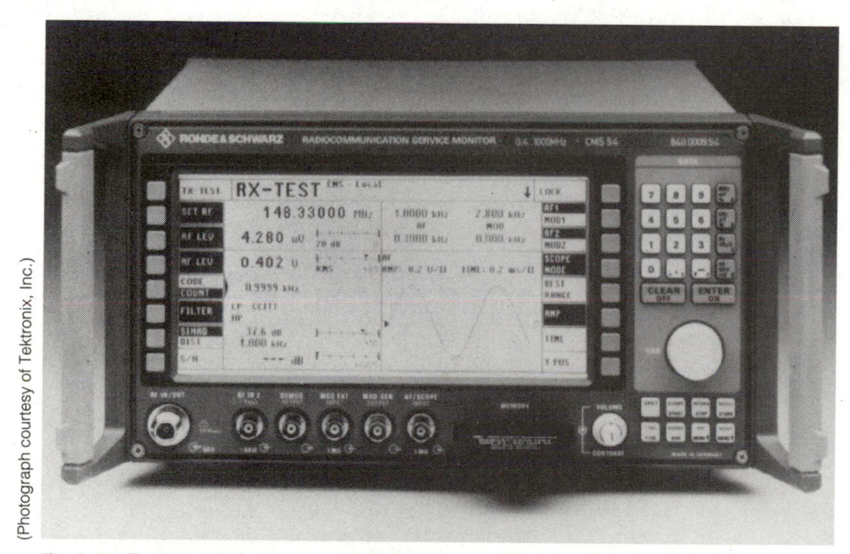

(Photograph courtesy of Tektronix, Inc.)

■ **8-8** *Rohde & Schwarz CMS 54 radio communications service monitor.*

Preventative Maintenance

Although preventative maintenance is vital to the successful operation of any electronic system, it is often the most overlooked and underestimated. As with safety precautions, preventative maintenance procedures were developed in response to real problems and issues. It has been discovered and rediscovered over the years that timely preventative maintenance saves many times its costs in reduced corrective maintenance man-hours, parts, and equipment down time.

One very important preventative maintenance step that is often not given enough attention is equipment cleanliness. Many potential equipment problems are eliminated by simple cleaning and good housekeeping standards. If grease or oils are allowed to accumulate on electronic equipments, large amounts of dust are attracted and allowed to build up. Not just a visual problem, it is an eyesore that can lead to equipment failures as the dust can form a high resistance path to ground for current flow. The existence of these paths leads to component failure and equipment down time. Accumulated dust can also cause electronic components and subassemblies to run hotter. Heat is an enemy of sustained equipment operation, as it can directly lead to component failure. The hotter an electronic component operates, the shorter its operational life. Even if component failure does not result, overheated components can shift in value, changing circuit parameters. Locating a failed component in a circuit full of components that are changed in value is a challenge that adds considerable time and frustration to the task. Dust buildup in high-voltage circuits can lead to arc overs, leaving carbon trails. Once a trail has been burned into an insulated surface, the insulation must be replaced to prevent future arcs at a lower voltage.

Air filters are an important preventative maintenance check. Electronic systems produce a tremendous amount of heat—just stand behind a 486 personal computer. To provide cooling, electronic equipments require an unimpeded air flow. To minimize dust problems, filters are fitted to remove dust and particles from the moving air. The filters are important, because if they become blocked with dirt, air flow is decreased, increasing equipment internal temperature, leading to equipment down time. Equipment air filters come in two broad classes: disposable and nondisposable. If the disposable filter is dirty, just replace it. A dirty nondisposable filter must be cleaned with soapy water. After all the dirt and dust has been removed, rinse it in clear water, and then let it air dry. If time is a problem, use low-pressure air to blow the moisture out.

Equipment cooling systems require periodic attention as well. Most cooling systems are equipped with meters to track important system parameters such as pressure and temperature. By periodically recording the values, system operation is tracked. If a parameter changes, that indicates the need for more in-depth maintenance before damage results.

Liquid cooling systems, just like the air type, have filters to remove solid impurities. A filter that is blocked or only has partial flow increases equipment temperature, leading to component failure. To minimize cooling system failures, tap water should never be used in them. Water suitable for human consumption contains minerals that can cause deposits in small channels and in the filter system. RF cooling systems often have small channels for water that pass through the heat-producing components. Also, tap water can lead to biological growth, which can be a real problem. If this does occur, flush the system until the water is clean. You should never add chemical agents unless it is recommended by the manufacturer. That is because the interior of the cooling channels may be coated with a material that is susceptible to damage from the chemical agents you use.

As a minimum, on performing preventative maintenance you should follow what the manufacturer recommends. With experience you might find that local conditions change the frequency of some checks. A good example is that of air and water filters. Typically, both are a monthly check. One radar site that I was assigned to was so dusty that radar indicator air filters had to be changed on a weekly basis, rather than monthly. At another site, a portable x-ray machine was used on a cross-country pipeline. Due to the harsh conditions, the water filter had to be cleaned daily.

Transmitter Checks

Alignments and adjustments must also be verified and performed on a repetitive basis. Many times when an equipment is to be used for a demanding operation, that type of maintenance must be performed more frequently. Verification of equipment accuracy and operational checks are performed on a daily, weekly, monthly, and quarterly basis. Although seeming to be time-consuming, by adhering to the schedule, actual effort is reduced and equipment availability is increased.

A very important transmitter check is verifying the RF generator output frequency accuracy. For optimum system performance, an

RF generator is designed to produce an output within a very limited band of frequencies. This is because the radio frequency spectrum is a finite resource that is crowded with many different services. A transmitter that generates spurious frequencies or covers more than its assigned bandwidth will interfere with other systems. Also, as more bandwidth is occupied, system efficiency is degraded, resulting in a weaker signal and loss of range.

Other important checks include ensuring that modulation is within tolerances. Either an excessive or a low percent of modulation could lead to the loss of important intelligence. If the carrier frequency is overmodulated, additional sidebands are produced by the transmitter, causing extreme distortion.

Receiver Performance Tests

Efficient communications system performance requires a receiver that is aligned and operating to design specifications. The most important receiver maintenance checks are receiver sensitivity, receiver frequency, and amplification factor. The best way to ensure proper receiver operation is to follow the manufacturer's recommended alignments and adjustments.

Receiver sensitivity has a direct impact on the quality of the received signal. A decrease in receiver sensitivity has the same influence on detection range as decreased transmitter output power. As an example, say a receiver has an input sensitivity of 10 microvolts rather than a normal 5 microvolts. That would have the same effect on the quality of the received signal as a 3-dB decrease in transmitter output power. While that change in output power is easy to detect, it is not easy to find in the receiver. A simple misalignment is more than enough to induce such a loss in receiver sensitivity. Receiver sensitivity is a guage of its capability to detect weak signals.

Frequency accuracy of a receiver is very important in the overall quality of the communications link. The first step is to ensure that the receiver is accurately tuned and properly tracks the transmitter frequency. If there is a difference between the receiver and transmitter frequencies, then that indicates either a tuning problem or an equipment failure.

Corrective Maintenance

The first step to take in correcting a failure in any electronic system is knowing how it operates under optimum conditions. The time to learn how a system functions is before any failures occur. To fully understand it, it is advisable to begin with the control panels and equipment cabinets. Observe all meters, displays, and switch positions that pertain to the equipment and system.

Far more problems than you can realize are caused by lack of familiarity with an electronic system's control panels. I call the in-depth knowledge of equipment controls, meters, and status lights "KNOBOLOGY." Both maintenance and operational personnel must be familiar with the proper setting for all controls and front panel adjustments. Often, an equipment will have a local control panel with the equipment and a remote control panel in another location. As the two are often separated by a great distance, it is impossible to observe both sets of controls at the same time. Many times a control in the wrong position can give the appearance of a failed system. As can be seen from Figure 8-9, a control panel has numerous buttons, lights, and switches. This example is the local receiver control panel for an AN/GPN-27 airport surveillance radar system.

In summary, you must be aware of what each control does and its proper setting in each mode of operation. The resulting status lights must be just as familiar to you, as they can be an invaluable aid in fault analysis and correction.

Knobology directly relates to "KISS," the acronym for "Keep it simple, stupid." Always check the simple things first. If the complaint is that a receiver is inoperative, make sure that the patch panels for connection to the antenna are properly set up. If it is reported that a transmitter is down, make sure that it is selected and all controls are in the right place.

After you are familiar with the front panels, it is time to learn how the equipment is laid out. If the equipment consists of several cabinets or major subassemblies, learn where all the different functions are located. Which cabinet contains the power supplies? Is the RF section in the second or third cabinet? The time to learn is before the system is down and you are under pressure.

The next step is to use test equipment to observe voltages and signal waveforms throughout the system. Depending upon the complexity of the system, you might use just a multimeter to measure voltages. Larger systems could require an oscilloscope or a spectrum analyzer. By observing and recording correct input and output voltages, and waveforms for each chassis, you have a starting point when isolating failures.

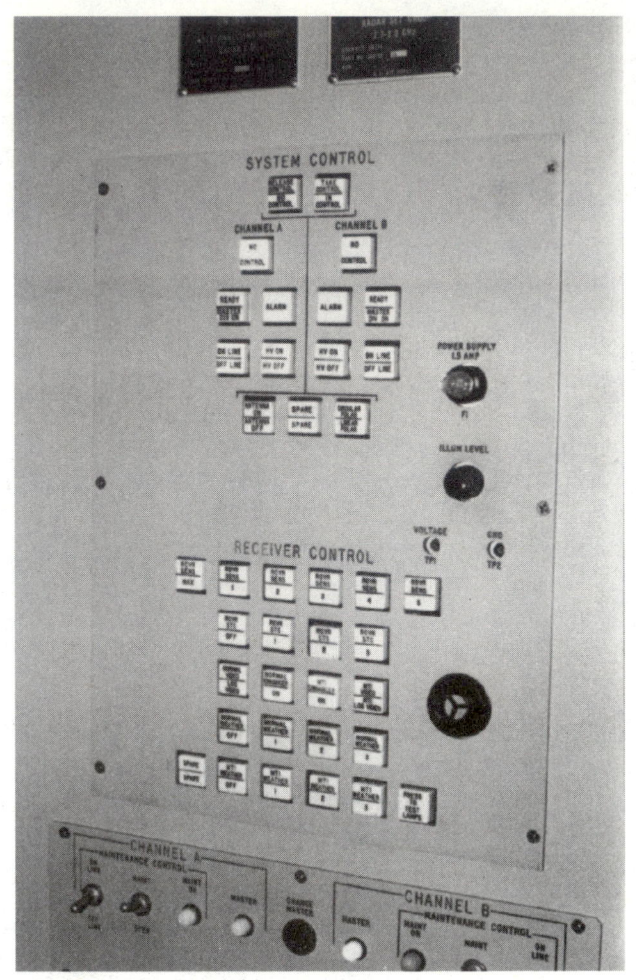

■ **8-9** *Receiver control panel.*

Failure analysis always begins with simple observation. This is not the time for test equipment, tools, or schematics. Before doing anything, just observe. If you are in the radar control room, what are the equipment status lights, meters, and controls telling you? How is the control panel set up? Which panel lights are on, which ones are off? Most failures can be isolated to one or two functions just from observation of front panel indications and controls.

Troubleshooting a failed piece of electronic equipment should be accomplished logically, regardless of the complexity or technological level you encounter. Approached correctly, the time required for corrective maintenance can be greatly reduced when you are methodical in your actions. Based upon years of experience, the best way to troubleshoot electronic equipment is called the *six-step troubleshooting procedure.* This will work with any electronic or mechanical device. The logical six steps are as follows:

1. Symptom recognition.
2. Symptom elaboration.
3. Identifying possible faulty functions.
4. Localizing the faulty function.
5. Localizing the faulty circuit.
6. Identifying the failed component.

This method was first taught to me when I attended my first technical school over 20 years ago, and it is still being stressed in both military and civilian technical schools. When I was an instructor in Memphis several years ago, we had several problems that drove home the need for a logical approach to troubleshooting. The problems were such that, unless you followed the six steps, a ten-minute analysis became over an hour of confusion and stress on the unwary. One of the more interesting ones involved the radar antenna. To simulate a fault, we disabled the antenna ON command in the remoting equipment. The result was that when the system was placed in remote control, there was no antenna motion. When the equipment was placed in local control, the antenna would scan normally. Many students failed to pick up on the effect on system symptoms when placing the unit in local control. What we soon learned was that the majority of the students who failed to be logical in the completion of the six steps were lost. The few that followed them faithfully found the failure very quickly. As a result, students referred to this problem as "The LA Freeway With No Off Ramps."

Symptom recognition is vital, because in order to repair a piece of equipment you must first know if it is functioning normally. A piece of equipment is designed to perform a specific function, at all times. Often with today's complex and capable devices, a series of events must occur in order. If it fails in any way, then it is in need of maintenance. By common definition, a trouble symptom is a sign of a malfunction in an electronic system. Symptom recognition is the action of the technician recognizing a malfunction. Symptoms can be obvious, or very subtle. That is why it is imperative that you are familiar with the normal operational characteristics of a device you are maintaining.

The easiest symptom to recognize is total equipment failure. When a very obvious failure occurs, a major function or the entire system is not performing. A good example would be the loss of main power. When that happens, the entire system or equipment is not functioning. Panel lights are out, cooling fans inoperative, and meters are motionless. Degraded performance is a little more difficult to ascertain. Low transmitter power or a decreased receiver sensitivity are examples of that type of problem. The equipment still appears normally functional, but something does not seem right. The receiver might not pick up as many other stations as usual. A transmitter used on a common link receives complaints of distortion or a weak signal. In order to perform this step, you must know the normal operational capabilities of your equipment.

Symptom elaboration is the second step. At this point, you are still observing the operation of the equipment without using tools, test equipment, or schematics. Are all transmitter or receiver front panel meters normal? Which status lights are illuminated? Are the indicator lights normal or abnormal? This is the step where you use front panel meters and built-in test equipment to verify system operation. This is the point where you check front control panel switches. Is the receiver defective, or is the BFO switch just turned to the OFF position? By carefully following this step, you can narrow the possible faulty functions to only one or two circuits within the communications system.

The third step is the action of identifying a list of possible faulty functions. This is a direct result of how well the first two steps were performed. If all equipment symptoms were correctly identified, then the situation is progressing well and should end successfully. This step is applicable to equipments that contain more than one function or operation. Examples of functions within a piece of communications equipment would be power supplies, a modulator, RF amplifiers, a detector circuit, and remoting equipment.

As can be seen, the first three steps did not rely on the use of test equipment or tools. Rather, you must use your senses and knowledge of the equipment to narrow the likely area of the failure to as few circuits as possible. As time-consuming as it might seem, it is much faster in achieving the goal of finding the failed component.

In the fourth step, you determine which function is at fault. This is the first step where you actually use test equipment to check equipment parameters. There are three factors to consider in this step. Choose measurements that will eliminate as many functional units as possible. Another point is the accessibility of test points. If the unit must be disassembled to check a test point, then you might want to skip it at this time. The final consideration is past experience with the equipment. If you have seen the exact symptoms before, then that may lead you directly to the problem circuit.

The fifth step is localizing the failed circuit. How you isolate the failed circuit depends upon the arrangement of circuits within the function. Figure 8-10 illustrates several possibilities of how circuits

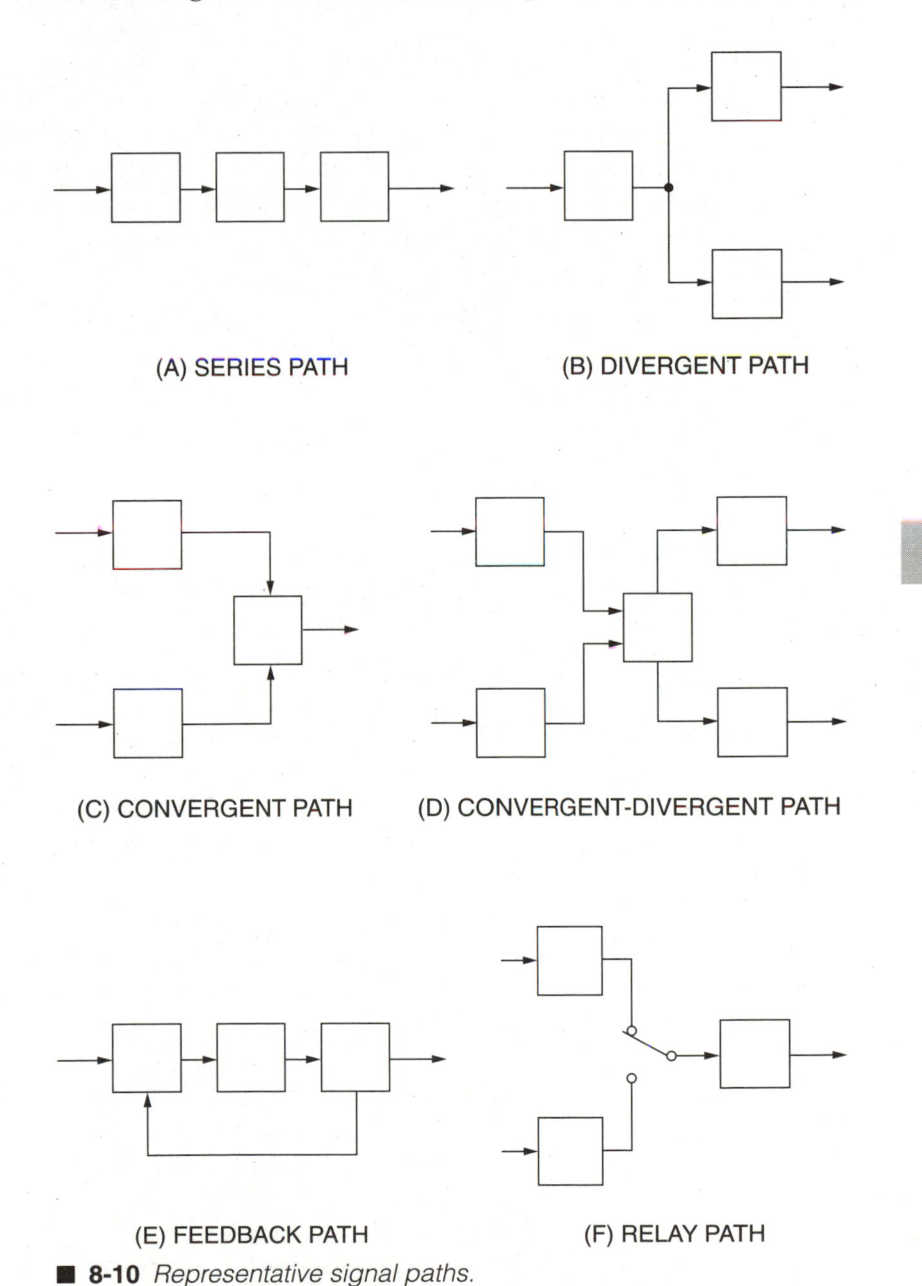

■ **8-10** *Representative signal paths.*

are interconnected. If the failed function has 20 different circuits connected in series, then it is known as a linear path. Because of the number of circuits, you would want to minimize the number of checks that you have to take. The most logical approach is to split the number of circuits under test in half. If the output of number ten is defective, then you know that the failure is between one and ten. The next check would be to split the circuits in half again by checking number five. In this manner only a few checks should isolate the failed circuit, rather than 20, saving a great deal of time. This type of troubleshooting approach is known as bracketing. It you failed to use this method and you checked the circuits sequentially and started with number one, it would take 20 separate checks if the last circuit was the inoperative one.

If the function has a divergent path, then there is no most logical first check. In cases such as this, personal preference would determine where you began your observations. In a convergent path arrangement, begin with the output circuit. In the convergent-divergent arrangement, begin with the common circuit. The most challenging of all is the feedback path. In this type of layout, check the input on the first stage. This type of problem is the most time consuming. To find the exact circuit, you must break the feedback loop where it is routed to the input stage. That action places the function in a static state. The final path is the switching path. This example is nothing more than a relay circuit.

A point to remember is that the output of one circuit might go to a number of circuits located in a different chassis or circuit boards. Cables, wires, circuit board pins, and connectors can fail. Just because the output of one circuit is normal does not mean that the input of the following circuit is good. A good idea is to test the output of the circuit board and the input of another where the signal is used.

After you have definitely identified the failed circuit, you are ready for Step 6. In this step the goal is to find the faulty component and review the steps you took to get there. In this step you will use detailed schematics and test equipment to complete the task. At this point you should be checking the operation of a single circuit consisting of an active device and support components. An active component would be a transistor, vacuum tube, or integrated circuit. You start with voltage checks, and can use either a multimeter or an oscilloscope. That tells you if the device is saturated, cut off, or distorted. The voltage checks will lead you to the component that has caused the problem. A quick rule on voltage checks is that within a circuit if you measure V_{cc}, that indicates a short above or an open below in the current path. If you measure a low voltage, then that indicates an open above, or a short below.

Review is an important part of the troubleshooting process. Through review you learn how to improve your failure analysis skills so that you can become a better electronic technician. Table 8-1 is a review of the six-step troubleshooting method. Each step lists its goal and which tools you are to use in its completion. When developing your style of troubleshooting, be consistent. The idea of the six-step method is to give you a common approach to problems that you use every time. Through equipment knowledge and consistency you will become a skilled troubleshooter. Troubleshooting is as much a science as it is an art.

Table 8-1 Six-step troubleshooting method

Step	Goal	Tools to Use in Its Completion
Step 1	Symptom recognition	Sight Sound Technical knowledge
Step 2	Symptom elaboration	Sight Sound Technical knowledge Equipment controls Equipment status lights Equipment meters
Step 3	Possible faulty functions	Sight Sound Technical knowledge Equipment status lights Equipment meters Equipment block diagram Equipment history
Step 4	Localize faulty function	Sight Sound Technical knowledge Equipment functional Diagrams Test equipment
Step 5	Localizing faulty circuit	Sight Sound Technical knowledge Equipment schematics Test equipment
Step 6	Localizing faulty component	Sight Technical knowledge Equipment schematics Test equipment

General maintenance

In electronic maintenance there are general points that pertain to all systems. To ease maintenance tasks, you should have a set of nonmagnetic tools for working around RF generators equipped with permanent magnets. It is a challenge to properly tighten bolts and screws when the magnet keeps pulling you toward it.

I have found it helpful to maintain my own set of schematic diagrams. That allows you to personalize them with your own notes and observed circuit parameters. You can mark each test point with voltages and waveforms to speed analysis. Another aid that I use is equipment history. If you compile a history of every failure that an electronic system experiences, you develop a powerful troubleshooting aid for yourself and anyone that might follow you. With a written history, you will find that many problems are repetitive in nature. As an example, I have one piece of portable equipment that burns out a particular transformer on a two-year cycle. It also helps with the rare-but-difficult problems by giving others a starting point, as many symptoms are similar with a number of failed components.

Spare parts are an area in which you should be very interested. You should try to maintain a sufficient number of parts to repair common and repetitive problems. A good rule of thumb is to not maintain more than a year's supply of parts. Many mechanical and all electronic components have a finite shelf life. Components such as capacitors might last only five years until the electrolyte dries out. A defective spare part is worse than having no part, because it provides a false sense of security. A bad habit to get into is excessive board and component swapping. If you suspect a failed circuit board, but the spare does not cure the problem, ensure that any alignments required are completed. It that still does not cure the problem, reinstall the original board and recheck your six-step method. As many circuit boards contain adjustments, if you swap several boards at the same time, the equipment could be inoperative due to the misalignments. If you are fortunate enough to have a spare set of boards, verify that they are operational by installing them in the equipment. After verification, store them in the original packing material in an approved cabinet.

Modern components require proper installation techniques. Attending a soldering course will save you time, effort, and mistakes. Also, many components today are static-sensitive. As such, they must have special handling, storage, and installation steps to ensure normal operation.

Figure 8-11 is what is considered to be a hybrid circuit board from an electronic test instrument. Notice that it consists of a high-voltage transformer in the upper right-hand side. This particular transformer boosts 115 VAC to 1800 VAC to operate a vacuum circuit. The other components are much lower-rated, in the range of 12 vdc to 18 vdc. This type of hybrid circuitry drives a requirement for more attention. In most sections of the board, the voltages and currents are low enough so as to not be a hazard. While within the same board, a high-voltage hazard exists.

Another problem is that the components are small and mounted close to the surface of the board. That calls for caution when taking voltage and waveform checks. This is an easier board to work on in one respect; most examples manufactured today are tightly packed, calling for extreme caution when taking observations with test equipment. The slip of a test probe can short out two leads,

■ **8-11** *Hybrid circuit board.*

destroying an integrated circuit or transistor. Another example is the subassembly from an AN/TPX-42 IFF device; this subassembly is pictured in Figure 8-12. In this example, the subassembly is mounted on a card extender to facilitate circuit analysis with test equipment. If you look in the lower left you will see a rail. Unless the assembly is locked in place, it can easily slide into the chassis, causing damage to open access ports and test probes. Other types of boards can be placed on extenders for ease of testing. If it is left in place while the drawer or assembly is reinstalled, damage results. This is an all-too-frequent occurrence. A technician will forget and close the drawer with the circuit board still in the extended position. Now, instead of having a single failed component, the board is destroyed.

Receiver and Transmitter Troubleshooting

Receiver troubleshooting is as straightforward as you make it. The following exercise of a component failure in a receiver will be solved using the six-step troubleshooting method. This is an example of how you might go through the various steps to correct a failed communications receiver. Being told that the unit provides no audio output, the first step will be to examine the receiver. Is it turned on? What are the readings of any front panel meters or displays and how are the controls set? It could be as simple as a volume control turned down, or the unit being placed in standby.

After determining that the front panel is satisfactorily set up, and the unit is still down, it is time to examine the interior. After ensuring that the unit is de-energized, open it up and inspect the interior. You are looking for obvious signs of failure. Burned components are an excellent and obvious sign of component failure. Are components burned, transformers leaking a tar-like substance, capacitors split open, resistors cracked, or signs of overheating on the board surface? At this time, also use your sense of smell. A burned component leaves a lingering odor that is easy to detect.

The next step is to begin using test equipment to isolate the problem. To effectively localize a failed component, adequate technical documentation is vital. As a technician you must be able to use a functional block diagram, servicing block diagram, and a detailed schematic. You are already familiar with the functional blocks. The servicing block diagram, as illustrated in Figure 8-13, breaks an equipment down into stages and functions. As can be seen in the diagram, this receiver consists of five major functions: the RF, mixer-LO stage, IF stage, the audio amplifier stage, and the power

■ **8-12** *Electronic equipment stack with extended chassis.*

supply. The functions are further broken down into 11 stages. Notice that no components are shown. As this point, you are just interested in the input and output signals of each stage.

The first check before proceeding would be the power supply fuses. In complex devices a series of power supplies might be used. The failure could be just a blown fuse caused by a primary power surge. The next step would be to signal-trace the equipment. A good way to troubleshoot a receiver is to inject a known-good modulated test signal. If an IF frequency were injected at the input of the second IF amplifier, you would listen for audio. If the unit produces sound, then the problem is in the first five stages. If not, then the problem is in the second five stages. The next step would be to split the possible failed stages in half. Keep this up until you have isolated the problem to one stage. You will not know you are there until you inject the signal

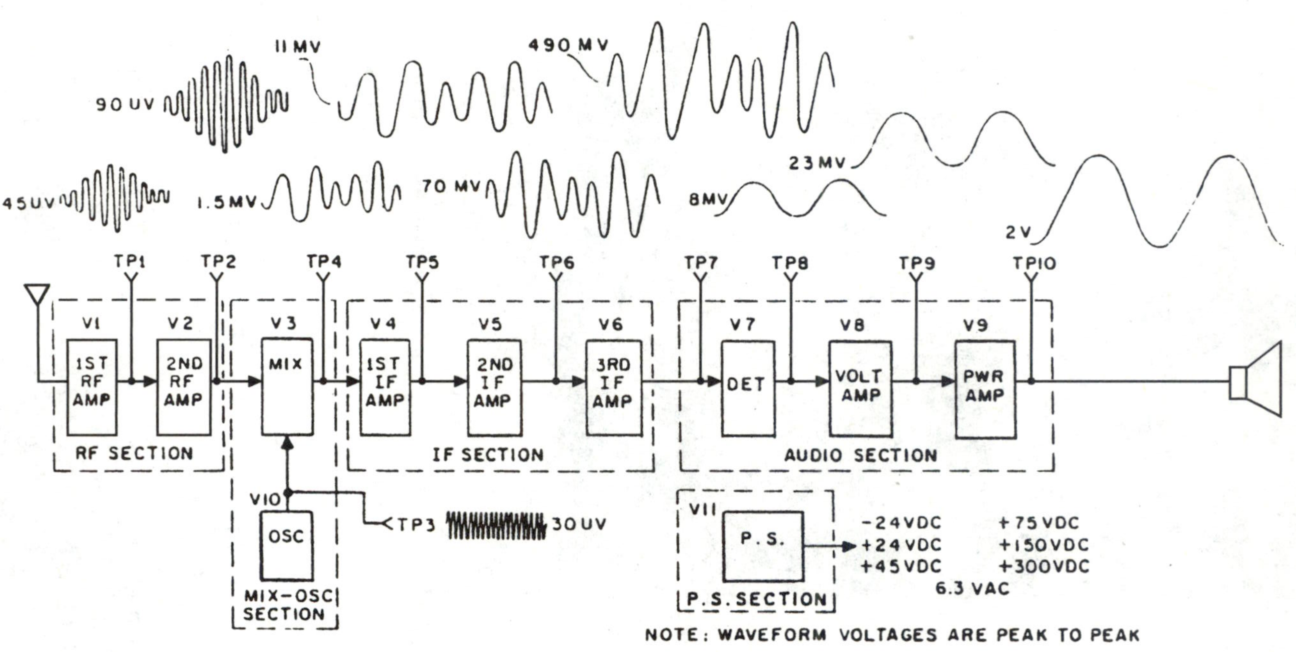

■ **8-13** *Receiver servicing block diagram.*

on the output of the defective stage and the system works. Signal injection on the input of the same stage will result in a receiver with no audio output. With this action, you now know that this is the defective stage.

At this point it is time to shift to a detailed schematic diagram, as shown in Figure 8-14. This drawing will have all components shown. Using this, you will measure the voltages and observe the waveforms around the active component. As an example, if the audio amplifier was not working and you measured no collector/plate voltage, then that would indicate an open load resistor.

A slightly more challenging problem would be a distorted output. In this case you would use a signal generator, your ear, and possibly an oscilloscope. From your background you would know that the most likely cause would be an open bypass capacitor. By signal injection at logical points throughout the receiver, you would soon have it isolated to a single stage.

Transmitter troubleshooting is more challenging due to the voltages and currents that you will encounter. Because of this, safety is a paramount conclusion. Failure to adhere to acceptable safety standards will result in an accident. Figure 8-15 is a picture of the RF section of a transmitter. Transmitter troubleshooting calls for a greater reliance on front panel meters and switches. Often, the final output stage has a meter wired in so that you can observe the operation of the tubes/transistors. Through the use of a chart in the technical manual, you can rapidly determine how the final stages have failed.

For personal and equipment safety, it is recommended that you use the front panel meters and lights to their fullest extent. Knowing what the indications are telling you will greatly diminish your troubleshooting efforts. The following are several examples that illustrate what the meters can tell you.

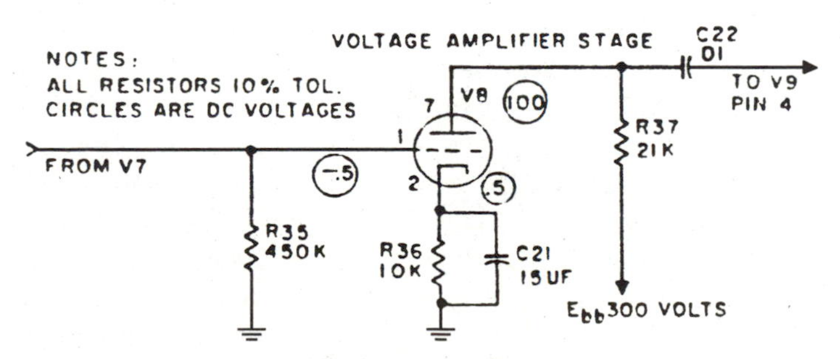

■ **8-14** *Receiver detailed schematic.*

A very important meter often encountered in transmitters is the *plate current meter.* If it reads lower than normal, that could indicate several different problems. First, if it reads zero or almost zero, that could be a dead power supply. It could also indicate improper tuning—a recurring problem. You should check the other front panel meters to see if the bias voltage is high, the filament voltage low, or the screen voltage low.

A high plate current reading could once again be improper tuning. Somewhat less common would be an excessive coupling problem between the stages. Finally, check the other meters to see if the bias voltage is low or the plate voltage is high.

The oscillator stage is also often equipped with front panel meters to allow you to safely observe system operation. In this stage, if the plate current is high, that would indicate a lack of oscillations. That could be caused by detuning or a shorted tube. If the plate current is low, then that would indicate that the following stage is defective and not drawing any current. A zero reading on the meter could be an open oscillator tube.

With vacuum tube equipment you will find that 80% of your problems are tubes. Elements can open or short together. Tube sockets, due to heat, can develop open pins and shorts to the chassis. Be aware of any arcs across insulating material or components. Capacitors can age or dry out, causing a dramatic shift in value. An overvoltage condition can cause the dielectric to short out and arc to ground. With all components, look for burns and signs of heating. Resistors are prone to shifting value due to age and heat. Transformers can short out due to damp conditions. That is because moisture results in oxidation, causing the insulation to break down, leading to arcs.

If the problem is not in the final stages, then as with the receiver you would use a functional block diagram and schematic to isolate the failed area. In this type of equipment, arcs are a definite possibility. Visual inspection becomes very important. If a complaint is a noisy signal or periodic arcing, observing transmitter operation is a good choice. Sometimes high-voltage cables are too close together. When that happens, arcing can occur between cables. Another problem is the insulation of the high-voltage cables breaking down. This would call for observing the equipment observation with the lights off—from a safe distance. The action of the insulation breaking down allows the development of a bluish corona around the damaged cable. This is not a repairable item; just replace it. Another problem associated with high-voltage sections is when

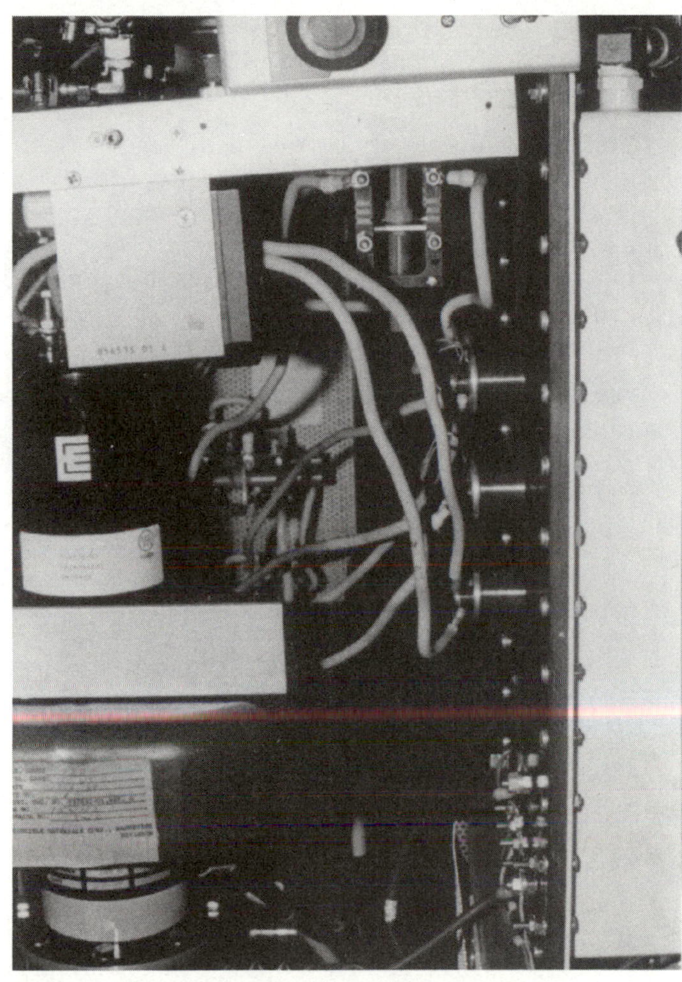

■ **8-15** *RF equipment cabinet interior.*

an arc burns a track in an insulated covering. You now have a high-resistance path to ground that will lead to further arcs in the future. The best idea is to replace it.

Antenna Maintenance and Troubleshooting

Antennas are typically a low-maintenance item. The two most common problems will be cleanliness and deterioration. If the insulators are allowed to have a dirt buildup, then a high-resistance path to ground will develop. In the case of a receiving antenna, you will notice a signal loss. If severe enough, only the strongest signals will be readable. It will cause a more serious problem with a transmitter. The most obvious symptom is when the Plate Current

Meter in the RFA or final stage is reading much higher than normal or pegged. If you detect this problem—shut down. Damage to the transmitter could result.

This problem is easily corrected through periodic preventative maintenance. Simply keep the insulators clean. The best method is with water and a sponge. Another part of preventative maintenance is to monitor the VSWR of the antenna system. If it begins to increase, then you have a problem in the making. It could be caused by dirty insulators, a cable breaking down, a corroded antenna, or defective connectors.

Cable connectors must be made up properly and sealed. If moisture is allowed to enter the assembly, then pins and jacks will corrode, increasing the overall impedance. Standing waves result. If the connectors are not repaired, then total failure follows, resulting in a downed system.

Summary

Effective electronic equipment maintenance is the result of careful preparation. You must know your basic theory, equipment capabilities, layout, and technical documentation. It is not necessary for you to know all the details, just where to find the information when you need it. As long as you have a good foundation in safety, the basic electronic principles, troubleshooting techniques, and situational awareness, training on individual devices is not required. The ideal technician is an individual who is knowledgeable, flexible, and who has a broad technical background. Due to the proliferation of communications equipment and technologies, the future is expanding in this interesting field of electronics.

9 Equipment Installations

Installation Considerations

A COMMUNICATIONS SYSTEM CAN RANGE FROM A SMALL, handheld transceiver to a high-powered, state-of-the-art, computer-controlled HF intercontinental transmitter with receivers around the world, or a global cellular phone network. The invisible web of communications links that bind us together are established and maintained by the US government, commercial communications corporations, internal corporate systems, and private organizations. News networks require virtually instantaneous communications from any point in the world, the financial centers must be accurately linked, and international transfer of data, funds, and entertainment depend upon a bewildering array of communications systems. Due to the explosion of advanced technology and the resulting low-cost communications applications, there are probably as many communications systems as there are reasons to talk, be entertained, and exchange data.

For the casual user, a personal communications system is very easy to install. All a handheld VHF or UHF unit requires is a high spot, not blocked by structures and hills. The greatest concern is ensuring a supply of fresh batteries. A mobile unit mounted in a vehicle is a little more complex. For best service, the antenna must be mounted so as not to be blocked by the metal body of the vehicle. As an example, if a UHF or VHF antenna were mounted on the back of a large camper, then some range would be lost in the forward direction due to the bulk of the vehicle blocking the radiated energy. The most effective mounting position is in the center of the roof. In this type of installation, the large metal surface acts as a ground plane, providing the best antenna radiation pattern. The downside is that a hole must be drilled through the roof to provide an entrance to the vehicle for the cable. If that installation is not practical, then the trunk lid is the next best location. The least satisfactory places are either a window mount, or on the rear bumper.

As with any communications system, a good ground is essential for proper equipment operation. An improperly grounded antenna can allow it to function as a diode, producing spurious radiated signals that will interfere with other two-way systems. After the antenna, the transceiver must be mounted inside the vehicle. Position it so that all controls are accessible while you are driving. Another good point is to anchor the unit. If it is allowed to vibrate due to a loose installation, then that will lead to component failures. A solid mount is highly desirable, as you do not want the radio becoming a missile hazard in the event of an accident. To complete a professional installation, follow all instructions found in the manual for best effect. If you don't have one, then numerous excellent books are available from communications organizations such as the ARRL, REACT, and *CQ* magazine.

A home base station installation is very simple with the new systems available today. Even a fairly high-powered ham HF system would require only a 220 VAC service, which can be pulled in any home. If a 220 VAC line is required, unless you are qualified to perform the work, it is best to have the modification performed by a licenced electrician. This ensures that the job is completed correctly and according to all applicable local and national codes. Any communications unit, including a small base-station CB transceiver, needs a little planning before starting. The area designated for its location should be uncluttered, with excellent air circulation. Many electronic systems, such as a communications transceiver or home computer, are capable of generating a great deal of heat. It should be placed close to where the antenna transmission line will enter the building. The shorter the transmission line, the lower the losses you will experience. The entry point for the cable should then be sealed to block moisture.

Any home personal communications system will require an outside antenna, and this is the problem area. Do it right and do it safely. Currently, many homes that you see are equipped with antennas for satellite TV, CB radio, ham radio, or a shortwave radio. Regardless of the type of antenna, there are many considerations that are common to all, whether a home, a military, or a commercial installation.

The first and most important point of antenna installation is safety. Accidents resulting from installing antennas, masts, or antenna towers are an all-too-frequent occurrence that can be easily avoided with a few easy-to-follow tips. To begin with, you must survey the area where it is to be placed. Trees, nearby buildings, telephone lines, cable TV lines, and power lines can all cause difficulties. The most important consideration is to ensure that there are no nearby power lines. Even if the clearance does appear to be sufficient, take into account the effects of high winds on antenna mast and power line motion. Another point to remember is to give yourself enough clearance to perform any hookups or maintenance on the antenna. All parts of a communications system, including the antenna, require a certain level of periodic maintenance to ensure proper operation. If the antenna or tower is above a certain length, then you must use guy wires to provide the needed stability. Therefore, you must ensure that there is enough space for the guy wires as well. Guy wires must be installed properly. At a minimum, guy wires should be evenly spaced in at least three equidistant points around the mast. It has been determined that a 45-degree angle between the surface of the ground and the guy wire provides the best support with the least amount of strain on both mast and wire. Also, there should be guy wires for every 10 feet of mast. As the wind will have a great effect on a tower, the guy wires must be securely anchored to prevent the wires from pulling loose and subsequent tower collapse. The simplest type to use is an anchor that screws into the ground. It may be familiar to you as the type used to anchor a dog's chain. The physical size of the tower will determine the size and depth of the anchor. Another type of anchor is to set the guy wire in the ground with concrete. After digging a hole to the desired depth, wrap the wire around a section of pipe or pass it through a concrete block. Then fill the hole to within 2 or 3 inches of the top with concrete. That will allow enough soil for grass to grow over the concrete.

To provide for the proper tightness of the guys, turnbuckles are used. If more than one set of guys is required, tighten the lower ones first, working your way up the mast or tower. It is vital that the mast is vertical. If the wires are overtightened, or tightened in the wrong order, then the mast could bend out of vertical, leading to collapse. As guy wires are metal, insulation is needed. That is because if the length of the guy wire is anywhere close to the required antenna length of your transmitter, it will intercept and reradiate the RF energy. That can have a detrimental effect on the radiation pattern of the antenna. The best way to prevent this effect is to break the guy wire length up so that it is does not reradiate RF energy as an antenna. One way is to insert a ceramic insulator 1 foot from where it connects to the mast, and another 1 foot above the ground. *If the guy is long enough, then insert additional insulators so that no one section of the guy is divisible by 16 or 22.*

While installing an antenna or mast is not difficult, a tower is a different story. Due to the cost of the metal structure, physical length, and difficulty in handling this large an object, it is not for the uninitiated. Many localities will require a building permit before installation can begin. Also, due to local conditions, there may be a height restriction on towers and masts. As concrete footings are needed to provide a stable base, having a professional do the job is preferable. Proper installation techniques can be found in many books; however, there is a great deal of practice and skill involved. If in doubt, contact an amateur communications organization or a professional for assistance.

If you have never installed an antenna before, do not learn by installing a mast or large array by yourself. There is no loss of honor in asking for help or obtaining the services of a professional. The cost of a professional job will be far less than what would be incurred by an accident due to lack of skill. Therefore, an antenna tower installation should not be attempted unless you have sufficient knowledge and tools. Any ladders or scaffolding used should be appropriate for the job. If power lines are close, then wood is preferable to metal. Antennas that require some assembly should have as much of it as possible completed on the ground. Carrying it up the ladder might be a challenge in balance. Additionally, any stiff breeze can create a safety hazard. Carrying an antenna assembly up a ladder in these conditions requires the actions of at least two individuals: one on the ground and one on the ladder. It is preferable to first get yourself into position on the ladder, securely anchored with a safety harness, then haul the array up with a rope or block and tackle. If a power line is nearby, do not use a metal ladder, as striking the power line, even a glancing blow, will lead to a severe shock hazard. If part of the antenna or guy wires does come into contact with a power line, do not touch it or attempt to remove it in any way. Call the power company to correct the problem.

An outside antenna attracts the desired electromagnetic radiation, and possibly the unwanted attention of lightning. An outdoor antenna, whether mounted on the roof, a mast, or a tower, can attract lightning because, as it is one of the highest points around, it acts as a lightning rod. A lightning strike on an antenna system causes discharge current to travel down the mast and the lead-in cable to the electronic device. To prevent damage, the mast or tower must have a good ground. A metal ground rod, designed and recommended for the purpose, should be used for the earth ground connection point. Unless you are familiar with the proper procedures for establishing an earth ground, have a licenced electrician perform the task, as it might not be as simple as just driving a metal rod into the ground. First, national electric code recommends a depth of at least eight feet. There are locations where that depth might prove to be insufficient. The depth to which the ground rod must be sunk is determined by the conductivity of the earth. It can vary greatly from one place to another. As an example, Kennedy Space Center and Cape Canaveral Air Force Station are located on the same island. On the KSC side of the installation, a good ground can be obtained by using 8 feet of ground rod. Ten miles away, on the Air Force Station, the ground in some locations must be at least 20 feet deep.

The conductor that is used to connect the antenna mast or tower to earth ground is very important. Aluminum wire is not recommended for a conductor, as it is not an efficient conductor and is subject to corrosion. The best solution is a copper strap rather than a wire or cable. It should be connected to the base of the mast, following the shortest path to the ground rod. When you pull the strap, there should be no sharp bends or complex paths, as the current resulting from a lightning strike could find a lower resistance path to ground by arcing from the strap to a part of a structure, or to other cables.

Even with a grounded mast, an antenna or tower lightning strike can still cause damage to installed equipment unless further protection in the form of a surge eliminator or static discharge component is used. Designed in different body styles and impedances, the devices are easy to install. Basically all it is is a pair of metal contacts separated by an air gap. One contact is inserted in the antenna transmission line, and forms part of the path for the RF energy to follow from the antenna to the electronic equipment. The other metal plate is connected to earth ground. As long as just RF energy is present on the line, the only path for conduction is between the antenna and the equipment. If there is a lightning strike, the air gap breaks down, providing a low-resistance path to ground for the massive surge of electrical energy. An added plus is that if there is a buildup of static electricity on the antenna for any reason, then before it reaches harmful levels, the air gap breaks down, conducting it to earth ground. Frequent inspections of the devices are required, as a lightning strike could possibly destroy it, rendering it useless.

The transmission line used to connect the antenna to the equipment is of prime consideration. You must use the proper size for the installation. RG-8/U or RG-213/U are two excellent choices for most applications. With transmitting equipment, do not forget to periodically check the VSWR. An increase in reflections would indicate antenna and transmission-line problems before they impact operation. Another maintenance consideration is the cleanliness of any insulators installed on antennas. If they become dirty, it provides a high-resistance path to ground, degrading performance.

Antenna installation practices for the home communications hobbyist can be found in several fine publications by McGraw-Hill, the ARRL, CQ Publications, and other sources. As antennas are typically ordered from a communications supply firm, instructions in the form of manuals and instructional videos are often available.

The past few years have seen a dramatic increase in the numbers of private boats and small craft in use. Any trip past a marina will reveal large numbers of small craft equipped with radar, cellular phones, advanced communications equipment, and navigation systems. As with any other installation, a small boat has its own set of problems.

Moisture is the first consideration. Do not use any equipment that has not been certified for marine applications. Conditions found in very large lakes and the oceans are harsh and unforgiving. Documentation with the systems will provide installations requirements—follow them. Equipment cases should be at least drip proof, if not waterproof. Due to the damp conditions and possibility of coming in contact with water, you need to minimize shock hazards. Moisture will find the smallest opening, entering equipment and causing problems.

Another often-overlooked consideration is electrical power. To begin with, ground fault interrupters are very important for safety. That is because in the marine environment, any fault can place AC on exposed metal surfaces. Another consideration is the installation of an isolation transformer in steel-hulled vessels. This is useful when taking shore power from a marina. If the craft is operated outside of U.S. waters, then you must also be able to connect to whatever type of power is available. Other countries have different values of AC and wiring color codes. As electronic equipment is prone to interference from electrically noisy generators and alternators, noise suppressors should be installed on all offending components.

With the proliferation of electronics aids and communications systems, the small craft must have sufficient generating capacity to operate the equipment. If the only way systems can be operated is through load sharing, that is unsatisfactory, as you will never know when the communications, navigation systems, and radar might have to be operated simultaneously. You must also consider the use of other items, such as cooking appliances and environmental systems, when figuring loads.

As with shore installations, the electrical ground is also of paramount consideration. You must follow all manufacturer's requirements and electrical code rules. As small craft have AC and DC power systems, there is one very important point to remember—never bond AC ground to the negative terminal of the DC system, AC neutral, RF ground, or a lightning-protection system. That is because if there is an AC fault, then the ground could float, causing shock hazards off of metal fittings, the hull, and the mast. If anyone does advise you to connect DC negative to AC ground, it is a potential safety hazard and should never be done.

The RF ground is very important for the proper operation of all electronic systems. The keel makes an excellent ground point, but it is physically long and quite expensive to run copper that length. Ground shoes mounted on the bow and stern are good points. With fiberglass craft, copper plates can be pressed between the layers. Another type of ground is made by pulling a copper strap from the antenna mast to the ground plate.

In aviation and marine installations, antenna placement is often a challenge of monumental proportions. As many craft are equipped with HF, VHF, UHF, and cellular communications systems as well as radar and global positioning system (GPS), a small craft can be a forest of antennas. The GPS is the most sensitive system. The antenna must be placed outside the beam of any installed navigational radars. It must be clear of any HF and VHF communications antennas to prevent interactions.

Marine antenna placement might be a series of trade-offs. On sailboats, the masts, due to metal construction and rigging, can cause RF losses. Transmission lines passing through metal decks can suffer from induction losses unless through-deck insulators are used. As with any antenna installation, the VSWR should be checked periodically. Antenna insulator cleanliness is of paramount consideration. Any dirt or salt buildup will provide a high-resistance path to ground, degrading system performance.

Shipboard Installations

The greatest problem encountered in shipboard installation is antenna position. With the proliferation of high-quality, low-cost electronics, even small pleasure craft are not immune from this RF interference problem. The most electronically crowded installations are warships. The most obvious antennas on ships are radars that are used for navigation, air traffic control, threat detection, and weapons direction. Inserted among this virtual metal forest are the communications antennas consisting of long wires, whips, UHF arrays, VHF arrays, and GPS antennas. Installation problems arise, as ships are very crowded and become more so with each passing year. With new advances in technology comes more electronic equipment required for basic operation. In addition to the numerous communications antennas, ship's masts, cranes, and rigging all vie for a share of the limited space.

Since the beginning of WWII, military ships have become floating forests of antennas. Figure 9-1 illustrates a typical large auxiliary ship, a U.S. Navy oiler. Built to commercial standards, the ship must be equipped with a range of electronic systems. Although the vessel is large, designers actually were working within space constrictions to properly position rigging, booms, cranes, and all antennas, including radar, electronics warfare, satellite navigation, and communications antennas. An additional concern is that separate antennas were sometimes called for in reception and transmission, so as to minimize interference with other systems. Care must also be taken to ensure that radiated energy isn't picked up by rigging and masts through placement and bonding. If rigging, masts, and other metal items are not properly grounded, then they can act as a receiving antenna and become an electrical safety hazard to personnel.

Typically, search radar antennas are mounted on the highest mast or superstructure to give the greatest radar horizon. In Figure 9-1 the antenna is located at the highest point and looks like a horizontal orange peel. With this location, it is provided with a complete 360-degree field of view. On the horizontal bar just below the radar antenna are several UHF antennas. Just below and forward of the mast is a block-like antenna. A second one is mounted just to the left and below it. That is the satellite communications antenna for data transfer. To complicate matters, two radar-directed automatic cannons are fitted to the ship for self-defense. One, visible on the bow, is shaped like a white dome or can. As it is equipped with a small radar for direction, it provides an additional radiation and interference problem that must be considered. Scattered about the ship are several 32-foot whip antennas that are used for HF communications. As can be seen from the photograph, antennas are mounted in virtually every conceivable location. Antenna tuners are used to tune the whips to the exact electrical length to provide for optimum

■ **9-1** *Antenna installation on a U.S. Navy oiler.*

impedance matching. This aspect of ship design is an exact science, performed to minimize electronic interference. Modern military and commercial maritime activities require that any ship must be capable of exploiting virtually every portion of the electromagnetic spectrum. Advances in satellite communications and data links indicate that more, not less, communications equipment will be found installed on ships in the near future. Additionally, in military electronics, current trends indicate that infrared and ultraviolet portions of the spectrum will be used to a greater degree, resulting in more equipment and antennas.

In shipboard installations, there is a great deal of tradeoff between equipment and antenna placement. Ideally, the electronic equipment would be installed as close to its applicable antenna as possible to keep transmission lines as short as possible. The shorter the transmission lines, the lower the inherent line losses. Longer cable runs are a problem in ships, as they represent weight and cost that reduces other commodities that can be carried as payload. However, the antennas often must be positioned in any available space that does not cause RF interference with another system. As you can imagine, equipment and antenna locations and the resulting interactions are an art practiced by ship designers.

Space is at a premium throughout the ship. It is a constant effort to provide sufficient space for equipment to fit and still allow access for maintenance personnel to service and remove large subassemblies. Also, space for cable runs, waveguides, power panels, and ventilation must be accounted for. Poor ventilation has caused many needless problems due to excessive heat and humidity, contributing to shortened component life and added maintenance hours. Excess humidity, a leading cause of corrosion, damages switches and relay contacts. Equipment cabinets must be mounted so as to not block exits, access to other devices, and power panels (surprisingly, such blockage is a fairly common occurrence). In any aspect of installing equipment, preplanning will eliminate many future problems, such as having to move equipment cabinets in order to reach other devices for equipment adjustments.

Typical military shipboard installations are spread among several different rooms or spaces. To the casual observer, the only external sign of an electronic communications or radar installation is the antenna system. On surface ships, such as cruisers, destroyers, amphibious ships, and auxiliaries, the communications transmitters, receivers, patch panels, antenna tuners, and antennas are located in several different spaces. Radio central is the control room that provides the switches and panels to make it all work. High-powered HF transmitters are in a separate space to provide for electromagnetic isolation and noise reduction. The more high-powered equipment in a space, the more noise and heat is generated. Receivers are often located in a second space. If encryption equipment is involved in an installation, it will be isolated for security purposes. Any teletypes for fax and data are usually maintained in radio central.

Shore Installations

Although you might find this odd, a major communications system installed on land can be almost as challenging as one on board a ship or aircraft. First are the environmental concerns for both humans and the equipment. Although it hasn't been completely proven, it appears possible that exposure to low-level radio-frequency energy emitted by high-powered radar and communications systems can cause health problems. Because of this concern, major sites of electromagnetic radiation are as isolated as possible. Also, terrain has an impact on signal propagation. To provide unobstructed coverage in the desired direction(s), the antennas shouldn't be placed behind hills or in low spots. A hill blocking the radiation pattern from an antenna would result in a blind spot.

For a high-powered commercial or military communications system, possibly the most important consideration is the antenna location. A service is established to provide reliable coverage over a specific geographic area. To provide for proper propagation of the radiated signal, the antenna must have a clear and unobstructed view that is not blocked by structures or topographical features. The radiation patterns and frequencies of any other electronically close systems must also be taken into consideration. In AM and FM broadcast radio, the term *clear channel* is sometimes used. What this refers to is the primary and secondary service areas of a broadcast station. The primary service area is the geographical region where reception is primarily through the ground wave. The secondary service area is where reception is through the sky wave. When a station is given a clear channel, that means that in the primary and secondary service areas it will be the only broadcast on that and adjacent frequencies to prevent interference.

To provide the required service area, antenna height is vital. In a transmitting antenna, the effective antenna height is defined as the center of the radiation pattern above the surrounding average terrain. To accomplish this, the elevation above mean sea level is measured at the 0-, 2-, 4-, 6-, 8-, and 10-mile points. This is repeated for eight radials originating at the antenna. The radials are separated by 45 degrees. The resulting 48 altitude measurements are then averaged to obtain one value. The effective height is then stated to be the altitude of the center of the radiation pattern minus the average height.

With transmitting antennas, ground is very important. As an example, a half-wave Marconi antenna uses a ground as the other half of the antenna's effective length. It has been found that the best fabricated grounds are copper radials one-quarter wavelength long, spaced every 5 to 15 degrees. However, the finest possible ground is a naturally occurring salt marsh.

It is not uncommon for the transmitter site, receiver site, and the communications control site to be separated by several miles. This is to provide for a minimum of interference between the activities. Airports, both civilian and military, are very challenging. The tower, transmitter site, and receiver site typically are on almost opposite sides of the installation. This ensures that there will not be any RF interference between the transmitters, receivers, and other electronic systems. As the installation may comprise over 20 transmitters and 20 receivers, the chance for interaction if the devices are in close proximity are quite good. Newer aviation installations are often equipped with transceivers rather than separate receivers and transmitters. In this case, the HF, UHF, and VHF systems are often separated to prevent possible electromagnetic interactions. As the communications links are used by the air traffic controllers in the tower, all devices are interconnected by a web of long-distance cables that run under the field. If the distance are great enough, then microwave interconnection systems could be used. Other electronic systems are scattered around the field, ensuring sufficient separation.

If the equipment is installed in a manufacturer-supplied and approved shelter, then there will be far fewer challenges and problems. As many devices are installed in permanent buildings, then planning and care must be taken on the part of those doing the planning. Manufacturers provide all the information required to complete a practical installation. Conduit installation for power, control, and possibly transmission cables are a problem area. When buildings are constructed, there is pressure to keep costs down. Many times the number and sizes of installed conduits are marginal, or are sized for the current equipment installation. As you have noticed, due to technological advances, the number of devices in any type of installation appears to be expanding. New requirements or capabilities are added, placing a strain on the structure. Although some money might be saved on construction costs, insufficient conduit space can make future installations prohibitive, as the conduit is often encased in the concrete floors. If another conduit is required, then the floor must be torn up, tubing replaced, and the floor repaired. One facility that I worked in had its allowance of equipment increased. As a result, the conduit was barely large enough to allow all the cables to pass through. The last one we pulled through was a challenge that took far longer than expected.

Ventilation and cooling are critical in any electronic installation. I have personal experience fighting a corrosion problem caused by humidity for four years. I suffered through intermittent problems, chattering relays, disintegrating metal parts, and degraded equipment performance due to uncontrollable rust. Upon investigation, an environmental health inspection revealed that the building had only half of the fresh air flow that it required. As we could do little in that regard, the addition of a dehumidifier cured the problem.

Another consideration is to keep transmission line lengths as short as possible. These lines represent a loss to both output power and received energy. Any cables should be installed in such a manner as to eliminate their use as handholds, to prevent any resulting strain that could lead to conductor failure. Cable runs protect the cabling and give a much neater appearance. If a system operates at a high enough frequency, then waveguides can be used to conduct RF energy rather than transmission lines. As with the transmission lines, to reduce unwanted losses, lengths should be as short and straight as possible. Bends and twists in a waveguide can cause signal reflections, which result in losses that will reduce the power output of the system. Also, waveguides should be installed so that personnel can't use them for handholds and supports. It is also helpful to paint "DO NOT STRIKE" on waveguides, as paint crews might want to use a chipper to remove dead paint before refinishing. Any resulting dents and small bends can cause reflections, which in turn leads to reduced operation.

There are general considerations for all electronic equipment installations. Although they are addressed in the national electric code and numerous equipment manuals and electronics books, often they are not followed. All interconnecting cables must be routed to prevent physical interference with any other systems, doorways, vents, and panels. This will prevent chafing and wear on the outer sheathing. You must always use the properly rated cabling for primary AC power. Underrated cable leads to many problems, including major electrical and fire safety hazards. It is essential that all cables for both power and signals are secured with cable trays, stringers, hangers, or conduit. In addition to a neater installation, it prevents premature cable wear. Any interconnecting cables should be routed to prevent their use as handholds. Also, cables and conduit are not to be used as supports for other cables or pieces of equipment.

An often-overlooked point with any installation is the use of cable labels. Many advanced electronic systems are composed of several different subassemblies, chassis, or components. Factory-assembled cables are usually marked or coded so that the technician can tell what they are to interconnect. With age and use, labels are lost or illegible. Also, cables do fail, and have to be replaced. Ensure that all cables are properly marked, and when making new cables or replacing missing labels, use the identical markings as shown on the schematics. If none are indicated on the prints, then use a generic type that any technician can read and understand. A good example would be the signal cable connecting the receiver to a recorder or teletype. On the receiver end, mark it *AUDIO TO RECORDER.* On the recorder end, mark it *AUDIO FROM RECEIVER.*

Wiring diagrams of how the system is actually interconnected are very important to failure isolation. If you must add or remove cables due to a design change or capability addition, mark it on the schematics. First, you might not always be there, due to time off or a career change. Secondly, your memory will have a tendency to fail you when you need to remember the exact change that you made three years previous.

Often you will have to make your own cables for audio, video, and system interconnections. When fabricating your own cables, the soldered type are superior to the solderless. The mechanical strength of the soldered connector is much better, resulting in fewer future failures. When you do fabricate cables, perform a quick continuity check before installing. That will rapidly detect an improperly assembled connector before any time has been expended in installation.

Finally, and most importantly, the system must have a proper ground. Any equipment without an adequate ground is a safety hazard to personnel. Missing or poor grounds can cause power supplies to float, leading to component damage. Other circuits can fail to function, with odd symptoms that are time-consuming to correct.

It is safe to say that there are no two identical communications installations. The rapid replacement of equipment, changing technologies, and requirements are the driving factors. However, there are many common points that apply to the operation of all communications installations, such as safety, professionalism, and technical competence. With the continuing changes, it will be a challenging profession for years to come.

Personal Communications and the Future

10

Personal Communications

IN EXAMINING ALL THE COMMUNICATIONS ADVANCES OVER the past 40 years, one might feel as if we have reached a plateau. Nothing could be further from the facts. What would have been considered as science fiction on the early "Star Trek" shows is becoming science fact. Digital technology, miniaturized components, and feature-laden communications packages have fueled a revolution in portable and affordable personal communications. This complex electronic web that now interconnects us all will continue to expand for years to come, providing satisfying careers for many technicians. All communications services can be placed in one of two categories. The first is one-way broadcasts. That would be TV, AM radio, FM radio, and shortwave broadcasters. The second category is two-way services. That would include private land mobile, public safety, emergency, industrial, business, and personal communications. Here are a few of the more significant forms of personal communications now in use.

CB Radio

Citizens' Band, the first widespread form of a radio-based personal communications system, has been popular since 1959. Initially consisting of 23 channels, it operates in the frequency range of 26.965 MHz to 27.405 MHz, the old radio amateur 11-meter band. Utilizing AM voice only, with a maximum transmitter output power of four watts, it was meant to fill a void in inexpensive personal communications that anyone could use. Equipment was limited to transceivers designed for handheld, mobile, and base-station operation. In the early years, the FCC had a licencing requirement that was meant to keep track of the users by issuing call letters. With the low output power, it was intended for local HF communications links only. Growing very slowly, it finally exploded in the early-to-mid seventies. Its almost universal popularity was fed by movies, music, and millions of operators that permeated every part of our society. As useful as the communications link was, and is, the band space allotted was insufficient for the tremendous number of people who wanted to use it. Due to the unmanageable crowding of the channels, many lost interest in it, reducing the number of radios to a level that could comfortably handle the frequencies.

By 1982, in response to the limited band space, CB had been expanded from the original 23 AM channels to 40 channels for SSB operation. With the increase in channels, the SSB transceivers were allowed an output of 12 watts PEP. The equipment manufacturers responded by designing new SSB equipment with all the latest features. With SSB, the quality of the communications links improved dramatically. Dropping the requirement for a personal license, the service can now be used for business, personal communications, and emergency links. Regulations are held to a minimum for the group of communicators who use this band. The equipment must not be used to drive a high-powered amplifier that would increase output power above the legal maximum. Antenna height is limited to 20 feet above the building the equipment is installed in, or a maximum of 60 feet if a tower is used. For operations, the primary rule is that it cannot be used for illegal activities, entertainment services, advertising for businesses, or political purposes. Another requirement is that the two-way communications links are not supposed to be over distances greater than 155 miles or between nations. This last restriction is sometimes difficult to follow, because the frequencies allotted for CB communications are subject to skip conditions during periods of high sunspot activity. That results in reliable communications being possible in distances of several thousand miles.

Once more, CB radio appears to be enjoying a resurgence in popularity. With the latest technology incorporated into the new transceivers, it is easier than ever for the individual to begin operation. Currently, the installation and repair of CB equipment does not require an FCC license or certificate.

49 MHz

Just a few MHz up the radio spectrum is the 49-MHz communications band. This could be considered to be the little brother of CB radio. Other than power and frequency, there are almost no restrictions on this service. Meant for personal, business, or short-distance links, it is limited to equipment that is very small, lightweight, and inexpensive. With a maximum output power of .1 watt, it has a very limited range of about ¼ mile.

Due to its low power, small size, and low cost, it is ideal for large commercial activities for use as an internal communications system. Personal uses would include hikers, campers, and other outdoor enthusiasts. The radios are designed to operate on five channels in the frequency range of 49.83 MHz to 49.89 MHz.

Amateur Radio

For decades, amateur radio (or *ham radio*) has provided an outlet for the technically-inclined individual. Hobbyists were known for building their own equipment from salvaged parts and refurbishing ex-military equipment. Ham radio has roots going back as far as the beginning of radio itself. In fact, many of the technologies, equipments, frequencies, and procedures we now take for granted were first developed and exploited by hams the world over. Almost from the beginning, hams occupied radio spectrum space that was thought to have marginal commercial use. As a result, hams were the pioneers who helped electromagnetic communications grow. A difference between this and other communications services is that the operators have to be licensed to certify a knowledge of regulations, electronics theory, and Morse code. Because of the licensing requirements and the technical interests of the individuals, hams can transmit in Morse code, AM voice, SSB voice, UHF FM, television, and satellite communications.

Due to pressure brought about from the popularity of CB radio, the FCC relaxed the Morse code requirement for ham radio in the 1990s. The code requirement has been dropped for the entry level ticket, but remains for the higher-level licenses. Testing is now handled by local ham clubs, and several different organizations offer low-cost study guides. The basic license is the Technician class, which allows two-way communications on frequencies above 30 MHz. Three higher license grades open up the HF bands, which offer the possibilities of intercontinental communications. Amateur radio stations are limited to 1 kW in output power, and the communications links can be used only for personal and emergency uses.

As with all other aspects of electronics, ham radio has benefited from the electronics revolution. Figure 10-1 is a 2-meter mobile ham transceiver that illustrates the advances. Very small in size, it

(Photograph Courtesy of Tandy Corporation)

■ **10-1** *Radio Shack 2-meter radio amateur mobile transceiver.*

is still packed with features that were unheard of 20 years ago. A Radio Shack product, it is designed for ease of installation, flexible operation, and years of faithful service. It is a fine example of what is in use today.

Business Radio

Found in two separate frequency bands, 150 MHz and 460 MHz, the business radio service is set up for business purposes. This type of communications link is licensed by the FCC. The exact frequencies are 151.625 MHz, 154.57 MHz, 154.6 MHz, 464.5 MHz, 464.55 MHz, 469.5 MHz, and 469.55 MHz. Although it is labeled for commercial uses, many part-time businesses use it for their communications needs. Very compact and easy to use, the units are ideal for landscaping, golf courses, farmers, internal business communications, and other small firms.

General Mobile Radio Service

The 462-MHz band is the home of two very interesting communications services, the first of which is the *general mobile radio service* (GMRS). In its original form, when it was known as Class A

CB, it was meant to be an FM version of CB radio in the UHF band. Gradually, due to changes in communications needs, it evolved into a small commercial service, a communications net for public safety communications organizations, and an outlet for informational notification groups. At this time, there are several thousand licensed repeaters in operation in these frequencies throughout the United States. With the use of repeaters, the low-power, short-ranged equipment can reach greater distances. A repeater is an automatic RF amplifier that is used to boost the level of weak signals and retransmit them to obtain a much greater range. For this to be possible, the repeaters are mounted on towers and masts; they can greatly extend the horizon of short-range, line-of-sight devices.

With a growing, more mobile, and more technically oriented population, there is pressure for additional personal communications systems. In response to this need, the GMRS band is slowly evolving to support another personal communications service that is so new that the equipment has just gone on sale. As the new service is unlicensed and the GMRS is licensed, there could be some conflict.

GMRS, unlike CB radio, must operate on a single frequency or repeater frequency. If required, due to business traffic, a licensee can request a second frequency, and it will be granted only if one is available in the area. There is one frequency, 462.675 MHz, which is reserved as a monitored travelers' aid service. When the service was first established, space was allotted between each active channel to eliminate the possibility of interference. With the advances in technology, the frequency space between the channels is being allotted to a new service, The Family Radio Service. Any currently licensed GMRS stations are going to be allowed to operate without change for the foreseeable future. However, no new licenses are going to be granted.

Family Radio Service

The *family radio service* (FRS) is an unlicensed, low-powered, direct communications system using voice communications in the 462-MHz range. Its frequency allocations are between those channels assigned to the GMRS. The handheld transceivers are limited to a maximum power output of .5 watts, which should be sufficient for its intended uses. The idea behind the service is to give casual users a low-powered UHF voice communications link to support outdoor activities. One of the few regulations concerning FRS is the fact that the little transceivers must have a permanently attached antenna. Without a means of installing a high-gain external antenna, the range should remain limited. Also, that would prohibit the use of a high-powered RF amplifier attached to the antenna jack. As it is currently envisioned, it will be a very useful service for many individuals.

Radio Broadcast Data System

Radio broadcast data system (RBDS) is a one-way service that allows commercial stations to transmit text data to displays that are an integral part of a radio receiver. This new technology has been active in Europe since 1985, and is currently used by over 400 U.S. stations, which is less than 5% of the total. Many experts in the field are predicting that it will become one of the more significant advances in radio technology.

The *radio data system* (RDS) allows an FM station to electronically tag a conventional RF signal with a data stream through the addition of a digital encoder. Suitably equipped receivers decode the data, displaying it on an alphanumeric readout format. Currently, RBDS receivers are capable of providing information such as the current station's call letters, song title, artist, and any other digital information. An added feature is that it can allow your receiver to automatically seek and tune to another station that is of a similar music or talk format. On a more serious note, emergency broadcasts, traffic reports, weather, news bulletins, and paging information can be provided as well. In short, it converts the humble radio into a complete information source.

There are nonlistener uses for the emerging capability. It can be used to activate remote-controlled billboards, saving energy and time. Another important use that will gain in popularity in the future is the ability to transmit GPS error-correction information. As more vehicles become GPS-capable, this service will become important. For the non-radio listener it can also function as a paging service and direct important data to a personal computer. In effect, anything that can appear as an alphanumeric text can be handled and directed by the system. As the system is independent of any other entertainment medium, such as CDS or cassettes, the message still appears on the face of the receiver, providing a convenient warning system for traffic, weather, and other emergency problems.

Although it is a very useful service, broadcasters and receiver manufacturers are going to have to market the technology to make it universal. To the north, Canada is allowing foreign investment to purchase shares in stations that it will upgrade to standards that can use this feature. As it is a low-cost, versatile technology adding only a few dollars to the cost of a radio, in time it will reach its full potential.

Cellular Telephone

Cellular telephone service is a personal communications system in the truest sense of the word. Due to the way it operates, it is a hybrid of the telephone and short-range, line-of-sight communications. If you have not realized it yet, people love to talk. Under current estimates, the cellular market in the United States alone is expanding by almost 50% a year. Another country, Israel, is now considered to be the most cellular nation, with the little communications wonders seen everywhere. With one of these tiny electronic marvels, one is always in contact with the rest of the world. From the first beginnings of the car radio telephone, we now have the virtually wristwatch-sized units that provide excellent service to many people.

The first mobile telephone system was based upon one centrally located high-powered transmitter that linked up with all mobile units that operated within the assigned service area. Based on early technology, the equipment was very bulky and had to be permanently installed in a vehicle. Not only expensive to purchase, it was expensive to use. Improvements in technology have led to the amazing system that we have today, called *cellular mobile radio telephone service* (CMRS), it was the developmental product of Bell Laboratories. As it is an offshoot of the telephone system, it supports all the features associated with a conventional land-line telephone. The cell phones in use today are fully portable and small enough to fit inside a shirt pocket.

Each cell site is an area of up to 24 miles in diameter that has its own centrally located low-power transmitter, receiver, and control system. Situated in the 850-MHz band, it consists of 832 separate channels per cell. Transmitter power is limited so that it can reliably link with a cell phone at any point within the cell, but not interfere with any adjacent cells. An additional safeguard against interference is that channel frequencies are different for neighboring cells. Cell phones are very low-powered units, currently 3 watts or less. The low power requirement is an advantage, as the devices can be in smaller and smaller packages, adding to consumer convenience.

What makes the CMRS system function efficiently is the action of frequency reuse. All cells in a region are interconnected by computer to a *mobile telecommunications switching office* (MTSO) or *network control switch* (NCS). The function of the computer is to monitor the signal strength of the each cell phone within its area of control. As a cell phone moves from one cell to another, the channel frequency is shifted to one in the next cell. As the channel shift is accomplished in microseconds, there is no noticeable interruption in the communications link. If the call is between two cell phones, then the cell site receives the call, and then routes it back out through another transmitter. A call to a conventional land-line telephone is routed by the central computer from the cell phone to the local telephone switching circuits, and then to the destination phone. With current technology, cell phones are effective offshore to a distance of about 20 miles. However, with the advances in electronics, satellites, and computers, this is about to change. In the not-too-distant future, cellular phone service will be available that will have the capability of calling from any location in the world.

Each cell site consists of an antenna farm and the base site building housing the receivers and transmitters. The completion of a cellular telephone call requires the use of three antennas. One antenna is used by the transmitter and is capable of supporting 16 channels simultaneously. The transmitting antennas are high-gain, omnidirectional, and vertically polarized. Due to the frequencies involved they are fairly small, only 13 feet in height. The two receiving antennas can be configured for either omnidirectional or directional coverage. The reason why a cell phone can be so small is because all the required control circuitry is shared by all users and is mounted in one central location.

Satellite Communications

Satellite communications and cellular telephones are about to merge, to the benefit of many consumers who are not in a land-based service area. Highlighted in the Persian Gulf War in 1991, the satellite phone is about to become accessible to everyone. A military program in the beginning, it has grown to interconnect our modern world. Figure 10-2 illustrates how the armed forces uses satellites to provide complete communications. It was discovered early in the research that a satellite communications system would offer unparalleled flexibility and security. Through the application of the technology, ground units, ships, aircraft, and shore bases can be in constant and instantaneous communications.

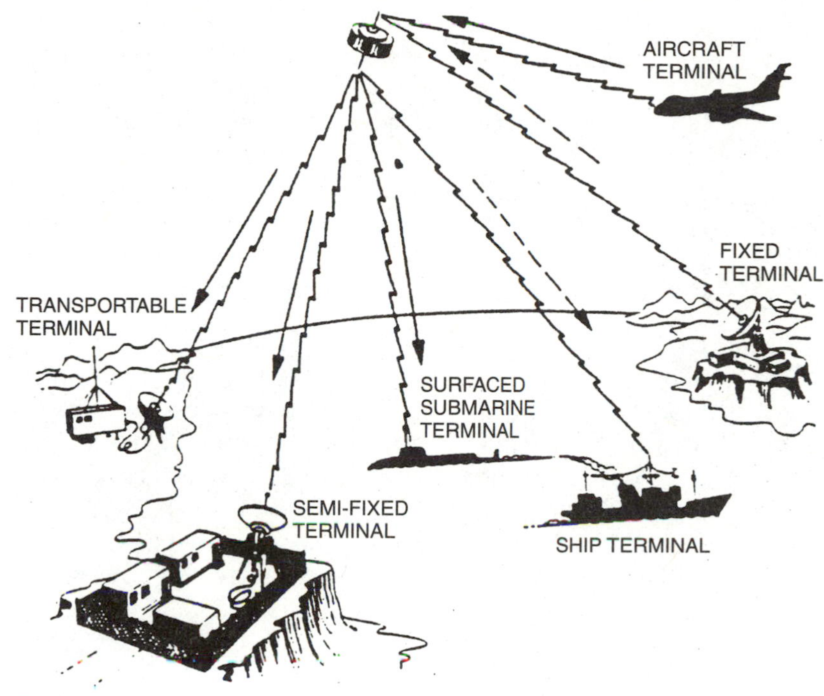

■ **10-2** *Representative U.S. Navy satellite communications network.*

Here is a brief history of satellite communications: The first communications satellite was the moon itself. Experiments by the U.S. Naval Research Laboratory in 1954 established an experimental communications link between Washington and Hawaii using the moon as a large reflector. Capable of handling voice and data information, it was used for priority traffic by the U.S. Navy. The first artificial communications satellite was launched in January of 1959. With an elliptical orbit ranging from 112 miles to 926 miles, it was a viable communications link for only 13 days, reentering the atmosphere on 21 January 1958. It served its purpose, as it moved satellite communications from science fiction to science fact.

Technology marched rapidly on, and experimental birds such as the Echo series pushed the envelope of understanding and capabilities. The first successful communications bird was launched in July of 1962; it was Telstar I. It was a 170-pound active communications satellite that functioned by using microwave frequencies to complete communications links with earthbound stations. Advanced for its time, it combined the emerging technologies of transistors, rechargeable NiCad batteries, and solar cells. Capable of voice, data, and video transmissions, it captured the imagination of the world.

The first geostationary communications satellite, Syncom 2, was operational in July of 1963. Weighing in at only 86 pounds, it was the first communications bird to support transpacific television links, the first of which took place in August of 1964. A viable commercial communications satellite was not placed in orbit until April, 1964, when IntelSat 1 was operational. From the halting first steps, the commercial communications market has heated up, with the U.S., France, Britain, Japan, and other countries establishing satellite-based communications links.

By using high frequencies that have short wavelengths, small, lightweight high-gain antennas can be used for the satellite communications systems. Early communications and navigation satellites had conventional orbits. That meant that the satellite was useable for only the length of time that it was in line with the ground station. The result was gaps in capability as the satellites moved out of range. As a stationary satellite could provide reliable and constant coverage, the geostationary orbit became the technological high ground, at an altitude of 22,000 nautical miles. Because the geostationary orbits are so high, a combination of high transmitter power and very sensitive receivers is required to ensure reliable communications links under all conditions. As the satellite transponders became more powerful, coverage expanded. With the satellites, ground stations, and shipboard installations in operation, communication from any point in the world to any other point is possible, but there is a weight and cost penalty. Earth installations to complete the voice and data links weigh about 200 pounds, are over four feet high, and are four feet in diameter. The cost of the ship-mounted equipment ranges from $18,000 to more than $30,000 depending upon the service and the type of intelligence.

Currently, there are four types of satellite communications systems that have a worldwide capability. Satcomm A is an older, analog system. It provides the user with voice, telex, data, and fax formats. An updated digital version is called Satcomm B. It provides all the services associated with Satcomm A, and the addition of high- and low-speed fax capabilities. A lower-cost system is Satcomm C. Using small equipment packages and antennas, it provides digital data and telex links. The final system is Satcomm M, which is considered to be a supplement to A and C. This version features voice and fax modes. It is a low-cost, physically small unit, but it is slower and has poorer voice quality. What these four satellite communications systems offer is the full range of communications possibilities, but they are clearly not in the price and size range of the average individual. However, for marine applications and anyone who has a need for worldwide telecommunications, they are effective. In the near future, improved versions of these services are about to be available.

Personal Pagers

Pagers are everywhere. The little plastic packages have become so low in cost that even children are now using them. The range of capabilities is amazing. The least complex ones alert you with a tone or vibration, and have you call a number to retrieve your message. The more advanced ones feature audio and alphanumeric readouts. So pervasive have these items become that pager notification systems are now in operation. These provide expanded services such as sports scores, emergency alerts, and stock quotes. It has reached the point where paging services are now in operation that specialize in alerting freelance photographers and news reporters of disasters and other sensational news events. In a reverse technology transfer, the U.S. Army is procuring pagers to use in military operations. With the vibration mode of notification, it provides a secure and silent way to accurately convey information, orders, and alerts to troops in the field.

Communications is all about information. Many people choose to just listen and be informed. Two consumer products that open the communications world are shortwave radios and scanners. Figure 10-3 is a good example of a modern tabletop communications receiver. Marketed by Radio Shack, it is small, but packed with advanced features. Using advanced circuitry and design techniques, it has a continuous frequency coverage from 150 kHz to 29.999 MHz. The companion scanner, illustrated in Figure 10-4, has a continuous frequency range of 25 MHz to 1300 MHz. Both units have numerous memory channels, frequency scanning, frequency tuning, and direct frequency selection. These advanced features add to the flexiblity and usefulness of both devices.

Internet

The Internet has become a major communications link for many computer users. A still-evolving medium, even in today's early stages it is far more than chat rooms where exchanges occur through the typed word appearing on a monitor's screen. New adaptations are allowing the transfer of video and voice data via Internet links. Due to the rate of data transfer, the video is still a little jumpy and the voice links sound tinny. With widespread

(Photograph courtesy of Tandy Corporation)

■ **10-3** *Radio Shack shortwave receiver.*

(Photograph courtesy of Tandy Corporation)

■ **10-4** *Radio Shack communications scanner.*

adaptation of fiber optic cables and higher data transfer rates, these problems should disappear.

Conventional broadcasters are showing an interest in the Internet. Many stations now have web sites where you can listen to the broadcasts via your computer. Radio services available on the net cover all types of music, talk radio, and sports broadcasts. Not to be left behind, international shortwave broadcasters now have audio hookups via the Internet.

The Future

The future for electronic communications appears to be brighter than ever. For the past several decades, advances in communications applications have been technology-led. Inventions such as integrated circuits, coaxial cable, microwave techniques, and geosynchronous satellites have been important milestones. It has been stated that the next several decades will be software-led, with refinements of today's technology and techniques.

Fiber optics are going to provide improvements in wired communications links. As more of the telecommunications and cable TV networks are changed to fiber optics, e-mail, visual teleconferencing, fax, and voice phones will merge and improve in quality. The digital revolution will continue to expand at a rapid pace, changing the face of communications as we know it. Continued pressure on the finite and precious electromagnetic spectrum will lead to the end of over-the-air television broadcasting. With better and cheaper computers, TV might become entertainment on demand. Rather than waiting for the program you want, you will request it when you are ready. Magazines and newspapers are ripe to be digitized. Currently, many magazines are totally available on the Internet. The print medium of the future will be downloaded overnight to be available on your handheld viewer in the morning. Breaking news, stock quotes, weather, and political commentary will be updated throughout the day to keep you fully informed.

Personal communications will become more common than ever before. Currently, phone service that rings your office phone, home phone, cell phone, and pager is being offered in many areas. Taken one step further, when you are equipped with a home terminal with links to the office and pocket communicator, you will be in touch with the world via one number, within microseconds. Vehicular navigation systems, smart highways, and digital links will call for technicians who are comfortable with all forms of communications technology and software applications. With the end of the Cold War, military and civilian applications will merge, with far more technology than in the past being transferred back and forth.

Military Applications

In the past, many civilian applications had their beginnings from the military sector as spinoffs. As an example, the humble cassette tape was first developed for the Air Force's U-2 spy plane program. As the aircraft were to fly long-range missions, vast quantities of data would be collected. Packed with sensors, fuel, and a pilot, a compact analog data storage medium was required to save the information for future analysis. The answer was a very small tape recorder using a very thin, but rugged, magnetic recording tape. We know it today as the common audiocassette tape. Other spinoffs include many equipment features taken for granted today. Digital tuning receivers, handheld transceivers, compact equipment, and digital data transfer all began as military programs. Now, with the reduction in defense spending, more civilian technology will be adapted to military applications.

In the past, items produced for the defense sector had to meet strict standards known as *military specifications* (mil-specs). At one time, due to the vast differences between military and civilian markets, this made sense. To meet these standards, manufacturers had to provide volumes of information. As an example, one medium-sized proposal for a development program weighed 4000 pounds. It was so large that a 737 aircraft had to be leased for $25,000 to carry it. Only 10% of the documents covered the hardware and software information that comprised the program; the other 90% was required to ensure compliance with mil-specs. The cost to the manufacturers, the government, and the taxpayers was less equipment for more money.

There are some items that must meet standards far above those encountered in the civilian community. However, many items built to civilian standards are suitable for the armed forces without any changes. As a result, programs have evolved to reduce mil-specs to the level required to ensure a product's ability to stand up to the hazards of military operations without producing a paperwork blizzard.

Another program, called *civilian off-the-shelf* (COTS) was instituted to channel more civilian equipment into military applications. Development time for the military is measured in years, while in the civilian community it is measured in months. By buying more civilian hardware and software, costs are held down and procurement times are sped up. Many items, such as communications systems and computers, are ideal for COTS.

Continued advances are going to lead to improvements in antenna design, increasing system efficiency. The latest moves appear to be toward digital beam forming (DBF) antennas. Through digitization, the radiated beam from a transmitter can be tailored for different circumstances. An antenna could go from highly directional to omnidirectional in a matter of microseconds. It promises to reduce side lobes to near-nonexistence, providing more energy for the beam. It is possible through the use of DBF to use the skin of a platform as an antenna, rather than an actual dedicated antenna.

Currently, the military is dependent upon satellite communications for operations. If a war is fought with a technologically advanced opponent, they will be targeted. Within the atmosphere, aircraft such as AWACS, JSTARS, Hawkeyes, and other aircraft provide radar and communications services. To protect this valuable asset, research is underway to use blimps in these and other roles. Under the U.S. Navy-managed program called Zephyr, it is a promising approach. If a balloon is at an altitude of 70,000 feet, it has a radio horizon of 500 miles. With a balloon-based transponder, a low-powered handheld transceiver could have that 500-mile range.

Increase the balloon's altitude to 120,000 feet, and the range is dramatically increased. A fleet of only 24 of these low-tech balloons could provide worldwide coverage. With automation, the craft could be unmanned, with an expected endurance of one year. This is an attractive program, as the balloons are very difficult to locate visually or with radar. Once operational, this could be an alternative to a more expensive satellite. Unlike a satellite in space, the balloons could be retrieved, repaired, and refloated.

The military will continue to expand in the areas of satellite communications, computer control, and digital applications. Military platforms such as aircraft, ships, and armored vehicles are very expensive to buy. It is cheaper to upgrade older units with new electronics. Communications systems are going to become more common rather than less so. Currently, battlefield management is able to track individual platforms. In the future, communications systems will link individual soldiers with each other and with their commanders. HF and communications links will provide real-time tactical information to the highest levels of authority. Changes in the future will occur at a tremendous rate. Military contractors, losing business due to the reduced spending on defense programs, will seek more civilian market share. Even with the changes, military spending in procurement and research will continue.

Civilian Applications

Since prior to World War II, technology has usually been developed to fill a military application, and then has shifted to civilian applications. Two excellent examples are the Jeep and the walkie-talkie. The term jeep came from the military nomenclature—*Vehicle, General Purpose,* or VGP. In its original contract specification it was to be a low-cost, four-passenger, four-wheel-drive vehicle capable of surviving military operations. At the end of the war, thousands of surplus jeeps were sold to a product-starved civilian market at low prices. The little vehicle became so popular that it started a market that still exists today. The walkie-talkie, which was developed to provide lightweight communications for the infantry, was also sold as surplus. Their popularity has led to the many forms of personal communications that we see today. With the end of the Cold War, more advanced technology will be developed in the civilian world and then be provided for military applications.

Modern society is becoming an information society. Today, information is power. The goal of the communications industry is to provide more information on demand to more consumers. To aid in this change, more intelligence is digitized. That makes it computer-compatible. Once any type of intelligence is digitized, it can be manipulated by a computer. It becomes more flexible, which makes it more valuable. The increased flexibility allows multiple formats to be integrated, processed, and shifted on land lines. The goal is to make an information pipeline for all types of data and intelligence. The first step is the *integrated services digital network* (ISDN). Available in two basic formats, it is a conduit that different electronic media devices can be plugged into. The basic rate interface is two 64-Kbps channels for data, and one 16-Kbps channel for control information. The primary rate interface is twenty-three 64-Kbps channels for intelligence, and one 64-Kbps channel for control information.

As stated earlier, the different forms of communications are going to merge. Radio is now available on the Internet. Recently, the first televisions capable of surfing the Internet went on sale. Difficult to use and somewhat limited, the manufacturers have already improved the basic design and will market it soon.

As the electromagnetic spectrum is finite, more services are going to leave the air and be available on the Internet or cable. Thus, the freed space can be used for current and future communications links. To ease crowding, the first steps are being taken to establish an infrared laser relay system to convey video, voice, and data intelligence. It promises to be very cost effective with current technologies, as it will have a large capacity. An additional plus is that it offers a high degree of security.

The satellite communications field will be undergoing a major change to reduce costs and increase service. A Motorola-led consortium of electronics firms is proposing the Iridium System for worldwide cellular phone and digital data services. The system is named after the element iridium, with an atomic number of 77, as the original network was to have 77 satellites. Studies have indicated that the same worldwide level of service can be obtained from only 66 satellites. It was decided not to rename the system, as the element with an atomic number of 66 is dysprosium; it sounds more like a medical condition than a high-tech communications network. To reduce the two-way transit time from the earth to the satellites, they will be at an altitude of 420 miles in six different low earth orbits. Each bird will be capable of handling 48 separate communications beams. Each beam will reliably handle 230 calls or links.

This system will connect any suitably equipped cellular phone with any other phone in the world. To communicate from a ship at sea in the Pacific to an aircraft over the continental U.S., several links are required. The call originates from the ship and is picked up by the closest satellite. It is then automatically relayed satellite-to-satellite until it reaches one close the aircraft; at that point it is downlinked to the aircraft to complete the call.

Other low-earth-orbit systems are being proposed. Orbcom, for example, would use 26 satellites in low earth orbit. This system is designed to handle digitized data and faxes. The low-cost terminal would be the size and complexity of a notebook computer. Equipped with a printer port, weather charts and message printouts would be available.

Two other systems are also working their way through the design process. One being developed by TRW would consist of 12 satellites that would orbit at about 1800 miles. This system would provide voice, data, paging, and location information. With the locator, you would be able to accurately fix any suitably equipped cell phone anywhere in the world. The second system, Globestar, is proposed by Loral and Qualcomm. This one is envisioned to consist of eight satellites placed in a 750-mile orbit.

Much military hardware and many capabilities are going to shift into the civilian sector as more defense contractors respond to the decreased military procurement. A good example of this high-tech transfer is in the area of cellular phones. As useful as they are, there is one major shortcoming: When placing a "911" call, you need to know where you are. Currently, there is no method of locating a caller automatically.

A U.S. Navy program called Classic Outboard might provide this much-needed feature. It is a computerized communications system that allows for the precise location of a signal by listening from only one ship. Typically, it takes a minimum of two ships in different locations to provide accurate positioning data by listening to a received signal. By adapting this technology to the cell phone control sites, any time a 911 call is placed, an almost instantaneous readout of the caller's position will be available. In the first version, it will be accurate only to a building or section of road. In the second version, the accuracy is to be increased to the point that in a multistory building, the floor and possibly the location of the room will be available.

The communications field is going to change and advance so much that it will be almost unrecognizable in a few short years. This explosion of technology and advanced applications will make the communications career field more dynamic and interesting than ever before.

Appendix

Electromagnetic Communications Personalities

Because of the universal nature of research, it is difficult to mention all the individuals who have had an impact on electronics and communications. Often, similar research and conclusions took part in several different locations, at the same time, without knowledge of one another. However, there are many who will always be remembered for their efforts. This list of notable electronics inventors, experimenters, and pioneers is long, and consists of many distinguished intellectual explorers.

Howard Aiken (1906-1973) was a mathematician who is renowned for his research leading to the design of the forerunner to the modern digital computer. His efforts led to the computer and microprocessor becoming everyday tools of our society.

Andre-Marie Ampere (1775-1836) was a French physicist and mathematician considered by many to be a prodigy. During his career he served as the professor of chemistry and physics at Bourg University. As his studies progressed and he gained in intellectual stature, he later served as the mathematics professor at the *Ecole Polytechnique* in Paris. Because of his experiments, he is considered to be the founder of electrodynamics (or as it is known today, electromagnetism). His most important contributions are his experiments with currents and magnetism, which resulted in Ampere's Law. Ampere's Law is the mathematical description of the magnetic force that exists between two currents flowing in two separate conductors. His contributions to basic research are so important that the unit of electrical current is named in his honor—the *ampere,* or *amp.*

Alexander Graham Bell (1847-1922) is best remembered for his work in the invention and development of the telephone. His interest actually began as an educator for the deaf. Through his work with the hearing impaired, he became interested in the possibility of using electromagnetic energy to transmit sounds. After his successful experiments, he first demonstrated a useful telephone system in 1876. His other important inventions included the photophone for transmitting sound by modulated light, and the wax cylinder to record sounds.

Bell's simple telephone system consisted of a transmitter and a receiver. Each was constructed with a thin iron diaphragm mounted on an iron core surrounded by a coil of wire. During operation, the iron core would be magnetized by a direct current passing through the coil from a battery. Sound waves pushing against the diaphragm caused it to vibrate. That, in turn, would generate a variation in the magnetic field. Variations in the magnetic field would in turn induce a variation in the current flowing through the coil. The current variations would then be transmitted over wires that connected two telephones. The current variations felt by the distant receiver would be used to generate a variation in the magnetic field in the diaphragm, reproducing the sound waves. With a few improvements, all telephones up until the modern digital revolution were based on the same simple designs. To recognize Bell's contributions to science, the unit of signal gain, known as the bel, is named in his honor.

Lee De Forest (1873-1961) began his studies as a mechanical engineer, but became interested in the then newly discovered phenomenon of radio waves. His doctoral thesis was on the subject of developments in wireless telegraph. His first work, in the decreasing of the time required for wireless messages, was applied by the U.S. military and had its first use in reporting the Russo-Japanese War of 1904 (1904-1905). He was a very active inventor, holding more than 300 patents. His most important work was in the advancement of vacuum tube technology. His other areas of interest covered radio broadcasts, movies with sound, and television.

Michael Faraday's (1791-1867) contributions to the fields of physics and electronics cannot be overemphasized. A self-professed student of philosopher (the phenomena of nature), his studies and research into the properties of electromagnetism provided the basis for coils, transformers, motors, transmission lines, and even electronics itself. Because of his contributions to the electronics field, the electrical unit of capacitance, the *farad,* is named in his honor.

Karl Friedrich Gauss (1777-1855) was a German mathematician who was interested in the emerging study of electricity. He became the director of the Gottingen Observatory. In addition to his significant advances in mathematics in the areas of number theory and the theory of series, he made extensive contributions in electricity and magnetism. As his contributions provided the foundation for many researchers who followed him, the unit of magnetic flux, the *gauss,* is named in his honor.

Herman Von Helmholtz (1821-1894) was a German anatomist, physicist, and physiologist. Although his inclusion in this book might seem to be odd, his experiments on the speed of nerve impulses and animal heat ultimately led to the principle of the conservation of energy and the introduction of the concept of free energy.

Henrich Hertz (1875-1894) an eminent German scientist, earned his place in history as the first researcher to verify the theoretical work of many physicists, such as Maxwell, Helmholtz, and Faraday. Based on their theories, he demonstrated the wave characteristics of electromagnetic energy in both conductors and free space. His groundbreaking work had application not only in the propagation of high-frequency electromagnetic energy, but paved the way for much higher-frequency systems such as radar, satellite communications, and VHF/UHF systems. Because his efforts were so valuable, he has been immortalized by having the unit of frequency, the *hertz,* named for him.

James P. Joule (1818-1889) was a British physicist noted for his research in electricity and thermodynamics. Based on his studies and observations, he deduced Joule's Law. The law states that the amount of heat produced each second in a conductor by the flow of electricity is proportional to the resistance of the conductor and the square of the current. Due to Joule's Law Of Electric Heating and other studies, he is honored by having a unit of energy, the *joule,* named after him.

Guglielmo Marconi (1874-1937) is best known as one of the early radio pioneers who commercially exploited the theories and experiments of those who preceded him. Although many learned scholars felt that long-range communications were an impossibility and used scientific studies to prove it, Marconi persevered and was the first individual to electronically bridge the Atlantic Ocean.

James Clerk Maxwell (1831-1879) used inductive reasoning and mathematics and determined the fundamental relationship that exists between electricity, magnetism, and electromagnetic wave propagation. So essential is Maxwell's work that his equations are still the basis for the design of electrical cables, microwave components, and antennas. He was the first scientist to compute the velocity of electromagnetic waves in free space. As a result of his tremendous contributions to science, the unit for magnetic flux, the *maxwell,* is named after him.

Georg Simon Ohm (1789-1854) a student of mathematics and physics, gained immortal fame as a physicist. His extensive studies resulted in a book that established the relationship between electrical potential, current, and resistance in a circuit. That little formula is the first one learned by a student of electronics:

$$E = I \times R$$

Although the concept of the relationship between current, resistance, and voltage is considered basic knowledge today, 200 years ago it was a major step forward and was justifiably considered revolutionary. His place in history is assured by having the unit of electrical resistance, the *ohm,* named in his honor.

Wilhelm Roentgen (1845-1923) was a German physicist who is credited with expanding our knowledge of the full range of the electromagnetic spectrum. He was one of the first to discover and explore the nature of X-rays. They have since become a research tool used to discover flaws in metals, diagnose disease in humans, combat cancer, and explore the nature of atoms and the universe. He is honored by having the unit of exposure dose for x-rays and gamma rays, the *roentgen,* named after him.

Nikola Tesla (1856-1943) was an electrical engineer who is recognized as one of the foremost pioneers in the electric power field. Through his experiments, he was able in 1888 to design the first practical system of producing and distributing AC electrical power. Among his more important discoveries are high-frequency generators and the Tesla coil, a type of transformer with vital applications in electromagnetic communications.

William Thomson (1824-1907) was a scientist whose contributions included studies in thermodynamics, magnetism, electricity, and mathematics. His experiments were considered to be so important that he was knighted as Lord Kelvin and had the Kelvin temperature scale named in his honor.

Alessandro Volta (1745-1827) a noted Italian physicist, is remembered as the inventor of the first electric battery, the voltaic pile. His research also resulted in the construction of the charge-accumulating machine, which applied the principle of electrophoresis. Its concept serves as the basis of condensers and other devices that operate based on the properties of static electricity. Due to his many contributions, he is memorialized with the symbol for the unit of electrical potential—the *volt.*

Glossary

acceptor An impurity that increases the number of holes, or spaces for electrons, in the valence shell of a semiconductor material. As each of the holes is the equivalent of a unit of positive charge, the resulting material is called *P-type* material.

active antenna An antenna used by a transmitter to radiate electromagnetic energy into free space.

alphanumeric Consisting of both numbers and letters.

amplitude The vertical dimension or height of a signal's characteristic waveform on an oscilloscope screen; this characteristic corresponds to the signal's strength or "loudness."

amplitude modulation Varying the amplitude of a signal to transmit intelligence.

analog The continuous representation of a physical phenomenon that is the amplitude of a signal as compared to time.

antenna An electronic device that radiates or intercepts radio waves.

aperiodic antenna An aperiodic antenna is an antenna that is capable of operating within a wide range of frequencies.

apparent power The power obtained in an AC circuit that is obtained by multiplying the effective values of current and voltage. The resulting value is expressed in voltamperes.

array A group of electrical elements arranged to provided directional characteristics to the radiated RF energy. The elements can be antennas, reflectors, or directors.

attenuation A decrease in power, either intentional or unintentional.

audio frequency A frequency that is capable of being detected by the human ear.

automatic frequency control (AFC) An AFC circuit provides a receiver with the automatic ability to rapidly compensate for small changes in frequency.

automatic gain control (AGC) AGC is the ability of a receiver to automatically compensate for small changes in signal strength.

band A range of frequencies between an upper limit and a lower limit is known as a band.

bandpass filter A filter that passes a band of frequencies between two limits without attenuation while blocking all frequencies that are outside of the limits is known as a bandpass filter.

bandstop filter A filter that attenuates all frequencies between two limits, while passing those frequencies that are outside of the limits without attenuation.

bandwidth The difference between the lowest and highest frequencies that can be used by a communications system.

base station A fixed, land-based communications installation.

beam antenna An antenna with focused directional characteristics.

beamwidth The angular measurement of a radiated beam of energy.

capacitance A property that exists when two conductors are separated by a nonconductive material; it allows the storage of electrical energy in an electrostatic field.

capacitive reactance The opposition to alternating current flow presented by a capacitor. It is measured in ohms and is determined by the value of the capacitor and the frequency of the applied voltage.

capacitor A component consisting of two or more conductive plates separated by a nonconducting insulator. The device stores energy in the electrostatic field that exists in the insulator.

carrier A frequency that is modulated by an intelligence frequency; it "carries" the data of the intelligence frequency.

carrier frequency The frequency of a carrier signal before modulation.

channel A circuit or means of information transmission between two locations.

characteristic impedance When an impedance is connected to the output terminals of a transmission line, the impedance value that makes the line appear to be infinitely long is the characteristic impedance. At the characteristic impedance there are no standing waves appearing on the line; all the RF energy is being transmitted through the line.

circulating current This is the current that flows between the reactive components within an oscillating tank circuit.

clear channel A broadcast channel in which one dominant station has a wide coverage zone. When a station is issued a clear channel, it will not have another station operating on either the same frequency or an adjacent frequency; this is to prevent interference in the primary service area, or a large portion of the secondary service area.

coaxial cable A transmission line that consists of one conductor in the center, surrounded by an insulating material, then a conducting shield, and finally an outer protecting jacket.

coherer A coherer is a vacuum tube containing two metal plugs filled with metal filings. When placed under the influence of an electromagnetic wave, the metal filings *cohere* or stick together, forming a low-resistance path for current flow. It functions as a very simple detector.

comparator An active electronic device that compares two outputs and then generates an output based on their relationship to each other.

converter An electronic device that can change the frequency of an AC signal, convert DC to AC, or change an analog signal into a digital signal.

counter-electromotive force (CEMF) CEMF is the voltage produced by an inductive circuit under conditions of changing circuit current.

critical angle In terms of antenna radiation, it is the smallest angle from the vertical that will still be reflected back to earth by the ionosphere.

critical coupling The degree of coupling between two resonant circuits that transfers the maximum amount of energy.

critical frequency The highest radiated radio frequency that will be reflected back to earth by the ionosphere. Frequencies above the critical frequency pass through the atmosphere to outer space.

cross modulation When the carrier of a desired RF signal is modulated by the intelligence of another, unwanted signal. This interference results in preventing the intelligence of the wanted signal from being received.

data Information; it can be analog or digital in format.

decibel One tenth of a bel. The decibel is ten times the logarithm of the power ratio being measured. 1 dB is a power ratio of 1.259. 10 dB is a power ratio of 10.

degenerative feedback Degenerative feedback is when the output of a circuit is fed back to the input in such a polarity as to reduce the circuit gain.

demodulate When a high-frequency modulated carrier is converted to a form that can be used by an operator, it is said to be demodulated. The action is also known as *detection.*

detection The action of removing the lower-frequency intelligence from the higher-frequency carrier.

dielectric A dielectric is a material that functions as an insulator; a material with low conductivity.

diode A two-element electronic device that permits current flow in one direction and blocks it in the other.

dipole antenna An open antenna constructed from two straight conductors mounted in line, and end-to-end. The connection to the transmitter or receiver is made at the point where the two conductors meet. Maximum radiation is produced in the plane perpendicular to its axis.

direct wave A radiated wave of RF energy that is propagated directly through space. It is not reflected from the ground or atmosphere.

director A parasitic antenna element located ahead of the driven element on the main antenna beam used to increase the radiation and directivity of the antenna in the direction of the main lobe.

discone antenna An antenna with a disc mounted symmetrically near the apex of the cone.

dish A reflector with a surface in the shape of a partial sphere or paraboloid.

distributed constants Any parameter that exists along a transmission line is known as a distributed constant as compared to unit length. Examples would include series resistance, series inductance, shunt-conductance, and shunt capacitance per unit length.

diversity A single received signal derived from the combination of or selection from a number of transmission paths.

donor An impurity with a free electron that is added to a semiconductor material to form N type material.

doping The manufacturing process of adding impurities to molten semiconductor material to form N and P type materials. After cooling, the doped semiconductor material forms a crystal structure.

double conversion A very high frequency superheterodyne receiver that has two intermediate frequencies. The first IF lowers the incoming RF which then mixes the resulting frequency with a second, lower local oscillator frequency. The technique is used in UHF receivers to obtain high gain, stability, selectivity, and image rejection.

double sideband The technique of suppressing the carrier and transmitting the two remaining sidebands.

driven element An antenna element connected directly to the transmission line. It is the major radiating element of a transmitting antenna.

duplex The action of data transmission in both directions between two transceiver sites simultaneously.

E field (electric field) In radiated RF energy, it is always associated with a magnetic field (H field). The E and H fields form the electromagnetic field that is the radiation of electrical energy into free space.

envelope When low-frequency intelligence is impressed upon a constant-amplitude higher-frequency carrier, the resulting varying shape is the signal envelope.

fidelity The degree of accuracy between the original modulating signal in the transmitted signal and the detected signal in the receiver.

field strength The value of electric field intensity (measured in volts per meter) produced at a known point by a radiating antenna.

filter An electronic circuit that passes some frequencies and blocks others.

folded dipole antenna An antenna constructed from two parallel, closely spaced dipole antennas. Both are connected at the ends, but only one is fed at the center.

free space The region that is high enough above surrounding terrain, buildings, and vegetation so that an antenna radiation pattern is not affected by it.

frequency The number of times a periodic phenomenon is repeated in a given unit of time.

frequency conversion The process of taking a received RF signal and converting it to a lower intermediate frequency by mixing it with a local oscillator frequency.

frequency diversity The communications link technique of modulating and then transmitting two carrier frequencies separated by a minimum of 500 Hz to compensate for signal fading. That works because fading is not a synchronous event on two separate frequencies.

frequency modulation (FM) A type of modulation in which the frequency of the modulated signal varies from the carrier frequency by a value determined by the modulation signal, or the intelligence.

frequency response A comparison between the frequency of a signal applied to a circuit and the degree of attenuation of that signal at the circuit's output.

frequency shift keying (FSK) A type of modulation in which different frequencies are used to represent the digital bits 1 and 0.

frequency synthesizer An electronic circuit or device that is capable of producing a wide range of precise frequencies.

ground wave An RF signal, the propagation of which is controlled and characterized by its close proximity to the surface of the earth.

H field The magnetic component of a radiated RF wave.

half duplex The transmission of data in both directions between two transceivers at separate times.

half-power points The two points on a frequency response curve that are at 70.7% of the maximum amplitude of the peak.

hertz The unit of measurement of a waveform's frequency; one hertz is equal to one complete cycle per second.

heterodyne The action of mixing two separate frequencies with the intent of obtaining an output consisting of the two original frequencies, the sum of the two frequencies, and the difference between the two frequencies.

high frequency (HF) The frequency range of 3 MHz to 30 MHz.

high-pass filter An electronic circuit that passes all frequencies above a value known as the *cutoff frequency.* All frequencies below the cutoff frequency are blocked, or attenuated.

horizontal polarization Horizontal polarization refers to the position of the E field in an RF signal in relation to the earth's surface. In horizontal polarization, the E field is parallel to the surface of the earth and the H field is perpendicular to the earth's surface.

image frequency In a superheterodyne receiver, it is an unwanted carrier frequency that is twice the intermediate frequency of the desired carrier signal. Receivers with poor selectivity pass both the desired signal and the image frequency, leading to interference.

impedance The measurement of the opposition to current flow in a circuit. It consists of two characteristic components (reactance and resistance) and is measured in ohms.

impurity Substances that are added to a semiconductor element to form N and P crystals.

inductance Inductance is the electrical characteristic of a circuit or component that opposes a change in circuit current flow. In a circuit containing inductance, the change in current lags the change in voltage.

inductive reactance Inductive reactance is the opposition to current flow presented by a circuit or component containing inductance. Measured in ohms, it is dependent upon the frequency of the applied signal.

inductor An inductor or coil is an electronic component constructed from a series of turns of wire wound in such a way as to produce a magnetic field when current passes through the wire.

integrated circuit An integrated circuit or chip is a *monolithic semiconductor device* that is composed of a combination of diodes, transistors, resistors, capacitors, and inductors; they function as a complete circuit.

intelligence Intelligence is the low-frequency voice, music, or data signal that is used to modulate a higher-frequency carrier for transmission through free space.

interference Any electromagnetic or electrical disturbance which causes an undesired response, malfunction, or reduced operation of electronic or electrical equipment. The disturbance can be lightning, undesired radiation from other equipment, or sunspots.

intermediate frequency The lower frequency to which a high-frequency carrier signal is converted so that the receiver can extract the intelligence.

low frequency (LF) The frequency range of 30 KHz to 300 KHz.

low-pass filter A filter that passes frequencies below a cutoff frequency and attenuates all frequencies above it.

lower sideband The lower sideband is the sideband below the carrier frequency that contains all of the intelligence of the modulating frequency.

lowest usable frequency (LUF) The LUF is the lowest frequency that can be propagated between two points and is determined by such factors as atmospheric absorption, transmitter power, antenna gain, and background noise.

magnetostrictive The property of changing physical dimensions exhibited by conductive material placed in the influence of a magnetic field.

maximum useable frequency (MUF) The MUF is the highest frequency that can be propagated between two points. It is determined by such factors as atmospheric absorption, transmitter power, antenna gain, and background noise.

medium frequency (MF) The frequency range of 300 KHZ to 3 MHz.

mixer An electronic device or circuit that has two or more inputs and produces a single output. In a superheterodyne receiver the intercepted RF signal is combined with the lower frequency local oscillator signal to produce four frequencies—the two original ones, the sum of the two, and the difference between the two.

modulation The process by which intelligence is impressed on a carrier frequency for transmission.

Morse code Developed by Samuel Morse, it was an easy to learn code that could be readily transmitted by telegraph or radio. It consists of a series of dots and dashes which can correspond to a keyed transmitter carrier or closed buzzer circuit.

noise Any undesired signal disturbance in an electronic system.

oscillator An oscillator is an electronic circuit that produces an alternating signal.

parabolic antenna A paraboloid reflector with the feed mounted at the geometric focus.

peak envelope power (PEP) The average level of power applied to an antenna, transmission line, or dummy load from a transmitter during one signal cycle. It is measured at the points of greatest deviation.

pentode A pentode is a vacuum tube that consists of five elements: cathode, control grid, screen grid, suppressor grid, and plate.

phase-locked loop (PLL) A phase-locked loop is an electronic circuit that compares the frequency of an applied reference signal to an internal one produced by a *voltage-controlled oscillator* (VCO). The PLL locks the phase of the VCO signal to the reference.

polarization Polarization refers to the orientation in space of the electromagnetic field that is radiated by a transmitting antenna.

primary service The primary service area is the zone that is covered by the ground wave radiated by a broadcast station, and that is not susceptible to fade or to interference from another station.

propagation Propagation is the movement of waves through a medium (such as electromagnetic waves through the atmosphere).

reactance Reactance is the opposition to current flow offered by a circuit containing inductors and capacitors. The degree of opposition is determined by the value of the components and the frequency of the applied signal, and is measured in ohms.

reflector A parasitic antenna element located opposite the general direction of the maximum field strength.

regenerative feedback Regenerative feedback is what you have when an amplifier stage's output signal is fed back, in phase, to the input of the stage.

repeater A repeater is an RF amplifier that boosts the power of weak signals so that they will cover a wider area.

secondary service The area served by a broadcast station by the radiated sky wave. By regulation, the signal is not to be subjected to interference from other broadcasters.

selectivity The ability of a receiver to discriminate between two or more frequencies.

semiconductor A semiconductor is an electronic device that functions as an open circuit under some circumstances and as a closed circuit under others.

sensitivity The operational characteristic of a receiver that determines the minimum useable input signal that can be processed. It is generally expressed in terms of signal-to-noise ratio.

sideband The range of frequencies immediately on either side of an AM signal's carrier frequency. The sideband frequencies contain all of the modulating intelligence signal.

signal strength A measure of the power of a signal produced by a transmitter at a known location. It is expressed in millivolts per meter.

signal-to-noise ratio The signal-to-noise ratio is the comparison of the amplitude of a desired signal to the amplitude of noise on the same frequency.

simplex A circuit that permits the transmission of signals in both directions between sites, but not at the same time.

simplex transmission Transmission of intelligence in one direction only. A good example would be a commercial radio or TV station.

single sideband A type of communications in which a signal's carrier frequency and one sideband are not transmitted.

sky wave The sky wave is the portion of a transmitted RF signal that is radiated up toward the atmosphere. Depending upon factors such as time of day, sunspots, frequency, power, and angle, it might be reflected back to earth.

space wave The space wave is the portion of a signal's radiated RF energy that travels directly from the transmitting antenna to the receiving antenna.

squelch Squelch is a technique that automatically reduces receiver gain to eliminate background noise when no signal is present.

stability The ability of a receiver to maintain an exact frequency.

standing wave A standing wave is when the ratio of the instantaneous voltage value at one finite point to that at another point along a signal conductor is constant with time. A standing wave is the product of two waves with the identical frequency traveling in opposite directions.

standing wave ratio (SWR) The SWR is the ratio of the minimum to the maximum values of a signal component along a transmission line.

syntony Syntony is when two separate oscillators have the same resonant frequency.

tank circuit A tank circuit is an electronic circuit that is capable of storing electrical energy in reactive components.

tetrode A tetrode is a four-element vacuum tube consisting of a cathode, plate, control grid, and screen grid.

transistor A transistor is a three-element semiconductor device consisting of an emitter, base, and collector. It corresponds functionally to a triode vacuum tube.

transmission line A transmission line is a conductor that is designed to efficiently transfer RF energy from one point to another. It can be a coaxial cable, twisted pair, two-wire ribbon, or waveguide.

triode A triode is a three-element vacuum tube consisting of a cathode, control grid, and plate. It corresponds functionally to a transistor.

tropospheric duct A tropospheric duct is an atmospheric anomaly that results in abnormally long-range RF propagation.

tropospheric scatter The propagation of radio waves by scattering from irregularities in conditions in the troposphere.

upper sideband The upper sideband is the sideband above the carrier frequency of an amplitude-modulated RF signal. It contains all of the intelligence of the modulating frequency.

vertical polarization Vertical polarization refers to the position of the E field in an RF signal in relation to the earth's surface. In vertical polarization, the E field is perpendicular to the surface of the earth, and the H field is parallel to the earth's surface.

wavelength The distance between two successive points of a periodic wave measured in the direction of propagation.

Bibliography

Adams, Charles K. 1984. *Basic Integrated Circuit Theory and Projects*. Blue Ridge Summit, PA: TAB Books.

Bray, John 1995. *The Communications Miracle*. New York, NY: Plenum Publishing Corporation.

Buchsbaum, Walter H. 1978. *Buchsbaum's Complete Handbook of Practical Electronics Reference Data*. New York, NY: Prentice Hall.

Davidson, Homer L. 1995. *Electronic Troubleshooting and Repair Handbook*. New York, NY: TAB/McGraw-Hill.

Dulin, John J., Veley, Victor F., and Gilbert, John. 1991. *Electronic Communications*. Blue Ridge Summit, PA: TAB Books.

Floyd, Thomas I. 1986. *Digital Fundamentals*. Columbus, OH: Charles E. Merrill Publishing Co.

Gibilisco, Stan. 1984. *Basic Transistor Course*. Blue Ridge Summit, PA: TAB Books.

Glass, Dick, and Crow, Ron. 1993. *CET Exam Book*. Blue Ridge Summit, PA: TAB Books.

Grob, Bernard. 1984. *Basic Electronics*. New York, NY: McGraw-Hill.

Horn, Delton T. 1987. *Oscillators Simplified with 61 Projects*. Blue Ridge Summit, PA: TAB Books

Horowitz, Mannie, and Horne, Delton T. 1988. *How to Design Solid State Circuits*. Blue Ridge Summit, PA: TAB Books

Husick, Charles B. 1996. "From Sputnik to Iridium: - A Brief History of Satellite Communications." *Southern Boating*. January 1996.

Kaufman, Milton, and Seidman, Arthur H. 1988. *Electronics Sourcebook for Technicians and Engineers*. New York, NY: McGraw-Hill.

Kim, John C., and Muehldorf, Eugen I. 1995. *Naval Shipboard Communications Systems*. Upper Saddle River, NJ: Prentice Hall PTR.

Lacy, Edward A. 1982. *The Handbook of Electronic Safety Procedures*. Blue Ridge Summit, PA: TAB Books.

Horowitz, Paul, and Hill, Winfield 1989. *The Art of Electronics*, 2d ed. New York, NY: Press Syndicate, The University Of Cambridge.

Marcus, John, and Sclater, Neil. 1994. *McGraw-Hill Electronics Dictionary*, 5th ed. New York, NY: McGraw-Hill.

Orr, William I. 1989. *Radio Handbook*. Carmel, IN: Howard W. Sams and Company.

Parker, Sybil P., ed. 1988. *McGraw-Hill Encyclopedia of Electronics and Computers*. New York, NY: McGraw-Hill.

Polmar, Norman. 1996. *Ships and Aircraft of the U.S. Fleet*, 16th ed. Annapolis, MD: The United States Naval Institute.

Potter, Capt. Michael C. 1995. *Electronic Greyhounds: The Spruance Class Destroyers*. Annapolis, MD: The United States Naval Institute.

Schniederman, Ron 1994. *Wireless Personal Communications*. Piscataway, NJ: IEEE Press.

United States Navy. 1986. *Basic Electronics,* Vol. 1. Washington, DC: Bureau of Naval Personnel.

United States Navy. 1986. *Navy Electricity and Electronics Training Series, Module 17*. Washington, DC: Naval Education and Training Program Development Center.

Index

A

Illustrations are in **boldface**.

B

C

D

E

F

N

O

P

Q

R

S

T

U

V

W

X

Y

Z

About the Author

Frederick L. Gould (Titusville, FL) is currently involved in the space program at Kennedy Space Center as an electronics technician. He has more than 25 years of experience as a maintenance technician and technical instructor for various military and civilian electronic systems. He has previously written *Radar for Technicians* (TAB Books), and has also developed technical educational materials for the U.S. armed forces, governmental agencies, and industry.